Preliminary Reconnaissance Report of the 2011 Tohoku-Chiho Taiheiyo-Oki Earthquake

GEOTECHNICAL, GEOLOGICAL AND EARTHQUAKE ENGINEERING

Volume 23

For further volumes:
http://www.springer.com/series/6011

Architectural Institute of Japan
Editor

Preliminary Reconnaissance Report of the 2011 Tohoku-Chiho Taiheiyo-Oki Earthquake

Springer

Editor
Architectural Institute of Japan
Shiba 5-26-20, Minato-ku
108-8414 Tokyo
Japan

ISBN 978-4-431-54096-0 ISBN 978-4-431-54097-7 (eBook)
DOI 10.1007/978-4-431-54097-7
Springer Tokyo Heidelberg New York Dordrecht London

Library of Congress Control Number: 2012946182

Printed on acid-free paper

Springer is part of Springer Science+Business Media (www.springer.com)

Foreword

The March 11, 2011, Tohoku-Chiho Taiheiyo-Oki earthquake of Mw 9.0 left enormous destruction in Japan where we had thought science, technology, and seismic engineering were well developed. The aftermath has not been localized only in east Japan but has extended over the entire country. The tsunami was one of the most serious events. The disaster left a number of lessons for many people who are involved with engineering and construction of buildings and infrastructures.

Buildings in Iwate, Miyagi, Fukushima, and Ibaraki, the prefectures closest to the epicenter, suffered some structural damages. However, buildings in wider areas experienced large numbers of nonstructural damages in inner and outer walls and ceilings. Nonstructural damage to buildings was so severe that it should be regarded as a serious problem. Also the liquefaction in Chiba Prefecture, which neighbors Tokyo, caused major difficulties to the people living in the region around Tokyo Bay. This earthquake shook buildings not only in east Japan but also those in Tokyo and Osaka, regions distant from the epicenter. Some areas experienced fairly strong ground motion.

The disaster at the Fukushima Daiichi nuclear power plants shocked people around the world. Such an incident in Japan, which produces a limited amount of energy from its own resources except for nuclear energy, is recognized not only as a nuclear disaster but also as a challenge of how we can develop the future of society in this country.

Besides the earthquake disaster, Japan already had been experiencing the compound effects of an aging society with fewer children, population concentration in large cities and depopulation in rural areas and provincial towns. To develop a stable and sustainable society, these issues must be considered. Society must be sustainable and it must maintain an edge in global economical competition. It should also accommodate citizens' lifestyles.

We have to prepare for future seismic events in Japan and other countries. We need to use our experience to develop strategic ideas to implement structural safety, nonstructural safety, and protection against tsunamis.

When we examine a regional area, we must keep in mind that a large earthquake is an event that may happen once in several hundred years. Consequently, it would be the first disaster experience for the people who reside in the region, and earthquake risk management may not be easy to implement. Thus it is critical to record the experiences of building responses and city earthquake sustainability in a report and to share the information with the public in many regions so as to reduce the risks of disaster.

This English version is based on the preliminary reconnaissance report issued in July 2011 in Japanese by the Architectural Institute of Japan (AIJ). I hereby express my gratitude to Professor Hisahiro Hiraishi, who heads the research committee on disaster of the AIJ, and to researchers and engineers who participated in reconnaissance and in the publication of the report.

Tokyo, Japan

Akira Wada

Preface

This report is on the disaster caused by the magnitude-9.0 gigantic subduction-zone earthquake that occurred off the coast of Miyagi Prefecture on March 11, 2011. It was the largest earthquake that has struck Japan in its recorded history.

The Disaster Committee of the Architectural Institute of Japan (AIJ) was responsible for coordinating AIJ members for reconnaissance and investigation of the disaster. The effort faced many difficulties as the AIJ headquarters office is located in Tokyo. For several days immediately after the earthquake, cell phone and Internet communications between Tokyo and Tohoku were interrupted, so there were significant problems in coordinating the reconnaissance. The search for people missing as a result of the tsunami and the concurrent rescue efforts went on for several months. Essential supplies, including such daily necessities as food, water, and fuel, were in short supply for weeks due to the failure of logistics systems in Sendai, Fukushima, and other cities and towns. Extensive damage to the infrastructure and transportation systems made it difficult to enter the affected areas. The accident at the Fukushima Daiichi nuclear power plants added worries about radioactive contamination. The Disaster Committee therefore decided to request AIJ members to voluntarily suspend reconnaissance efforts as well as trips from outside of the Tohoku area for a while.

On March 30, 2011, the Disaster Committee issued *The Guidelines for Damage Survey Method*. These guidelines requested AIJ members to undertake reconnaissance activities only in coordination with the Disaster Committee. It remained difficult, however, to enter the regions devastated by the tsunami and the area suffering from the effects of the nuclear accident. In late April, the main target in the devastated areas had shifted from the phase of searching for the missing to the recovery phase, and the clearance of debris finally started in many towns. Hence the Disaster Committee lifted the voluntarily suspension of reconnaissance efforts on April 25 through the amendment of *The Guidelines for Damage Survey Method*, for the sake of collecting valuable data. As the situation improved in the affected area, AIJ members, working with the Committee, could conduct a well-coordinated reconnaissance.

Considering the extent of the damaged area, the number of towns devastated by the tsunami, and the unprecedented nuclear accidents, the disaster taught us some important lessons and will directly and indirectly affect our future society significantly. This book in English is an abridged version of the Japanese-language interim report entitled *Preliminary Reconnaissance Report Issued on the 2011 Tohoku-Chiho Taiheiyo-Oki Earthquake* by the Disaster Committee of AIJ. We hope this report will offer readers some valuable lessons in providing a stronger backbone for countries that face the risk of earthquake and tsunami disasters.

The Disaster Committee members extend our deepest respect and gratitude to the AIJ members of the Tohoku Chapter and the Kanto Chapter for their dedicated and energetic reconnaissance activities; to other chapters of AIJ, including but not limited to the Tokai and Hokuriku chapters for their speedy reconnaissance activities; to the committee members of AIJ Research Steering Committees and the Academy Board for their comprehensive and systematic survey implementation; and to other AIJ members who worked very hard on this project. We also express our great appreciation for the cooperation of national and municipal governments and the various degrees of support from local citizens, without which we could not have made this report.

Finally, and most importantly, we express our sincere condolences to the victims of the disaster and pray for a speedy recovery of the devastated area.

Tokyo, Japan Hisahiro Hiraishi

Contents

Committee

Research Committee on Disaster

Chair	Hisahiro Hiraishi, Meiji University
Secretaries	Hiroshi Kawase, Kyoto University
	Hitoshi Shiohara, University of Tokyo
	Koichi Kusunoki, Yokohama National University
	Mikio Koshihara, University of Tokyo
	Osamu Murao, University of Tsukuba
Members	(not listed)

Working Group for the AIJ Reconnaissance Report on the 2011 Off the Pacific Coast of Tohoku Earthquake

Chair	Hisahiro Hiraishi, Meiji University
Vice Chair	Reiji Tanaka, Tohoku Institute of Technology
Secretaries	Hiroshi Kawase, Kyoto University
	Hitoshi Shiohara, University of Tokyo
	Koichi Kusunoki, Yokohama National University
	Masato Motosaka, Tohoku University
	Mikio Koshihara, University of Tokyo
Members	Akihiko Kawano, Kyushu University
	Eiichi Inai, Yamaguchi University
	Hiroshi Isoda, Shinshu University
	Kenji Kikuchi, Oita University
	Kohji Tokimatsu, Tokyo Institute of Technology
	Masahito Kobayashi, Meiji University

Contributing Researchers

Chapter 1 Outline of Investigation and Damage

1.2 Outline of damage

Akihiko Hokugo, Kobe University
Haruki Nishi, National Research Institute of Fire and Disaster
Hitoshi Shiohara, University of Tokyo
Koki Goto, Tohoku University
Masaomi Teshigawara, Nagoya University
Yoshio Inoue, UR Linkage Co., Ltd
Yoshio Takiuchi, East Japan Railway Company
Yu-You Li, Tohoku University

Chapter 2 Earthquake, Geology, and Tsunami

2.2 Ground Motion Records in Tohoku District

Akihiro Shibayama, Tohoku University
Kazuya Mitsuji, Yamagata University
Kenji Mori, Tohoku Institute of Technology
Koichi Suzuki, Yamashita Sekkei, Inc.
Makoto Homma, Tohoku University
Makoto Kamiyama, Tohoku Institute of Technology
Masaaki Saruta, Shimizu Corporation
Masayuki Hando, Sendai National College of Technology
Naoki Funaki, Tohoku Institute of Technology
Shin Koyama, Building Research Institute
Takaaki Ikeda, Tobishima Corporation
Toshihide Kashima, Building Research Institute
Wataru Goto, NTT Facilities Inc.
Yuta Sugawara, Tohoku Electric Power Co.

Chapter 3 Damage to Timber Buildings

Masahiro Inayama, University of Tokyo
Naoyuki Itagaki, Akita Prefectural University
Takafumi Nakagawa, Building Research Institute
Yasuhiro Araki, Building Research Institute

Chapter 4 Damages to Reinforced Concrete Buildings

Hisashi Takahashi, Nagoya Institute of Technology
Katsuhiko Nakano, Chiba Institute of Technology
Masaomi Teshigawara, Nagoya University
Susumu Kono, Kyoto University
Toshihiro Nakamura, Nagoya University
Yasushi Sanada, Toyohashi University of Technology

Chapter 5 Damage to Steel Reinforced Concrete Buildings

Hisatoshi Kashiwa, Osaka University
Juan Jose Castro, Osaka University
Koichi Minami, Fukuyama University
Masato Sakurai, Osaka University
Masayuki Handou, Sendai National College of Technology
Shintaro Matsuo, Kyushu University
Suguru Suzuki Osaka University
Tomomi Fujita, Sendai National College of Technology
Yo Kuratomi, Fukuoka University

Chapter 6 Damage to Reinforced Concrete Box-Shaped Wall Buildings

Tetsuya Nishida, Akita Prefectural University team
Kenji Kikuchi, Oita University team
Masayuki Kuroki, Oita University team
Koji Nishino, Oita University team
Hiroshi Kuramoto, Osaka University team
Juan Jose Castro, Osaka University team
Hisatoshi Kashiwa, Osaka University team
Masato Sakurai, Osaka University team
Taku Suzuki, Osaka University team
Yo Hibino, Tokyo Institute of Technology team
Yasushi Sanada, Toyohashi University of Technology team
Yuta Sato, Toyohashi University of Technology team
Sohei Matsubara, Toyohashi University of Technology team
Masaomi Teshigawara, Nagoya University team
Ippei Maruyama, Nagoya University team

Motoyuki Nakamura, Nagoya University team
Toshiyuki Masubuchi, Japan Prefabricated Construction Suppliers & Manufacturers Association team
Hiroki Yasuda, Japan Prefabricated Construction Suppliers & Manufacturers Association team
Shigemitsu Hatanaka, Mie University team
Naoki Mishima, Mie University team
Nanako Marubashi, Yamaguchi University team
Junji Ozaki, Yamaguchi University team
Ken Harayama, Yamaguchi University team
Shiro Suzuki, Urban Renaissance Agency team
Yoshio Inoue, UR Linkage Co., Ltd. Team
Motoyuki Tanaka, UR Linkage Co., Ltd. Team
Masumi Hishikura, UR Linkage Co., Ltd. Team
Takashi Kitahori, UR Linkage Co., Ltd. Team
Yasuhiko Kitano, UR Linkage Co., Ltd. Team
Yoichi Wakasugi, UR Linkage Co., Ltd. Team
Naoto Shiraishi, UR Linkage Co., Ltd. Team
Akira Tasai, Yokohama National University team
Koichi Kusunoki, Yokohama National University team
Yuichi Hatanaka, Yokohama National University team
Hidekazu Watanabe, Yokohama National University team
Masatsugu Iwata, Green Design Office Co., team
Kimiko Tamura, Green Design Office Co., team
Rie Kawahara, Green Design Office Co., team

Chapter 7 Damage to Masonry Buildings

Chikahiro Minowa, Tomoe Research & Development
Daiki Sato, Tokyo University of Science
Hisashi Yamasaki, Polytechnic University
Koji Nishino, TOHO
Kozo Tsumura, Hirosaki University

Chapter 8 Damage to Steel Buildings

Akinobu Takada, Osaka University
Atsushi Sato, Nagoya Institute of Technology
Bu-Sung Kong, Mie University
Hideki Idota, Nagoya Institute of Technology
Hironori Otomo, Hokkaido University
Hisashi Namba, Kobe University
Jun Kawaguchi, Mie University
Keiichiro Suita, Kyoto University

Kenzo Taga, Kobe University
Masanobu Sakashita, Kyoto University
Naoto Yamada, JFE Shoji Construction Materials Sales Co.
Norihisa Hirabayashi, Mie University
Seiji Mukaide, Osaka University
Seiko Tsuge, Mie University
Sho Watanabe, Osaka City University
Susumu Kuwahara, Osaka University
Taichiro Okazaki, Hokkaido University
Taisuke Muraki, Hokkaido University
Takamasa Yamamoto, Toyota National College of Technology
Tsuyoshi Tanaka, Kobe University
Yoshito Tomioka, Mie University
Yoshiya Taniguchi, Osaka City University
Yuichi Sato, Kyoto University
Yuji Koetaka, Kyoto University

Chapter 9 Damage to Non-Structural Elements

Hirozo Mihashi, Tohoku Institute of Technology
Manabu Kanematsu, Tokyo Univ. of Science
Masaki Tamura, Kogakuin University
Naoyuki Itagaki, Akita Prefectural University
Satoru Ishiyama, Akita Prefectural University
Takatsune Kikuta, Tohoku University
Tomo Inoue, Kyushu University
Tomoya Nishiwaki, Tohoku University
Toshiyuki Kanakubo, University of Tsukuba

Chapter 10 Damage to Soils and Foundation

Akio Abe, Tokyo Soil Research Co.
Amane Kuriki, Kyoto University
Chung Yulin, Kyoto University
Eisuke Nishikawa, Tokyo Institute of Technology
Jennifer Donahue, Geosyntec Consultants
Jonathan Stewart, University of California, Los Angeles
Kazuaki Hoki, Kyoto University
Kazuhiro Hayashi, Kyoto University
Kazushi Tohyama, Tokyo Institute of Technology
Kei Inamura, Tokyo Institute of Technology
Kodai Watanabe, Tokyo Institute of Technology
Kyle Rollins, Brigham Young University
Mai Ito, Kyoto University

Ross Boulanger, University of California, Davis
Ryuta Enokida, Kyoto University
Sachi Furukawa, Kyoto University
Scott Ashford, Oregon State University, Corvallis
Steve Kramer, University of Washington
Yoshiaki Ibaraki, Tokyo Institute of Technology
Youhao Zhou, Tokyo Institute of Technology
Yuki Takeda, Tokyo Institute of Technology
Yusuke Nakano, Tokyo Institute of Technology

Translators

Carmine Galasso, University of California, Irvine
H. Kit Miyamoto, Miyamoto International, Inc.
Hiroshi Kawase, Kyoto University
Hitoshi Shiohara, University of Tokyo
Kristijan Kolozvari, University of California, Los Angeles
Msasahiko Higashino, Takenaka Corporation
Taylor Rawlinson, Auburn University

Chapter 1
Outline of Investigation and Damage

Mikio Koshihara, Reiji Tanaka, Junji Ogawa, Ryota Yamaya, and Yukihiro Take

Keywords Damage to building equipment • Fire • Historic buildings • Oil facilities • Water facilities

1.1 Outline of Damage Investigation

On March 11, 2011, at 2:46 pm Japan standard time, a magnitude 9.0 earthquake struck off the eastern coast of Honshu, Japan's main island, approximately 129 km east of Sendai. The earthquake and tsunami produced severe building damage and over 15,000 people were killed due to tsunami. This earthquake was later named "The 2011 off the Pacific coast of Tohoku Earthquake" by the Japan Meteorological Agency (JMA).

Shortly after the earthquake, the Architectural Institute of Japan (AIJ) set up headquarters for the Great East Japan earthquake disaster investigation and

M. Koshihara (✉)
University of Tokyo, Tokyo, Japan
e-mail: kos@iis.u-tokyo.ac.jp

R. Tanaka • R. Yamaya
Tohoku Institute of Technology, Sendai, Japan
e-mail: rtanaka@tohtech.ac.jp; ryamaya1215@yahoo.co.jp

J. Ogawa
Akita Prefectural University, Akita, Japan
e-mail: ogawa_junji@ebony.plala.or.jp

Y. Take
Sendai Technical High School, Sendai, Japan
e-mail: yyyytake@nexyzbb.ne.jp

Architectural Institute of Japan (ed.), *Preliminary Reconnaissance Report of the 2011 Tohoku-Chiho Taiheiyo-Oki Earthquake*, Geotechnical, Geological and Earthquake Engineering 23, DOI 10.1007/978-4-431-54097-7_1, © Springer Japan 2012

rehabilitation and the Research Committee on Disaster conducted an investigation of the damage of structures. The reconnaissance teams were coordinated by four chapters of the AIJ: Tohoku, Kanto, Hokuriku, and Tokai, with support from the head office and other chapters of the AIJ. The internet working group in the Research Committee on Disaster started the web page "SAIGAI Wiki" to share the research status.

To make a detailed record of building damage, attention was paid to all types of building structures including wooden houses, reinforced concrete structures, steel-encased reinforced concrete structures, masonry structures, steel structures, non-engineered structures, and foundations.

The reconnaissance team consisting of four chapters and over 320 members surveyed the damaged area until April 5, 2011. Due to the large scale of the earthquake and tsunami damage and issues such as rescuing of the missing, serious damage of the nuclear power plant by the tsunami, fuel shortage, and damage to transportation infrastructure, the investigated area was limited.

Further detailed studies of the damage area have been planned and conducted by Research Committees of the AIJ after April 12, 2011. Over 282 groups and 2,500 members surveyed and detailed studies continuing until January, 2012.

1.2 Outline of Damage

1.2.1 Damage Statistics

1.2.1.1 Main Shock and the Aftershocks

A very strong earthquake of magnitude 9.0 in JMA scale occurred on March 11, 2011. It was named the off the pacific coast of Tohoku Earthquake by the Japan Meteorological Agency (JMA). The maximum seismic intensity of 7 was recorded at Tsukidate and Kurihara in Miyagi Prefecture. Four aftershocks with JMA intensity greater than 6 were; M6.7 of Nagano Prefecture on March 12 with maximum intensity of 6+, M6.4 of Eastern Shizuoka Prefecture on March 12 with maximum intensity of 6+, M7.4 of Miyagi-oki Oga peninsula on April 7 with maximum intensity of 6, and M7.0 of off shore Fukushima on April 11 with maximum intensity 6, as of June 30.

The characteristics of these events are summarized as: (1) The magnitude is extremely large and produced very high ground accelerations. The maximum observed base acceleration was 2,933 gal (all three components superimposed) recorded at Tsukidate, Kurihara city in Miyagi Prefecture. Peak ground acceleration higher than 1.0 G was recorded at 18 strong motion stations. (2) Two distinctive groups of strong waves were observed in the acceleration time history record and the latter part had the stronger intensity. (3) Soil failures including landslides, subsidence, liquefaction were observed over a wide area. (4) Extensive damage was

Table 1.1 Summary of human impact and damage to residential buildings by prefecture [1]

	Human impact			Damage to residential buildings				
Prefecture	Dead	Missing	Injured	Extensively damaged	Moderately damaged	Partially damaged	Fire	Extinguished
Hokkaido	1	–	3	–	–	3	–	–
Aomori	3	1	47	306	879	85	5	5
Iwate	4,709	2,233	186	20,998	3,174	3,498	26	26
Miyagi	9,194	4,639	3,477	65,462	48,684	76,785	163	163
Akita	–	–	7	–	–	4	1	1
Yamagata	3	–	29	–	1	37	–	–
Fukushima	1,709	245	236	15,897	29,250	93,013	11	11
Ibaraki	24	1	695	2,163	15,164	132,541	37	37
Tochigi	4	–	131	257	2,074	56,799	–	–
Gunma	1	–	41	–	6	16,145	2	2
Saitama	1	–	104	7	41	13,863	13	13
Chiba	20	2	248	771	8,056	27,714	13	13
Tokyo	7	–	111	11	128	2,893	33	33
Kanagawa	4	–	134	–	11	168	6	6
Niigata	–	–	48	34	218	1,837	–	–
Nagano	–	–	13	34	169	495	–	–
Shizuoka	–	–	54	–	–	523	1	1
Mie	–	–	1	–	–	–	–	–
Total	15,680	7,121	5,565	105,940	107,855	426,405	311	311
1995 Hansin-Awaji Earthquake[1]	6,434	3	43,792	104,906	144,274	390,506	–	–

Note: [1] Fire Department confirmed on May 19, 2006
(–) No description

by tsunami. (5) Deaths in large space facilities were caused by collapse of ceilings. (6) Critical accidents occurred at nuclear power plants.

1.2.1.2 Human Loss and Residential Buildings

The number of human losses and damaged residential buildings in Table 1.1 were provided by Fire Department Headquarters on June 30. The 99.5% of deaths and missing were found in Iwate, Miyagi and Fukushima Prefectures, whereas 58.6% of total deaths and 65% of total missing were found in Miyagi Prefecture.

1.2.1.3 Non-residential Buildings by Prefecture

Table 1.2 shows the number of damaged non-residential buildings by prefecture provided by Emergency Disaster Countermeasures Headquarters, National Police

Table 1.2 Number of damaged non-residential buildings by prefecture [2]

Prefecture	Number of damaged non-residential buildings
Hokkaido	470
Aomori	1,184
Iwate	1,538
Miyagi	17,315
Akita	3
Yamagata	N/A
Fukushima	1,015
Ibaraki	8,499
Tochigi	295
Gunma	195
Saitama	33
Chiba	708
Tokyo	20
Kanagawa	1
Niigata	7
Nagano	N/A
Shizuoka	N/A
Mie	9
Total	31,292

Agency (NPA) on June 28. It can be observed that the 55% of damaged buildings is found in Miyagi Prefecture, while 27% is found in Ibaraki Prefecture.

1.2.1.4 Results of Post Earthquake Inspection of Buildings

The results of post earthquake inspection of buildings performed by the Ministry of Land, Infrastructure, Transport and Tourism as of June 27 are listed in Table 1.3. It can be observed that out of 95,227 damaged buildings, 12.2% were unsafe, 24.3% demanded caution, and 63.5% were safe. The ratio of unsafe buildings to the number of inspected buildings is the largest in Gunma Prefecture (27.3%) and Fukushima (21.0%).

1.2.1.5 Damage to Buildings in Coastal Areas of Pacific Ocean

Overview of damage to buildings located on the coast of the Pacific Ocean is shown in Tables 1.4, 1.5, 1.6, 1.7, 1.8, and 1.9. It is considered that most of the buildings were damaged by the tsunami, and therefore buildings were not distinguished as being damaged directly by earthquake or by tsunami.

Table 1.3 Result of post-earthquake quick inspection of damaged buildings by prefecture [3]

Prefecture	Unsafe (red tagged)	Caution (yellow tagged)	Safe (green tagged)	Total
Iwate	168	445	459	1,072
Miyagi	5,088	7,511	37,968	50,567
Fukushima	3,314	6,718	5,775	15,807
Ibaraki	1,561	4,684	9,618	15,863
Tochigi	676	1,845	2,658	5,179
Gunma	30	61	19	110
Saitama	0	42	83	125
Chiba	677	1,625	3,213	5,515
Tokyo	59	137	252	448
Kanagawa	14	81	446	541
Total	11,587	23,149	60,491	95,227

Table 1.4 Overview of damage to buildings in coastal cities of Aomori Prefecture (both residential and non-residential buildings) [4]

Municipalities	Extensively damaged	Moderately damaged	Partially damaged	Total
Misawa City	96	45	28	169
Oirase Town	99	63	51	213
Hachinohe City	593	1,444	–	2,037
Hashikami Town	16	12	1	29

Table 1.5 Overview of building damage in coastal cities in Iwate Prefecture [5]

Municipalities	Number of extensively damaged buildings
Hirono	26
Kuji (Kuji Villages, Yamagata Village)	274
Noda Village	478
Fudai Village	0
Tanohata Village	270
Iwaizumi Town	197
Miyako City (Miyako, Taro, Shinzato Village, Kawai Village)	4,675
Yamada Town	3,184
Otsuchi Town	3,677
Kamaishi City	3,723
Ofunato City (Ofunato, Sanriku-cho)	3,629
Rikuzen Takata City	3,341
Total	23,474

Table 1.6 Overview of building damage in coastal cities in Miyagi Prefecture [6]

Coastal municipality	Extensively damaged	Moderately damaged
Kesennuma (Kesennuma, Karakuwa City, Yoshimoto)	8,383	1,861
Minamisanriku Town (Shizugawa City, Town of Utatsu)	3,877	Under investigation
Ishinomaki City (Ishinomaki, Kawakita, Ogatsu, Kanan, Momoo, Kitakami, Oga)	28,000	Under investigation
Onagawa-cho	3,021	46
Higashi Matsushima (Yamoto, Naruse)	4,791	4,410
Matsushima City	103	390
Rifu Town	12	84
Shiogama City	386	1,217
Shichigahama Town	667	381
Tagajo City	1,500	3,000
Sendai City	9,877	8,227
Natori City	2,676	773
Iwanuma City	699	1,057
Watari Town	2,369	823
Yamamoto Town	2,103	939
Total	68,464	23,208

Note: Names of the municipalities listed in parentheses are those before the consolidation of municipalities starting in 1999

Table 1.7 Overview of building damage in coastal cities in Fukushima Prefecture [6]

Coastal municipality	Extensively damaged	Moderately damaged
Shinchi Town	548	
Soma City	1,120	392
Minamisoma City (Haramachi, Odaka, Kashima-cho)	4,682	975
Namie Town	–	–
Futaba Town	58	5
Okuma Town	30	–
Tomioka City	–	–
Naraha Town	50	–
Hirono Town	102	38
Iwaki City	5,234	9,037
Total	11,824	10,447

Note: Names of the municipalities listed above are those before the consolidation of municipalities starting in 1999

Table 1.8 Overview of building damage in coastal cities in Ibaraki Prefecture [7]

Coastal municipalities	Extensively damaged	Moderately damaged	Partially damaged
Kitaibaraki City	217	1,035	3,580
Takahagi City	95	555	2,849
Hitachi City	336	2,374	6,732
Tokai Village	39	77	1,789
Hitachinaka City	70	421	3,257
Oarai Town	5	221	809
Hokota City	73	147	5,250
Kashima City	186	700	2,567
Kamisu City	128	1,175	1,882
Total	1,149	6,705	7,854

Table 1.9 Overview of building damage in coastal cities in Chiba Prefecture [8]

Coastal municipalities	Extensively damaged	Moderately damaged	Partially damaged
Choshi	28	69	–
Asahi	332	883	–
Sousa-shi	6	13	–
Yokoshibahikari	6	9	–
Yamatake	35	102	–
Kujukuri	–	2	–
Ohami shirasato-cho	1	–	–
Shirako-machi	1	–	–
Isumi-shi	–	1	–

Note: No description about building damage from Chosei Village, Ichinomiya Town Onjuku Town, Katsuura City, Kamogawa City, and Minami-boso City
(–) Not listed

1.2.1.6 Lifeline Damage

Tables 1.10, 1.11, 1.12, and 1.13 show the damage to infrastructures. Table 1.10 compares the summary of lifeline damage with 1995 Hanshin-Awaji Earthquake disaster.

1.2.1.7 Earthquake Damage from the Great East Japan Earthquake and Tsunami

Table 1.14 shows the cost of damage caused by the Great East Japan Earthquake and Tsunami estimated by the Cabinet Office.

Table 1.10 Summary of lifeline damage [9]

	2011 Great East Japan Earthquake and Tsunami disaster	1995 Hanshin-Awaji Earthquake disaster
Electricity	Tohoku electric power network: Approximately 446 million units in power outage (March 11)	About 2.6 million power outages in households
	TEPCO network: approximately 405 million units in power outage (March 11)	
Gas	Iwate, Miyagi, Fukushima Prefecture	About 84.5 million gas outages
	City gas: about 42 million units in outage (March 11)	
	LP gas: about 166 million units in outage (March 11)	
Water	Water supply of 2,290,000,000 households in 19 prefectures outages	About 1.27 million units with limited water
Sewer, etc.	Sewer: 48 damaged sewage treatment facilities	Approximately 260 km of damaged sewer length
	Approximately 96 km of damaged culvert	
	Plant: 78 water pumping station out of order	
	Drainage: 11,403 affected districts (in Tokyo and 10 Prefectures)	
Tele-communication	Landlines: about one million lines interrupted (March 13)	Due to telephone switchboard: about 28.5 million lines
	Wireless network: about 14,800 stations out of order (March 12)	Due to phone line: about 19.3 million lines

1.2.2 Overview of Damage in Tohoku District

1.2.2.1 Damage to Buildings

Many buildings designed in accordance with old seismic codes suffered damage from structural problems such as a lack of strength, shear failures in captive columns due to spandrel walls, and eccentric plans with non-uniform lateral resistance. Buildings with seismic retrofit generally performed well and proved the effectiveness of such countermeasures. Buildings designed by current seismic codes showed good performance in the earthquake resulting in very little damage to the main structural elements. Only minor cracks and concrete spalling on structural members can be observed proving that current seismic design codes are very effective. However, significant damage to the nonstructural elements (nonstructural

Table 1.11 Damage to lifelines [9]

Lifeline	Summary of damage
Wholesale markets and retailers	Wholesale markets: facilities of central wholesale markets in Sendai City, Fukushima City and Iwaki City have been damaged Retailer: about 30% of supermarkets and over 40% of convenience stores stopped working causing many markets to shut down their businesses
Fuel	Refineries: six of nine refineries were shut down in the northeastern Kanto region Gas stations: about 53% of total 1,834 gas stations remained operational in three northeast prefectures (as of March 20)
Banks	For operating 2,700 stores in 72 financial institutions headquartered in Ibaraki Prefecture and the Tohoku region 6 were disabled, which corresponds to about 10% 280 offices are closed (as of March 14) In total 280 branches of 70 banks were closed; which are approximately 10% of all the braches, in Aomori, Iwate, Akita, Miyagi, Yamagata, Fukushima and Ibaraki were closed
Postal service	Iwate, Miyagi and Fukushima Prefectures Post Offices: 583 of 1,103 stations stopped working after March 14 (about 53%) Postal services: delivery was suspended in 44 areas, which is 15% of all the areas as of March 14
Courier	Iwate, Miyagi and Fukushima Prefectures: Out of service for about 1 week after the earthquake throughout the whole regions.
Radio and TV broadcasting	TV: 120 relay stations out of order Radio: up to four relay stations out of order

walls, ceilings), as well as damage to building equipment (air conditioning, lighting, etc.) compromised the functionality o many buildings. This suggests that it is necessary to improve the performance of buildings in earthquakes, not only from the structural point of view, but also from a view point of resilience.

1.2.2.2 Non-structural Elements

One of the major characteristics of this earthquake was nonstructural damage. Many cases of damaged non-structural walls, ceilings, and glass-tiled roofs were reported. In many large space buildings, the ceilings were damaged regardless of the purpose of the building. According to eyewitnesses, the falling of the ceiling lights occurred. The most typical damage to nonstructural elements was the fallen ceramic tiles and mortar used for cladding of reinforced concrete buildings and the fallen autoclaved lightweight panels for cladding of steel buildings.

Table 1.12 Damage to transportation [9]

Type to transportation	Summary of damage
Roads	Roads closed for traffic: Route 15 Highway, the road directly under section 69, section 102 of National Road Administration prefectures, prefectural roads, etc. In total, 539 damaged locations
Railway	22 Railroad companies could not operate the trains normally
	64 Railroad lines including: Tohoku Shinkansen, Yamagata Shinkansen, and Akita Shinkansen, were suspended as of March 13
	Tohoku Shinkansen: 1,200 damaged locations
	Conventional JR lines: 4,400 damaged location (not including seven lines suffered from tsunami damage)
	7 lines damaged by tsunami: 23 station buildings swept away, while the length of swept away or buried railroad was 60 km
	101 Railroad bridge girders were also swept away or buried
Bus	196 Damaged buses (collective bus : 62, chartered bus 134)
	115 Damaged buildings of bus companies (severely damaged 30, moderately damaged 85)
Airlines	Sendai Airport was disabled by tsunami
	Fallen ceiling of terminal buildings in Hanamaki Airport and Ibaraki Airport
Harbors	An international hub port, the other 14 import, and 19 local ports out of service (Hachinohe Port, Kuji Port, Miyako Port, Kamaishi Port, Ofunato Port, Ishinomaki Port, Sendai Shiogama Port (Sigma and Sendai), Souma Port, Onahama Port, Ibaraki Port (Hitachi, Hitatinaka, Oharai and Kashima))
Islands route	Four island routes were suspended connecting Kesennuma/Ohshima, Onagawa/Ejima, Ishinomaki/Nagato and Shiogama/Bokujima due to ship landing and due to damage to wharves
Ferry	Loading and unloading of ferry boats were disabled due to damage of wharves at Hachinohe Port, Sendai Shiogama Port, Ibaraki Port, while a number of routes were suspended (Hachinohe/Tomakomai, Nagoya/Sendai/Tomakomai and Oharai/Tomakomai)

In addition, the glass breakage and the spalling of mortar hurt and killed many people, suggesting that design and construction of non-structural elements has to be considered with more attention in the future.

1.2.2.3 Historic Buildings

In the five prefectures in Tohoku, a damage survey of historic buildings was carried out. The severe damage was observed in lowland river basin in Miyagi Prefecture. In some parts of Akita and Iwate Prefectures, the effect of snowfall may have caused extensive damage. In the tsunami-inundated area, the deterioration of the buildings due to salt damage was a serious threat to the buildings, which may not have had severe damage otherwise. In Sendai, homestead wood surrounding a building called an "igune," may have reduced the destructive power of the tsunami.

Table 1.13 Damage to other infrastructure and other [9]

Other infrastructure facilities, etc.	Summary of damage
River	National River Administration: embankment collapse occurred at 2,115 locations due to river outflow
	Prefecture City Administration: embankment collapse occurred at 1,360 locations due to river outflow
Coastal	Iwate, Miyagi and Fukushima Prefectures: approximately 300 km of coastal embankments were affected from which 190 km were completely destroyed
Fishing port	Iwate, Miyagi and Fukushima Prefectures: almost all the fishing ports (about 260) were heavily damaged by the tsunami. The damage was estimated to 594.4 billion yen in these three prefectures
Farms, etc.	Iwate, Miyagi and Fukushima Prefectures: estimated are of land affected by the tsunami is approximately 2.3 million hectares, which represents 5.2% of arable land. About 7,400 agricultural facilities were affected
Educational facilities	76 National institutions, 6,414 public schools and 2,928 educational facilities were damaged
	The primary damage was manifested in collapsed and partially destroyed buildings due to the earthquake and tsunami
Medical facilities	Iwate, Miyagi and Fukushima Prefectures: 11 collapsed hospitals, 296 partially damaged hospitals (total 381 hospitals affected by the earthquake) where, partially damaged building varies from partially unfunctional to failure of building equipments
Debris	Iwate, Miyagi and Fukushima Prefectures: approximately 24.9 million tons of debris
	(Iwate: about 6 million tons; Miyagi: about 16 million tons; Fukushima: about 2.9 million tons)

1.2.2.4 Damage to Building Equipment

The level of damage to electrical equipment, plumbing, and air conditioning were surveyed, among which the highest level of damage was observed on electrical equipment. The most common damage was to the air conditioning equipment involved the falling of air vents, then overturning and damage of outdoor units, the fracture of joints, and the damage of the casing of the blowers.

Observed damage to plumbing equipment was the destruction of pipe holders, buckling and failures of stands for water tanks, as well as the destruction of the anchorage. Electrical equipment was suffered from the falling of lighting equipment, falling of the cable racks, damaged supports and electrical units (boxes), and broken bolts at the anchorage. Besides mentioned, damage to the lifelines was also investigated.

Table 1.14 Economic loss estimated [9]

		Amount of damage	Remarks	
2011 Great East Japan Earthquake and Tsunami disaster	Estimated loss	About 10.4 trillion yen in buildings About 1.3 trillion yen in lifeline facilities About 2.2 trillion yen in infrastructure facilities About 1.9 trillion yen in Forestry and Fisheries About 1.1 trillion yen the rest of the damage	Damage from each prefecture (buildings, lifeline facilities, infrastructure, etc.) Estimates are made based on provided information Changes are possible if new damage discovered Damage caused by nuclear disaster is not included	
1995 Hanshin-Awaji Earthquake disaster	Estimated by National Land Agency (February 14, 1995) Estimated by Hyogo Prefecture (April 05, 1995)	About 9 trillion 600 billion yen About 9.268 trillion yen	Budget for restoration	Government bond 5,020 billion JPY (from FY 1994 to FY 1999) Source: Feb. 20, 2000 (Prime Minister's Office, Report on restoration after Hanshin-Awaji Earthquake disaster)

1.2.2.5 Geotechnical Consequences

Numerous types of soil failures occurred as consequences of the earthquake such as soil subsidence, liquefaction, and soil failures. 866 locations in Miyagi Prefecture, 269 locations in Fukushima Prefecture, and 98 locations in Iwate Prefecture have been found to be hazardous after the investigation. Out of 866 hazardous locations in Miyagi Prefecture, 794 were located in Sendai. Land subsidence of about 76 cm was observed in Ofunato, Iwate Prefecture, while 56 cm of subsidence was reported in Kamaishi. In addition, 65 cm of subsidence is reported in Kesennuma in Miyagi Prefecture. The most typical areas with land subsidence are at the vicinity of the estuary of the Kitakami River and the Old-Kitakami River in Ishinomaki City. Hence the residential and business area in central part of Ishinomaki City sunk below sea level. Also, building structures in Onagawa were swept away and overturned by the tsunami.

1.2.2.6 Environmental Related Damage

For life environment related damage, status of shelter use, subsidence damage of housing, housing damage, and damage associated with urban high-rise housing were investigated. Damage to medical facilities was also investigated. Relationship of damage to educational facilities and teaching schedules, diversity of the damaged area, diversity of demography, and industrial structures were also investigated. Also, restoration plans were also examined in four prefectures in Tohoku.

1.2.2.7 Damage Caused by Tsunami

Tsunami damage in Aomori, Iwate, Miyagi, and Fukushima Prefectures was quite extensive and investigated thoroughly. Most of the damage in Fukushima Prefecture was surveyed with the exception of two locations due to the restriction of Fukushima Dai-ichi nuclear power plant. In the most cases, the tsunami destroyed wooden houses, while structures of reinforced concrete and steel frame structures remained stable. So far, those embankments intended for tsunami protection were not perfect and sometimes not effective at all. The tsunami effects should be considered and reflected to the design of buildings.

1.2.3 Overview of Damage in Kanto District

An overview of damage observed on March 11 in the region off the northeastern Pacific Ocean coast in Kanto district is presented herein.

1.2.3.1 Damage to Buildings

The ratio of number of collapsed buildings and partially damaged buildings to total number of damaged buildings are shown in Table 1.15. In addition, the total number of buildings destroyed and partially destroyed, are shown in Fig. 1.1 on the map of the rate ratio and some partially destroyed total collapse.

Most of the collapsed and significantly damaged buildings were masonry and wooden structures, while there are very few RC and S structures that collapsed. RC buildings designed by the old seismic code, before 1981, had structural damage such as shear failure of shear walls and columns. It has been observed that the damage in RC structures occurred as a result of columns shear failures. On the other hand, significant damage occurred in steel structures as a result of failure of braces and anchor bolts connections.

In the area close to the Ibaraki Prefecture, north, east central part of Chiba Prefecture and Tochigi Prefecture, damage in buildings was extensive. In the north of Ibaraki Prefecture 1.87% out of 6,504 buildings were significantly damaged. 1,328 houses were flooded due to the tsunami in that area.

In the area between Kasumigaura Lake and the Tone River in the Ibaraki Prefecture, 3,418 buildings were affected and 1.85% collapsed, mostly due to soil failures. The ratio of the buildings extensively and moderately damaged is 0.41% while that of the buildings with partial damage is 12.87%, the largest in Kanto.

In the northern part of Tochigi Prefecture, which shares the boundary with Fukushima Prefecture, the number of the extensively and moderately damaged buildings is 1,338, 0.39% of all the buildings in the area. The typical construction that suffered damage is old wooden buildings without braces and wooden houses with unreinforced masonry stone foundation beams.

In the west region of Chiba Prefecture, which is Tokyo Bay, 814 houses were significantly damaged due to soil subsidence caused by liquefaction, which is 0.13% of total number of houses in the area.

In the east region of Chiba Prefecture, which is watershed of the Tone River or the coastal area of Pacific Ocean, 1,343 buildings were extensively or moderately damaged due to tilting or settlement by soil subsidence caused by liquefaction, or tsunami inundation, which is 0.57% of total number of houses.

In southern Tochigi, Gunma, Saitama, Tokyo Metropolis, Kanagawa and southern Chiba Prefectures damage on buildings is relatively small. Common types of damage were the falling of roof tiles, spalling of wall's exterior cover, and destruction of masonry walls. Buildings located on soft soils and slopes were suffered from soil subsidence.

1.2.3.2 Distribution of Soil Deformation and Damage Caused by Tsunami

In the northeastern coast of Pacific Ocean of Ibaraki and northern coast of Chiba Prefectures, building damage caused by tsunami was very prominent. In the southern coast of Ibaraki Prefecture, where there are not many houses along the

Table 1.15 Damage statistics in Kanto area

Metropolice or Prefecture	Area	Number of damaged buildings: Complete collapse	Partial collapse	Complete or partial collapse	Partial damage	Damage of non-residential buildings	Total number of buildings	Ratio of complete or partial collapse (%)	Ratio of partial damage (%)
Tokyo	East of 23 special wards	10	12	22	586	20	714,370	0.003	0.08
	West of 23 special wards	1	3	4	714		1,008,607	0.000	0.07
	Tama	0	1	1	67		953,648	0.000	0.01
	Subtotal	11	16	27	1,367	20	2,676,625	0.001	0.05
Ibaraki	North	915	5,589	6,504	25,034	1,844	347,152	1.874	7.74
	Central	172	926	1,098	32,616	1,767	267,063	0.411	12.87
	Rokko	458	2,960	3,418	15,257	274	184,747	1.850	8.41
	South	174	770	944	20,926	1,303	484,673	0.195	4.59
	West	55	416	471	25,095	1,619	354,391	0.133	7.54
	Subtotal	1,774	10,661	12,435	118,928	6,807	1,638,026	0.759	7.68
Tochigi	North	179	1,159	1,338	12,340	295	341,373	0.392	3.61
	Central	64	684	748	33,207		422,152	0.177	7.87
	South	2	15	17	6,584		432,046	0.004	1.52
	Subtotal	245	1,858	2,103	52,131	295	1,195,571	0.176	4.39
Gunma		0	1	1	15,428	195	1,227,537	0.000	1.27
Saitama		1	9	10	10,351	33	–	–	–
Chiba	West	188	626	814	19,037	369	625,296	0.130	1.03
	South	42	113	155			458,503	0.034	
	East	401	942	1,343			237,440	0.566	
	Central	60	561	621			568,118	0.109	
	Subtotal	691	2,242	2,933	19,037	369	1,889,357	0.155	1.03
Kanagawa							2,301,979	–	–

(–) Unknown

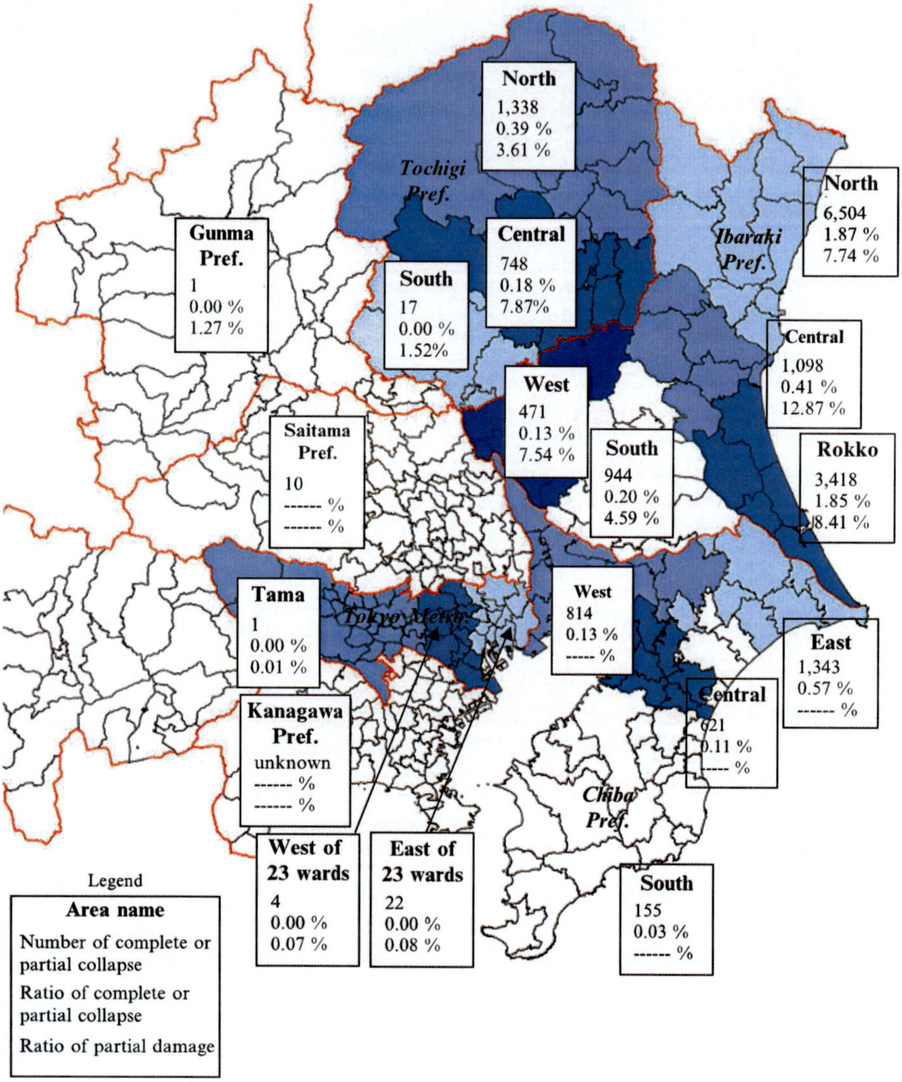

Fig. 1.1 Overview of damage in Kanto area

beach and little damage due to tsunami was reported. Soil liquefaction and subsidence were observed in many areas affected by tsunami, especially around Kasumigaura Lake and Kitaura Lake in Ibaraki Prefecture, in the basin of the Tone River (Ibaraki and Chiba Prefectures), around Imba-numa Lake, in Chiba Prefecture, along Tokyo bay landfill (Tokyo metropolis and Chiba Prefectures), the Arakawa and basin of the Sumida River (Tokyo metropolis).

1.2.3.3 Damage to Historical Structures

Agency for Cultural Affairs has reported that old buildings of cultural heritage suffered a lot of damage due of high seismic vulnerability. Typical damage is falling of roof and ridge tiles, spalling of wall plaster, dislocation and inclination of columns. However some important cultural heritages were significantly damaged. Such examples are Rokkakudo Building of Izura Art and Cultural Institute of Ibaraki University, which was swept away by tsunami, and stone walls of Edo Castle which were partially collapsed in Tokyo.

1.2.4 Other Damage

1.2.4.1 Fire

Earthquake that occurred off the northeastern coast of Pacific Ocean caused fire hazards in many buildings and facilities. The impact force due to tsunami caused eruption of gas from the tank trucks as well as land fixed tank facilities. Sparks due to clash and broken power lines ignited a number of fires in various places. In a few places, they became a large urban fires and wild fires. Evacuated tsunami victims needed to be evacuated again from fires.

According to the fire department 312 fires are reported in total. 200 of them are in Iwate, Miyagi, or Fukushima Prefecture and 158 fires are in the municipalities in the coastal area. The percentage of fires caused by tsunami-affected regions cannot be determined accurately due to lack of information. For example, six urban fires broke out in Yamada Town and Otsuchi Town in Iwate Prefecture, respectively, according to this investigation, while only two fires are reported by the Fire Department [10]. This means the capability of Fire Department in collecting information decreased at the area where were seriously damaged by tsunami.

Two typical causes of the fires are (1) tanks of hazardous material destroyed by tsunami being ignited and (2) cars were swept away by the tsunami, colliding into buildings and igniting spilled gas.

The former type of fire was observed in Kesennuma. Kerosene from the tank at Kesennuma Port spilled and flowed out into the water. The hazardous material reached to the debris of wooden houses and raft for marine farming and ignited.

The debris and raft material drifted from the harbor to fill between Ohshima Island and Kesennuma harbor. Then the fire spread to vessels, vehicles factories, and residential buildings as well as forest.

The latter type was observed to the cars parked in front of buildings. The cars were crashed into the building by the tsunami and ignited on the wall of the buildings, sometimes broke into the building through the opening then spread the flames to the inside of the buildings.

While the cause of the fire is unknown in some cases, burning houses, fishing boats, or propane tanks drifted to urban area or forest, spreading the fire. In this case, fire occurred in inundated area, which is very far from the coastline.

Table 1.16 Distribution of fire area, occurred around inundated area

City (Prefecture)	10–20 ha	5–10 ha	1–5 ha	0.3–1 ha	–0.3 ha
Noda Village (Iwate)	–	–	–	–	1
Miyako (Iwate)	–	–	–	1	–
Yamada (Iwate)	1	–	–	4	1
Otsuchi (Iwate)	–	1	2	3	–
Ofunato (Iwate)	–	–	–	2	–
Rikuzen-Takata (Iwate)	–	–	–	–	1
Kesennuma (Miyagi)	–	1	1	–	–
Minami Sanriku (Miyagi)	–	–	1	–	–
Ishinomaki (Miyagi)	–	1	–	1	1
Sendai (Miyagi)	–	–	–	2	1
Natori (Miyagi)	–	–	1	–	–
Watari (Miyagi)	–	–	–	–	2
Soma (Fukushima)	–	–	–	–	1
Iwaki (Fukushima)	–	–	1	–	–

Only large fire is counted. Unrecognized small fires are not necessarily included

The distribution of area affected by fire in the urban areas is shown in Table 1.16. Total area affected by fire exceeds 60 ha without considering fires that extended in the forest.

Many people were evacuated from the urban areas and areas under forest affected by fire, after the evacuation of the areas affected by tsunami. One example was reported at one elementary school in Ishinomaki City.

1.2.4.2 Water Facilities

Water facilities also suffered damage from the earthquake as well as tsunami. According to the announcement of the Ministry of Health, Labour and Welfare (MHLW), 187 cities suffered from the deficient water supply leaving 2.3 million households with water outage. Summary of the damage to water supply in Miyagi Prefecture is provided.

Sennan and Senen Water Supply System in Southern Miyagi Prefecture takes water from the Abukuma River at Shichikashuku Dam and supplies drinking water to 17 municipalities from a water purification plant in Shiroishi City. No significant damage was reported to these facilities and their major pipes.

Five water pipes were detached in main shock of March 11, 2011. As shown in Fig. 1.2, a high-pressure water pipe (φ2,400 mm) was completely torn out of the concrete block in the vicinity of the downstream of IC Shiroishi, only a few kilometers from the water treatment plant. In addition, low-pressure pipes (φ1,200 mm) were damaged in the similar way by the aftershock on April 7. Damage on the water pipe systems close to the water treatment plant interrupted water supply, prolonging water outage.

Fig. 1.2 Damaged steel pipe (φ2,400)

Ohsaki Water Supply System in Northern Miyagi Prefecture supplies to ten municipalities since 1980. The seismic joints are used in 64% of the water system among which the "K shaped joint" made of ductile iron are the most common type. Observed damage of this type of joint was fracture and leakage. Also, damage to water pipe bridges was observed due to the failure of supports, anchor bolts, leakage of flexible pipe and the leakage of air valves.

Sendai City Waterworks Bureau has four major water purification plants and they have no significant damage to the function of the plant. Water distribution plants suspended the operation at three of them, including An-youji (due to the collapse of waterflow control wall) and Koyodai (due to failure of artificial slope). In this water treatment plant, remote system monitoring was disabled due to the failure of the local power generator. This limited the operational time of the plant to a maximum of 24 h.

In Sendai City, which uses the 30% of the water from Sennan and Senen Water Supply System, 230,000 households lost water supply due to the trouble of the water pipes as stated before and leakage of distribution water pipes at approximately 900 locations in Sendai City. After the earthquake on March 11, it took 11 days (until March 22) for the water treatment plant system to be recovered. In the period from March 22 to March 29 restoration work was done on the supplying system, so that at the end of this period water treatment plant was able to supply water in the city of Sendai. Aftershocks on April 7 caused damage to the power system, which left 30,000 households without water supply. Water supply systems were recovered by April 11 in all affected areas except in the eastern part, which was flooded by tsunami.

Fig. 1.3 Damage to pipes caused by tsunami

After the earthquake on March 11; 75,673 households in Ishinomaki and Higashimatsushima were suffered from water outage. The cause of the damage is under investigation. It is confirmed that water pipes from the Kamata intake site located on the Ex-Kikami River to the Daikaido Water Purification Plant leaked (a joint of iron pipe with diameter of 500 mm). In addition, 2 out of 10,000 pipes attached to bridges on downstream side (ocean side) were damaged by tsunami (Fig. 1.3).

Some water supply systems near coastal area took on ground water inundated by tsunami and caused trouble in purification plants. Water supply to Miyako Island was cut for a long period of time due to the failure of the road and the bridge by tsunami near Nobiru Coast. The aftershock on April 7 caused damage again. Although the number of households that were not supplied with water decreased to 23,475 on April 6, it jumped up to 69,193 again on April 8.

A strong support for the tsunami inundated area by supplying more water for cleaning the area was necessary. However, the recovery construction was usually delayed due to the debris, land subsidence, flood due to high tidal waves.

1.2.4.3 Sewage Treatment Facilities

On May 31, 18 sewage treatment facilities in the eastern coastal area of the Pacific Ocean were out of operation. Summary of the damage is shown in Fig. 1.4. After the earthquake, sewage treatment plants immediately stopped working in inland areas

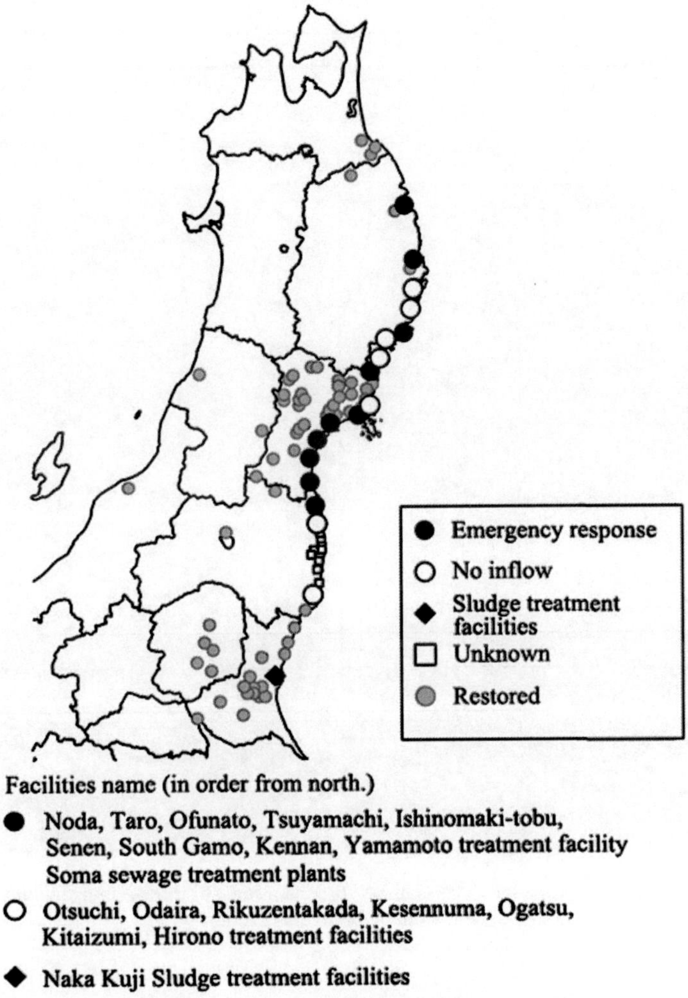

Fig. 1.4 Distribution of damage to sewage facilities (May 31, 2011)

of Kanto and Tohoku. Treatment plants were restored to be operational by June 10. Table 1.17 summarizes the report of the MLIT showing the number of daily changes after the sewage treatment plants stopped working. Six sewage treatment plants: one in Iwate, eight in Miyagi, three in Fukushima and one in Ibaraki are still out of service. In addition, there are many facilities in Fukushima Prefecture, which still haven't been investigated because of the effect of radiation.

South Gamo treatment facility has the ability to handle about 70% of sewage in Sendai with a daily inflow of 320,000 m^3 of water.

Located close to the shore, South Gamo water treatment facility has been damaged by the tsunami. After the earthquake, only the primary treatment and

Table 1.17 Recovery of sewage system

	–3/13	3/14	3/21	3/27	3/31	4/7	4/25	5/31
Unavailable	46	20	12	19	22	19	18	18
Unknown	–	–	–	14	–	–	11	9

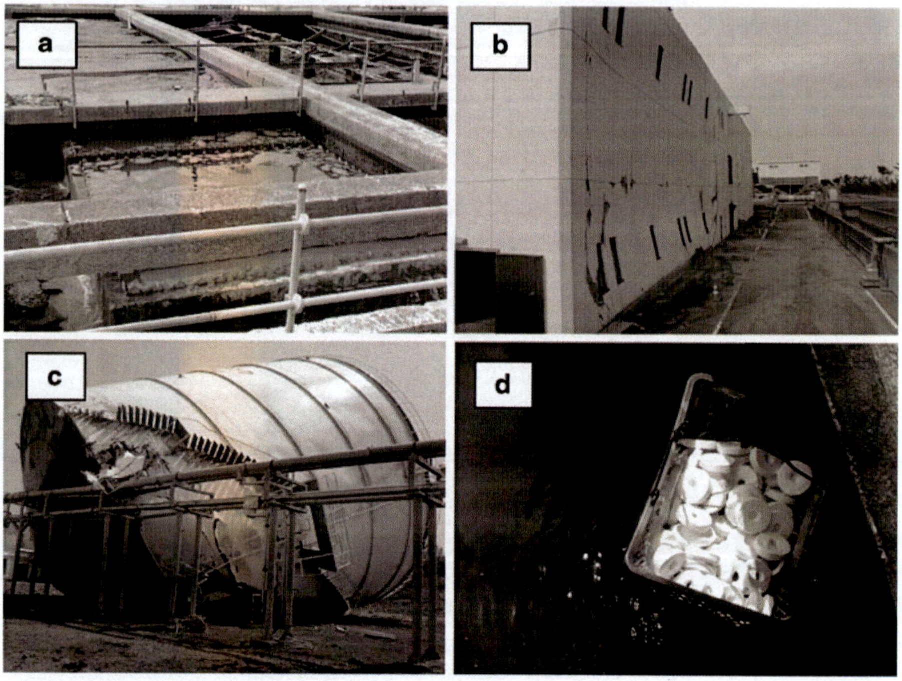

Fig. 1.5 Damage to sewage farm. (**a**) Gamo S: precipitation, (**b**) three pumps in current situation, (**c**) Senen: tank no. 2, and (**d**) Senen: solid chloride disinfection

disinfection with primary sedimentation tank were operated (Fig. 1.5a, b) At the first stage of operation, sludge in settling pond was vacuumed and stored at the final settling pond. A temporary centrifuge was used to dehydrate the sludge to reduce the volume.

Senen purification center is the largest treatment plant in Miyagi Prefecture. Since sewage pumps were installed under the ground, they became unavailable due to tsunami inundation. This accounts for the accident observed in Tagajo City in which sewage water blew out from manholes just after the main shock. Sludge treatment facilities were destroyed. Land subsidence and liquefaction caused various troubles such as failure of tubes for bypass of sewage water and the cave-in of roads.

For the recovery of the sewage treatment function after the earthquake, temporary pumps were usually installed with priority on exclusion function. In other word, all of the settling ponds and reaction tanks were diverted to the first stage settling ponds and water was disinfected with high density chlorine gas of the density of 11–13 mg/L and discharged (Fig. 1.5d).

Water pumps were needed to transfer the wastewater to sewage systems. To recover the function of the first-stage treatment, the recovery of the function of treatment of sludge is also important. However, due to the damage of the pipeline systems, transportation of sludge and its removal were significantly omitted. Particularly, reconstruction of the south sewage treatment facility in Sendai Gamo is estimated to be very expensive and requires several years.

1.2.4.4 Oil Facilities

The earthquake on March 11 followed by tsunami caused significant damage to the petroleum facilities. Investigation pointed out that some differences in the types of damage exist between the damage observed on Pacific Coast and Japan SeaSide. Features of damage on the Pacific Ocean side are (1) tsunami washed away the soil under surface pipes causing damage to the piping system, (2) sloshing due to seismic excitation with longer fundamental period and as a result the damage to floating roof are were evident, and (3) buckling of the side-walls due to short-period ground motion caused the uplift and damage were not observed. Whereas the features of the damage on the Japan Sea Side are (4) damage to floating roof due to sloshing and the spills of contents on decks.

An overview of damage on critical facilities is presented in Figs. 1.6, 1.7, and 1.8. Figure 1.6 shows oil spills from the refinery. It can be observed from Fig. 1.6: (a) tsunami reached about 3.5 m above the tank bottom plate. Although they were empty, no floating, uplift or movement were observed; (b) and (c), bending of pipes and oil spills; (d) damaged embankment wall (e) sloshing wave about 1 m height caused oil spills to occur on the tank roof.

Figure 1.6 shows oil spills from ten outdoor storage tanks in Kesennuma. The amount of spilled oil, kerosene, diesel and gasoline is 11,521 kL.

In Miyagi Prefecture, earthquake caused fire in the refinery that burned down a storage tank, while the dam and the foundations were carried away by the tsunami, causing spills of asphalt and sulfur.

Also, the sag tank in Yamagata Prefecture was badly damaged by sloshing of waves from the Japan Sea.

Observing damage from the each site individually it can be concluded that earthquake caused major damage to the critical facilities of hazardous materials. Stop valves for the piping system did not work as expected due to the power being down after the earthquake. As a result, tremendous hazardous material flew out from the cracks of the pipe damaged by tsunami action. Further investigation on the reason of the severe damage of floating cap inside of the tank is necessary.

Furthermore, in most cases the tsunami washed away storage tanks by fracture of the bottom part. However, some tanks were not washed away, even empty tanks. Some observed damages were caused by tsunami, which washed away the soil beneath the foundation, causing tilting and partial collapse of tanks.

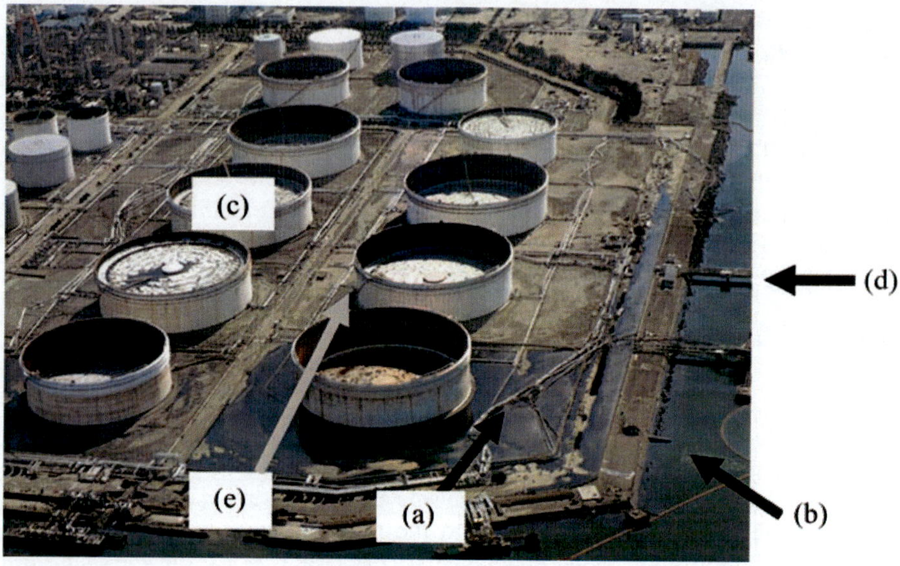

Fig. 1.6 Damage caused to the refinery. (**a**) tsunami reached about 3.5 m above the tank bottom plate. Although they were empty, no floating, uplift or movement were observed; (**b**) and (**c**), bending of pipes and oil spills; (**d**) damaged embankment wall (**e**) sloshing wave about 1 m height caused oil spills to occur on the tank roof

Fig. 1.7 Oil tank collapsed by tsunami

Fig. 1.8 Damage to aluminum lid

1.2.4.5 Railroads

Earthquake occurred on March 11, 2011 generated tsunami that caused catastrophically damages to the railroads of JR (Japan Railway Co.) train running along the coast of the Pacific Ocean. Figure 1.9 and Table 1.18 provides the information about the damage to the train stations and railroads caused by tsunami.

As shown in Table 1.18 tsunami affected total 42 stations, while 23 stations were damaged by water outflow. A great number of wooden structures were completely washed away from the foundation. In some stations, only RC structure remained with no finishing materials. Figure 1.10 shows a photograph of a train station in Onagawa before and after the tsunami. The wooden station was completely washed away. Steel structures of the elevator shaft and public toilets have been significantly damaged, but not washed away.

About 19 stations were not washed away but they were damaged by rubble and mud flowing into them. Although structures of these stations were not heavily damaged, they are still not operational due to damages to the equipment.

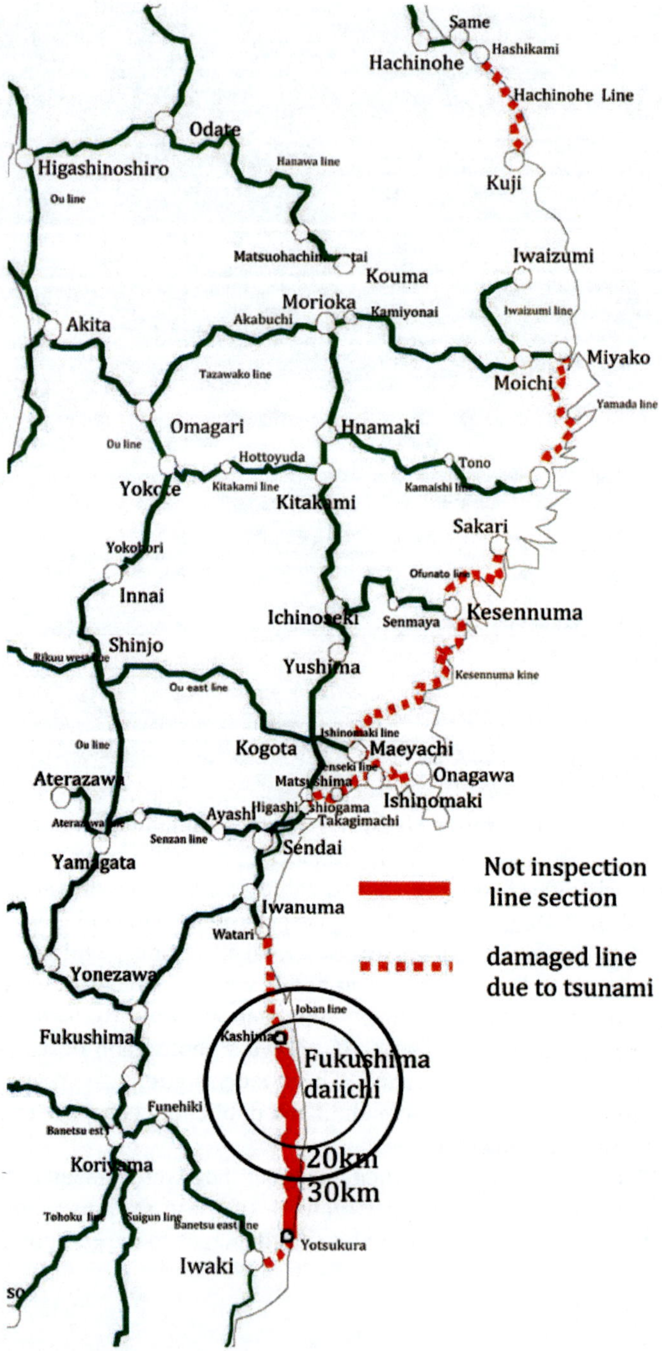

Fig. 1.9 Damage to railway lines due to tsunami

Table 1.18 Number of damaged station buildings due to tsunami

Name of train line	Section	No. of station	Tsunami no. of stations	Outflow no. of stations
Hachinohe	Kaijo–Kuji	12	2	0
Yamada	Miyako–Kamaishi	13	8	4
Ofunato	Sakari–Kesennuma	12	7	6
Kesennuma	Maeyachi–Kesennuma	21	11	9
Ishinomaki	Maeyachi–Onagawa	11	2	1
Senseki	Higashi Shiogama–Ishinomaki	16	7	0
Joban[a]	Watari–Iwaki	13	5	3
Total		98	42	23

[a]Within a 30 km radius from the Fukushima nuclear power plant (Tomioka station excluded)

Fig. 1.10 Onagawa station before and after tsunami

References

1. Fire Department Headquarters, June 30 2011
2. Police Emergency Headquarters, June 28 2011
3. Housing Bureau, Ministry of Land and Infrastructure, Transport and Tourism (MLIT), 27 June 2011
4. Aomori Prefecture Disaster Department, 20 June 2011
5. Iwate Prefecture Disaster Department, 1 July 2011
6. Fukushima Prefecture, "List of damages by Great East Japan Earthquake and Tsunami Disaster (May 31, 2011)"
7. Ibaraki Disaster Group Headquarters, 31 May 2011
8. Chiba Emergency Management Division, 1 May 2011
9. Eastern Conference Earthquake Reconstruction initiative, announced on 25 June 2011
10. Headquater of Disaster Management, Fire Department, 2011 Great East Japan Earthquake (No. 127), 2011

Chapter 2
Earthquake, Geology, and Tsunami

Hiroshi Kawase, Shinchi Matsushima, Baoyintu, Masato Motosaka, Susumu Ohno, Hidekazu Watanabe, Reiji Tanaka, Hitoshi Tanaka, Junji Ogawa, Ryota Yamaya, and Yukihiro Take

Abstract On March 11, 2011, a huge subduction zone earthquake occurred offshore of the Tohoku district, Japan, and created devastating damage, primarily due to the very high tsunami tide along the Pacific Coast. We first review several source studies and then summarize basic characteristics of observed strong motions with reference to the attenuation relationships, response spectra, linear site factors in the region, and nonlinearity. Fortunately we haven't seen heavy damage concentration even in the areas with high Peak Ground Accelerations, because Peak Ground Velocities were not high enough to make buildings heavily damaged or collapsed. The locally

---

H. Kawase (✉) • S. Matsushima • Baoyintu
Kyoto University, Kyoto, Japan
e-mail: kawase@zeisei.dpri.kyoto-u.ac.jp; matsushima@zeisei.dpri.kyoto-u.ac.jp; baoyintu@zeisei.dpri.kyoto-u.ac.jp

M. Motosaka • S. Ohno • H. Tanaka
Tohoku University, Sendai, Japan
e-mail: motosaka@struct.archi.tohoku.ac.jp; ohnos@saigai.str.archi.tohoku.ac.jp; tanaka@tsunami2.civil.tohoku.ac.jp

H. Watanabe
Hiroshima University, Hiroshima, Japan
e-mail: hidekazu-watanabe@hiroshima-u.ac.jp

R. Tanaka • R. Yamaya
Tohoku Institute of Technology, Sendai, Japan
e-mail: rtanaka@tohtech.ac.jp; ryamaya1215@yahoo.co.jp

J. Ogawa
Akita Prefectural University, Akita, Japan
e-mail: ogawa_junji@ebony.plala.or.jp

Y. Take
Sendai Technical High School, Sendai, Japan
e-mail: yyyytake@nexyzbb.ne.jp

Architectural Institute of Japan (ed.), *Preliminary Reconnaissance Report of the 2011 Tohoku-Chiho Taiheiyo-Oki Earthquake*, Geotechnical, Geological and Earthquake Engineering 23, DOI 10.1007/978-4-431-54097-7_2,

deployed network data as well as observed building responses are briefly summarized. Topographic maps and geological maps of Tohoku district and Kanto district will be shown. Finally, statistics of inundation survey, water level time histories, and examples of damage for tsunami in Tohoku district will be presented.

Keywords Building response • Damage potential • Inundation • Source process • Taro-Cho

2.1 Earthquake and Ground Motions

2.1.1 General Information on the Main Shock

A M9.0 mega-thrust earthquake occurred off the coast of Miyagi Prefecture (38°6.2′N, 142°51.6′E) at 14:46:18.1 on March 11, 2011, with the hypocenter being 24 km deep based on the information released by the Japan Meteorological Agency (JMA). JMA named this earthquake "the 2011 Off the Pacific Coast of Tohoku (Tohoku-Chiho Taiheiyo-Oki) earthquake". Since this earthquake was a low-angle reverse fault and aftershocks have been occurring along the upper surface of the subducting Pacific Plate, it was clear that the earthquake occurred on the boundary between the Pacific Plate and the North American Plate on which the Tohoku region of Japan rests. What was surprising at was the size of the earthquake, which created unprecedented heights of Tsunamis all along the coast of Tohoku region. This is a brief report on this earthquake and its observed strong motions, which seem to have smaller potential to structural damage despite of their quite high values of PGAs. Pervasively observed soil nonlinearity will also be briefly covered.

As mentioned above this earthquake was a low-angle reverse fault as shown by the analysis of teleseismic data in Fig. 2.1 [1]. Figure 2.2 shows the distribution of aftershocks that occurred on March 11 [2]. It can be seen that the aftershocks occurred widely in the Pacific Ocean between off-Sanriku and off-Ibaraki Prefecture. The area of these immediate aftershocks suggest about 200 km by 500 km of the

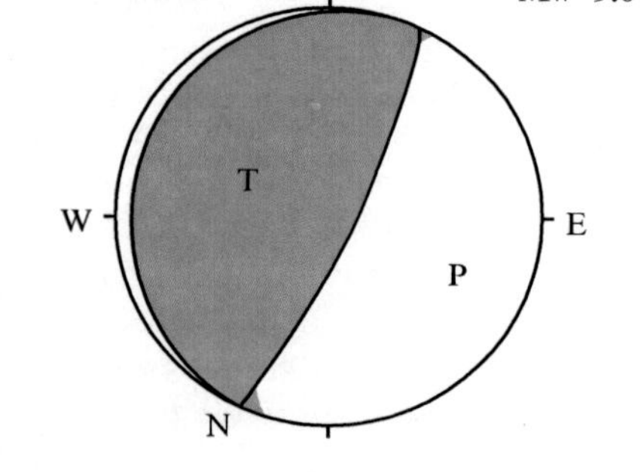

Fig. 2.1 Mechanism solution of the Off the Pacific Coast of Tohoku earthquake obtained by JMA [1]

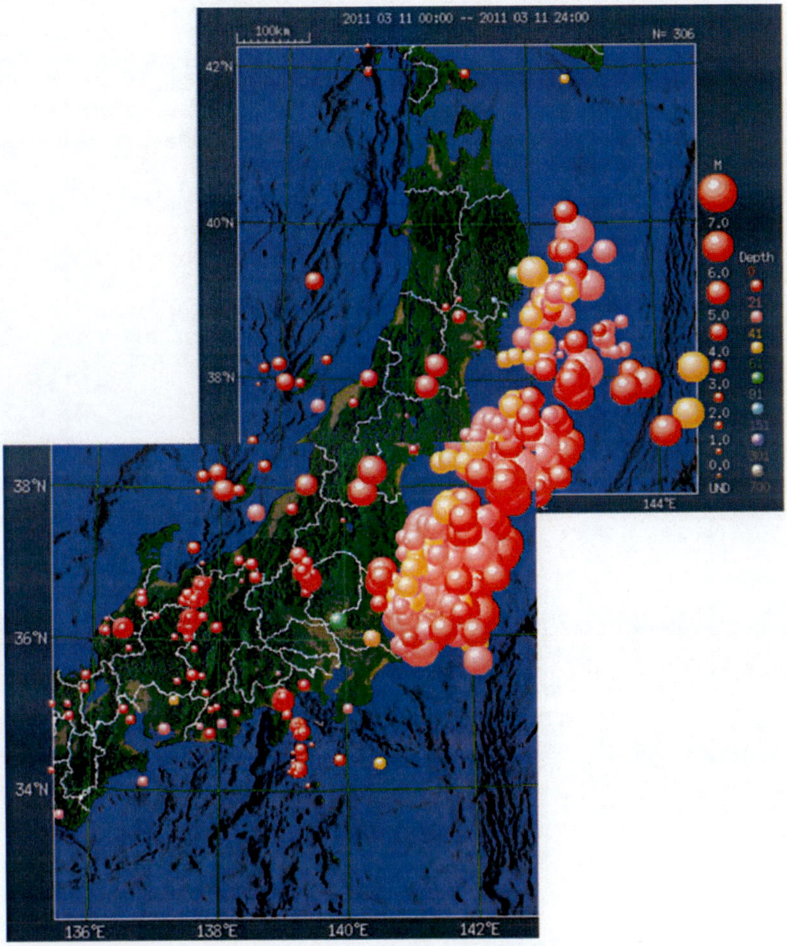

Fig. 2.2 Aftershocks and associated earthquakes on March 11 after the Off the Pacific Coast of Tohoku earthquake [2]

rupture area, which is about the size of M9 class earthquakes as the scaling low suggests.

Prior to this earthquake, an Mj7.2 earthquake occurred on August 16, 2005, off Miyagi Prefecture, west of the epicenter of the main shock. At first it was considered that this might be the recurrence of the Mj7.4 off Miyagi Prefecture (Miyagi-ken Oki) earthquake that occurred on June 12, 1978, causing significant damage in Sendai City. Subsequent studies, however, showed that only about half of the fault area ruptured during the 1978 event was ruptured at that time and the rest remained without slip. As a result of this study, the Headquarters for Earthquake Research Promotion estimated the possibility of an earthquake occurring off Miyagi Prefecture in the next 30 years at 99%.

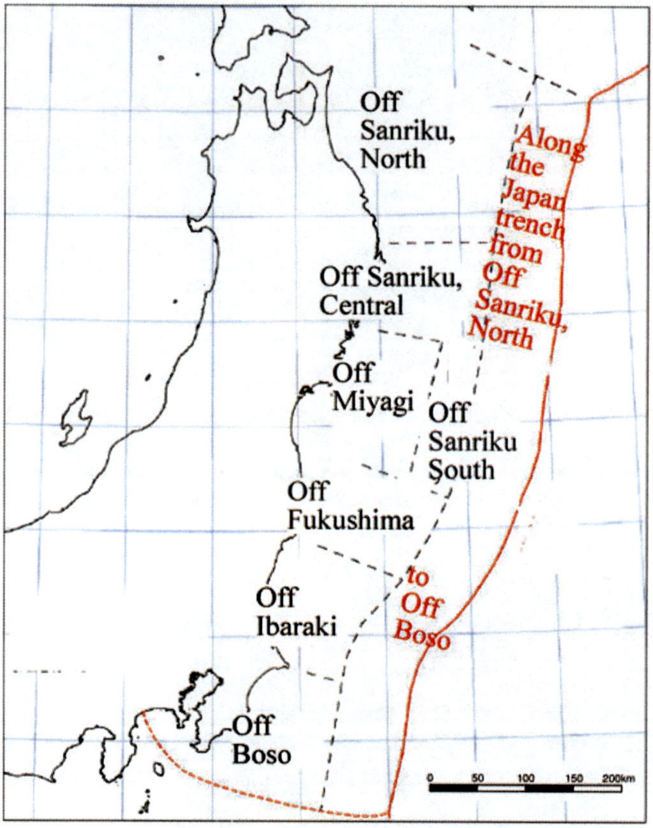

Fig. 2.3 Assumed segments of earthquake occurrence along the Pacific Coast of the Tohoku region [3]

Figure 2.3 shows the map of the predicted seismogenic zones (hypothesized segments) prepared prior to the occurrence of the main shock [3]. The segments shown in Fig. 2.3 assume their uncoupled behavior based on the historical records of earthquake occurrence, and this earthquake therefore raises the fundamental question whether such segmentation based on the past earthquake history over a relatively short period of time, along with the estimation of the possibility of earthquake occurrence for each segment, actually has any meaning at all or not.

In addition to the 2005 Miyagi-ken Oki earthquake, on March 9, 2011, shortly before the main shock, another M7.3 earthquake occurred off Southern Sanriku area (38°19.7′N, 143°16.7′E), at the same latitude but closer to the trench axis. The hypocenter of the main shock was located between the hypocenters of these two events, which suggests that the rupture of these two areas triggered the main shock. Figure 2.4 shows the distribution of the aftershocks from March 9th earthquake [4]. When JMA analyzed the b-value of these aftershocks after the main shock, JMA found that it was only 0.4, much smaller to the ordinary value of 0.9–1.0, suggesting that this was a foreshock.

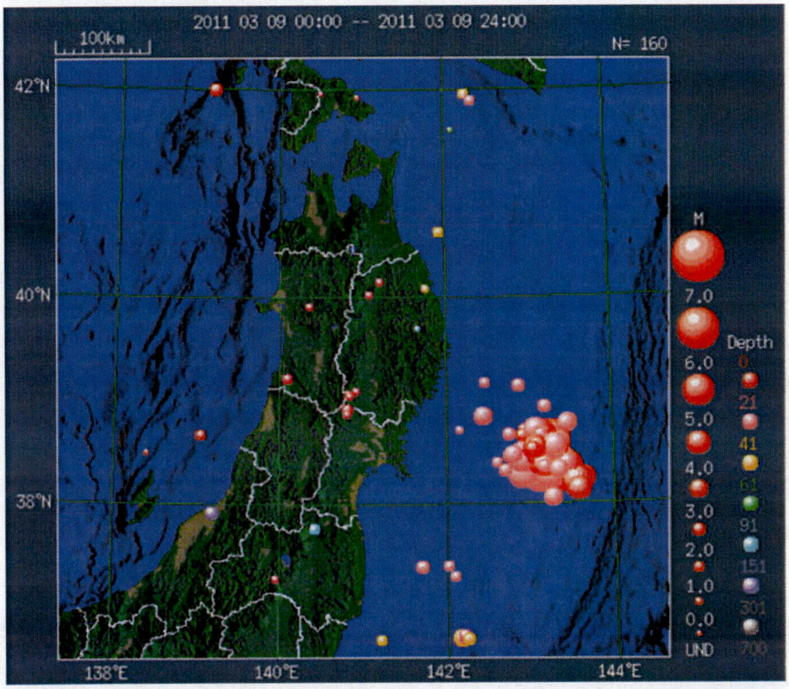

Fig. 2.4 Aftershocks of the Sanriku-Oki earthquake of March 9, 2011 during March 9, 2011 [4]

2.1.2 Source Process of the Main Shock

Several research institutes determined the source process, source region, and slip distribution using teleseismic data, tectonic deformation, tsunami height, strong-motion waveforms, and so on. Limited space does not allow us to present all of the results, and so we only present here typical inversion results for different constrained conditions.

First, Fig. 2.5 shows the crustal movement vectors (top: horizontal, bottom: vertical) determined by the Geospatial Information Authority of Japan associated with the main shock by using GPS data [5]. It can be seen that horizontal displacement of up to 5.3 m occurred at the tip of the Oshika Peninsula, and that the direction of the displacement was east–southeast, which is the same as the direction of the CMT slip shown in Fig. 2.1. Displacement in the vertical direction, without exception, shows subduction on land. As a co-seismic motion this is not consistent with the fact that before the earthquake these areas are also subducting continuously due to constant movement of the Pacific Plate.

The slip distribution was obtained first by determining the distribution of static displacement for the crustal movement data shown in Fig. 2.5. The area with high slip was found to be located near the hypocenter off Miyagi Prefecture, the maximum slip was estimated to be 24 m, and the width and lengths of the areas with

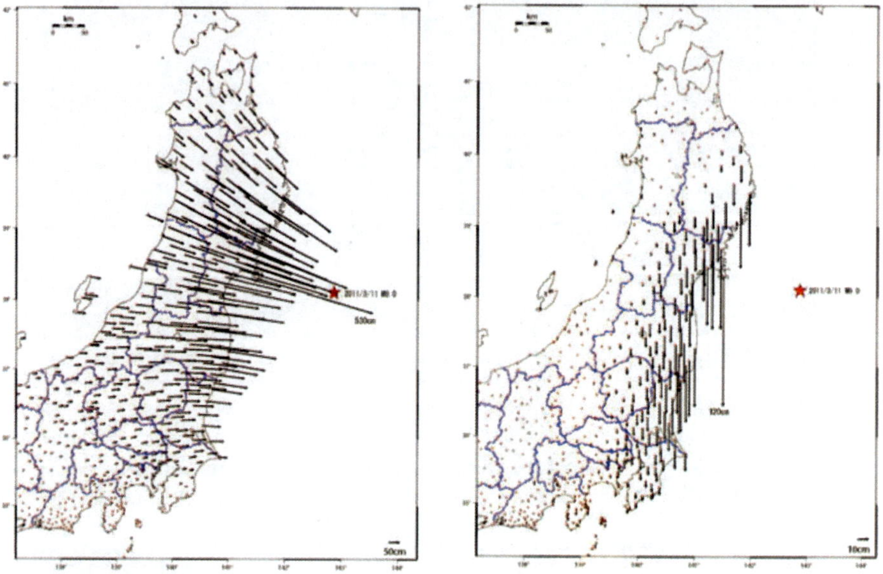

Fig. 2.5 Crustal Movement associated with the main shock (*left*: *horizontal*; *right*: *vertical*) by the GPS network of GSI [5]

great rupture were 200 and 400 km, respectively, reaching off Fukushima Prefecture in the south, but not any farther south than that, and reaching off Central Sanriku, but not any farther north than that.

Then, the Geospatial Information Authority of Japan (GSI) performed another inversion of slip distribution, this time including the ocean crustal movement data from the Japan Coast Guard [6]. The results are shown in Fig. 2.6. It was estimated that the maximum slip near the trench was 60 m, reflecting the ocean crust movement near the hypocenter being 24 m horizontally and 3 m vertically (upward). Based on this estimation, another area with a slip of around 10 m is located between off Southern Fukushima Prefecture and off Ibaraki Prefecture.

Figure 2.7 shows the distribution of slip for each fault element, determined based on tsunami waveforms by Fujii and his colleagues at the Building Research Institute [7]. Since the time constant of tsunami is in the order of tens of minutes, much longer than time scale of seismic waves, it is considered that this figure shows the final slip on the fault plane at the completion of the earthquake, which is the same as static crustal movement determined by GPS. The figure shows that slip became larger toward the east of the hypocenter, the area with a slip of 15 m or more was 150 km in width and 200 km in length, and that the maximum slip was over 26 m. The main rupture area was, as expected, southern off Central Sanriku and northern off Fukushima Prefecture, and does not extend to off Ibaraki Prefecture.

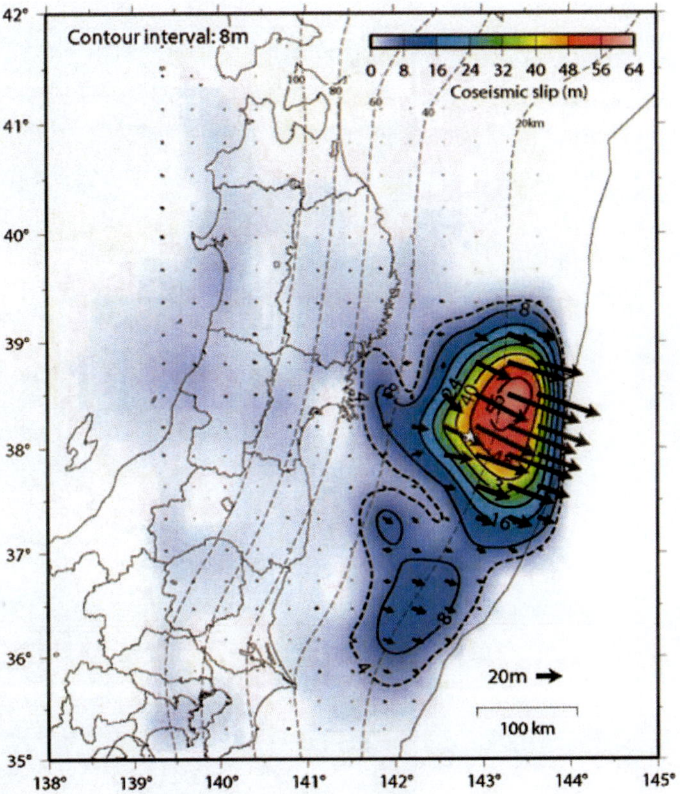

Fig. 2.6 Static slip inverted from crustal deformation from GPS and MSA ocean bottom measurement by GSI [6]

Figure 2.8 shows the distribution of slip inverted by the US Geological Survey [8] based on the teleseismic waveforms recorded by long-period (or broadband) seismographs located around the world. It is considered that this figure reflects the fault motion with the time constant from 20 s to several hundreds of seconds. This inversion from the teleseismic data also confirms the slip up to 30 m east of the hypocenter, closer to the trench, which is consistent with the results of the inversion using geodetic motions or tsunami waveforms (Figs. 2.6 and 2.7). The region with a slip of 10 m or more has the area of approximately 150 × 250 km, indicating again that a large slip occurred in a relatively compact region.

The inversion of source process using observed teleseismic waveforms was conducted by not only the USGS but researchers around the world, including Japanese researchers, and with minor differences in details, a consistent result was seen of a large slip of 25–30 m occurring over 200 km to the east of the hypocenter, near the Japan Trench. Please refer to the website of the Earthquake Research Institute, The University of Tokyo, for such a comparison of several results [9].

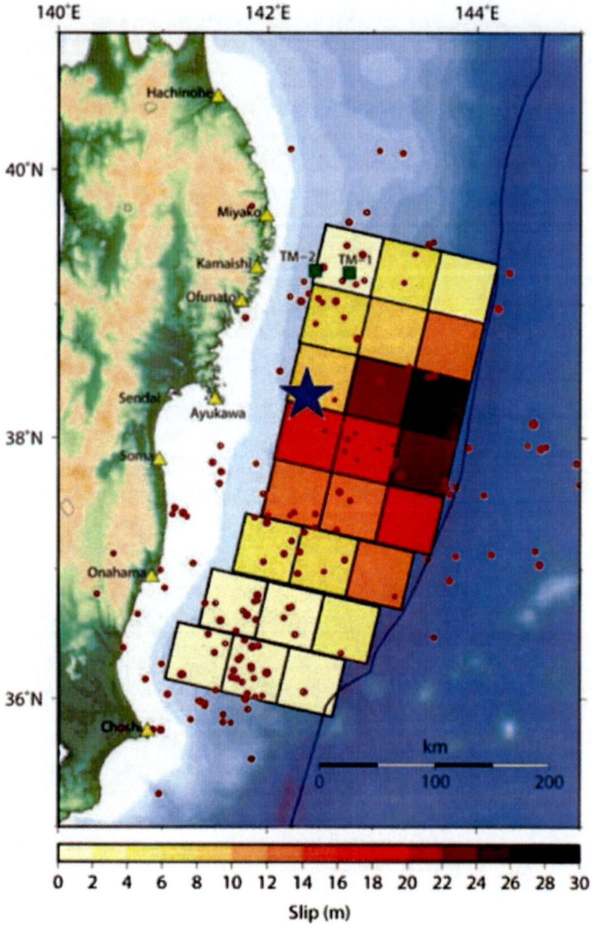

Fig. 2.7 Slip amount inverted from tsunami waveforms by BRI [7]

Figure 2.9 shows the results of the inversion, performed by Suzuki and his colleagues at the National Research Institute for Earth Science and Disaster Prevention (NIED), from long-period (8 s or longer) strong ground motions observed by K-NET and KiK-net [10]. The figure shows that large slips occurred near and to the east of the rupture initiation point (star), and between off Central Sanriku and off Miyagi Prefecture, and that the maximum slip was 25 m, consistent with the results of the other inversions shown here. However, in these results, the region of large slip (10 m or larger) extends along the Japan Trench to off Ibaraki Prefecture. Considering that the model of static displacement including ocean crust movement produced a slip of approximately 8 m off Ibaraki Prefecture as shown in Fig. 2.6, it might need to be considered that rupture did indeed reach that far.

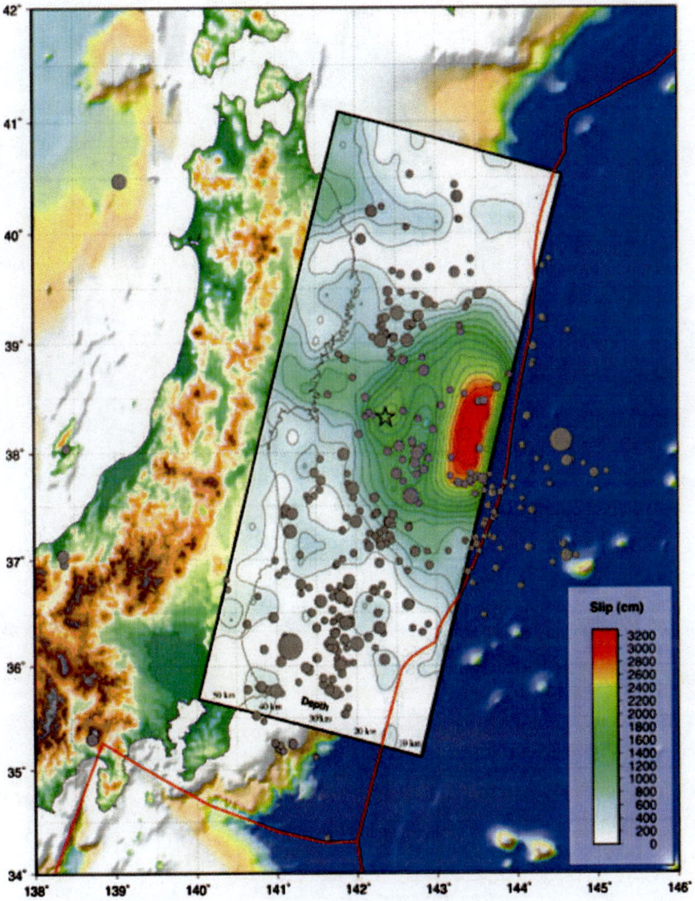

Fig. 2.8 Slip distribution inverted from teleseismic waveforms by USGS [8]

2.1.3 Observed Strong Motion Records

This earthquake generated a large number of strong motion records with high acceleration at many observation points, mainly in Miyagi and Ibaraki Prefectures. The distribution of seismic intensities observed or collected by JMA, which were broadcasted immediately after the main shock, shows that intensity 7 was recorded only at K-NET Tsukidate (MYG004) in Kurihara City 50 km north of Sendai, but that intensity 6 upper were recorded at 40 points in four prefectures.

Figure 2.10 shows the distribution of the peak ground acceleration (PGA) and velocity (PGV) of the strong motion records observed by K-NET and KiK-net of NIED [10], along with the empirical attenuation relation of Si and Midorikawa [11] with variations (red lines). In the attenuation plot, the horizontal axis indicates the shortest distance to the rectangular fault plane shown in Fig. 2.9. The magnitude of

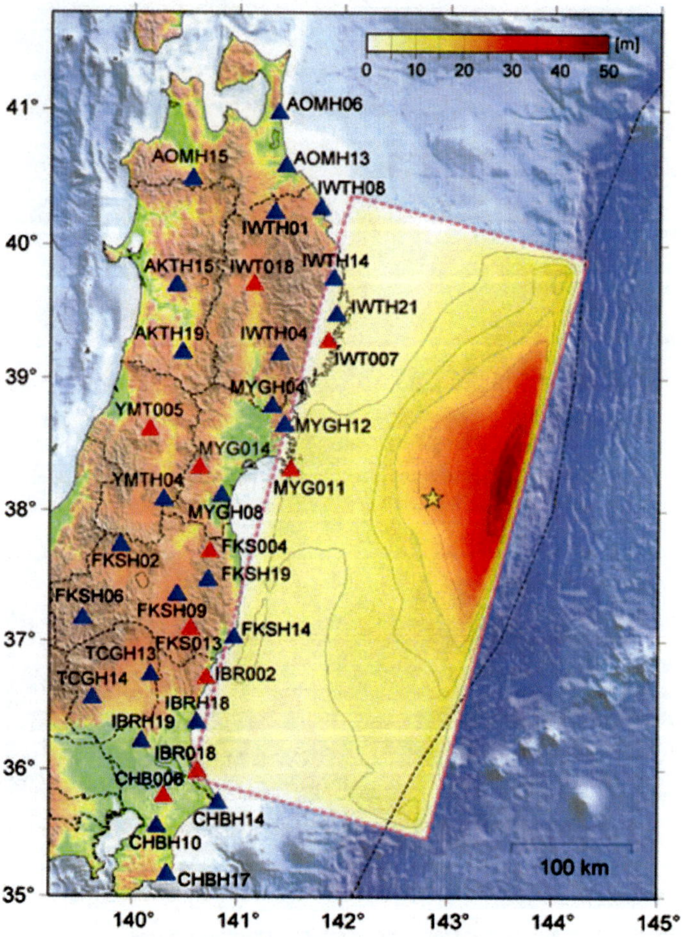

Fig. 2.9 Slip distribution inverted from long-period strong motions from K-NET and KiK-net of NIED [10]

the event M9.0 used to calculate this attenuation was apparently extrapolation for the formula since such a large event was not included in the regression data. It can be seen PGAs exceeded 500 Gal (cm/s^2) extensively along the coast from the Central Sanriku to Ibaraki Prefecture; however, PGVs in the area were lower than 80 cm/s (except for MYG004 where PGV is around 100 cm/s), and in most areas lower than 40 cm/s.

When compared with the attenuation formula, it seems that this earthquake produced strong motions of the average or slightly weaker level, except at points within 100 km that had higher PGAs. As previously mentioned, this earthquake was M9.0 and was one of the largest in scale in Japan's recorded history; however, the area with the largest slip was farthest from the coast, as shown in Figs. 2.6, 2.7, 2.8, and 2.9, and therefore even if that area with longer time constant of slip also produced short-period strong motions, these seismic waves might not have contributed

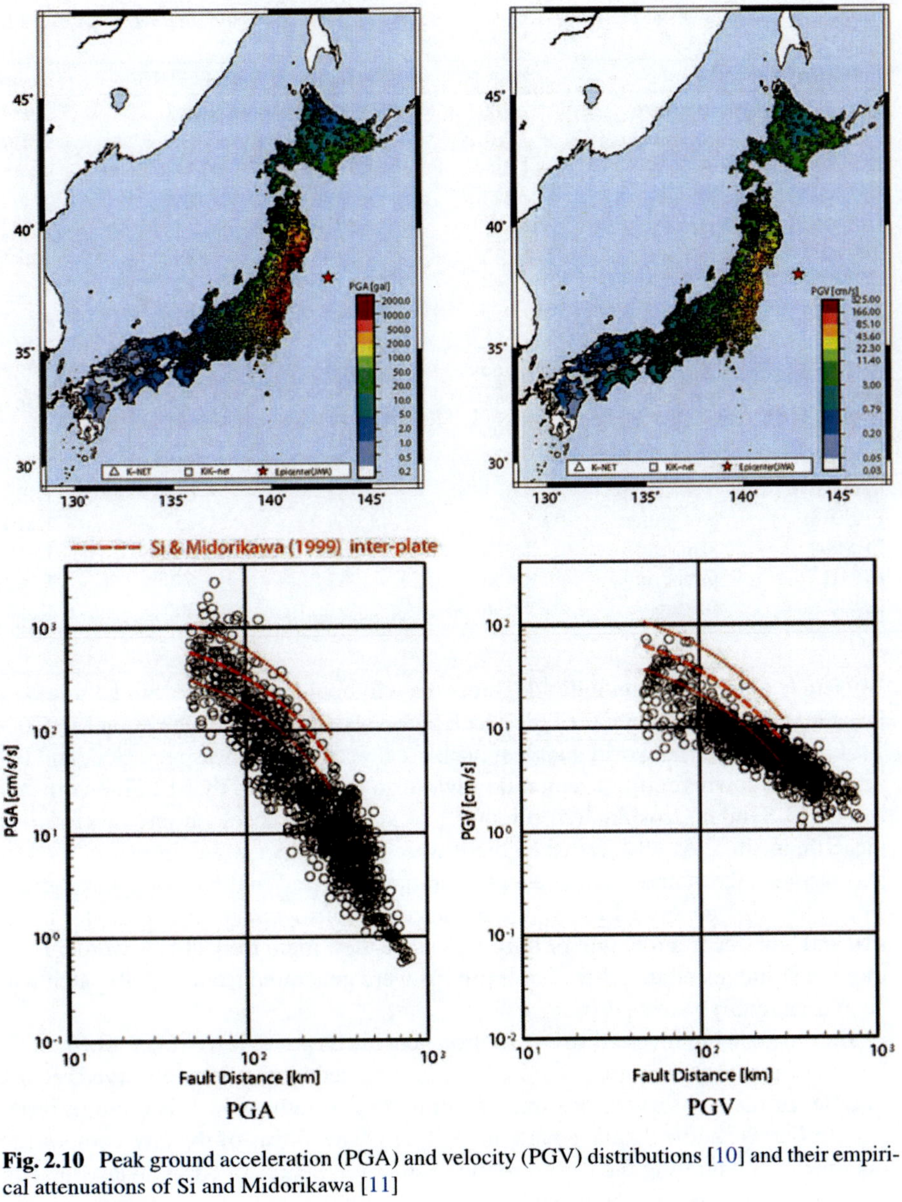

Fig. 2.10 Peak ground acceleration (PGA) and velocity (PGV) distributions [10] and their empirical attenuations of Si and Midorikawa [11]

significantly to the sites along the coast. The area that generated short-period seismic motions and their attenuation characteristics will be discussed later.

Table 2.1 shows the points where the synthetic maximum acceleration of strong motions observed by K-NET and KiK-net exceeded 1 g [10]. While many of the sites are located in Miyagi and Fukushima Prefectures, it is notable that Tochigi and Chiba Prefectures also have such points, even though they are distant from the main source area, off Miyagi Prefecture.

Table 2.1 PGA values at the K-NET and KiK-net stations where vector-summed PGA exceeded 1 g [10]

Site code	Name	PGA NS	PGA EW	PGA UD	Vector
MYG004	Tukidate	2,700	1,268	1,880	2,933
MYG012	Shiogama	758	1,969	501	2,019
IBR003	Hitachi	1,598	1,186	1,166	1,845
MYG013	Sendai	1,517	982	290	1,808
IBR013	Hokota	1,355	1,070	811	1,762
TCG009	Imaichi	1,017	1,186	493	1,444
FKS016	Shirakawa	1,295	949	441	1,425
FKSH10	Saigo	1,062	768	1,016	1,335
IBR004	Oomiya	1,283	1,007	775	1,312
TCGH16	Haga	799	1,197	808	1,305
TCG014	Mogi	711	1,205	494	1,291
IWT010	Ichinoseki	998	852	353	1,226
IBRH11	Iwase	815	827	815	1,224
MYGH10	Yamamoto	871	853	622	1,137
FKS018	Koriyama	745	1,069	457	1,110
FKS008	Funabiki	1,012	736	327	1,069
IBRH15	Omaeyama	606	781	640	1,062
CHB007	Sakura	1,036	491	200	1,054

Detailed discussion on individual records will not be made here, but let us examine at least three components of the acceleration waveform at three sites; at MYG004 Tsukidate, where the world's largest horizontal acceleration was recorded, and two nearby points where high acceleration was also recorded, MYG012 Shiogama and MYG013 Sendai Oroshimachi. Figure 2.11 shows the acceleration waveforms at these three sites. At all three sites there were two distinct wave groups in the first and second halves, and the highest acceleration occurred in the second wave group. The pulses that generated the maximum accelerations were not just a single shot or two and have a certain length of duration; since their main frequencies were 5 Hz or more, it is suggested that these wave groups were generated from a small patch with an extremely high stress drop.

Before showing inverted strong motion generation areas, it would be worth to see how the wave packets were generated and propagated by lining up observed records in order of their observed locations (from north to south), which is quite informative. In Fig. 2.12, the displacement and velocity waveforms of the EW components obtained by integrating the underground (borehole) records of KiK-net are lined up, considering the distance between the observation points to be the waveform interval and thereby synchronizing time. The figures show that seismic waves arrived earliest at MYGH08, which is located west of the epicenter, and it looks as if the seismic waves propagated from there to the north and south. Since the wave source was obviously near the hypocenter, not on land, they should not actually be straight lines, but they have been rendered as so in the figure for convenience.

Looking at the displacement waveform, there is a time gap of approximately 50 s before the arrival of the second peak from the first one. The peak level is three times

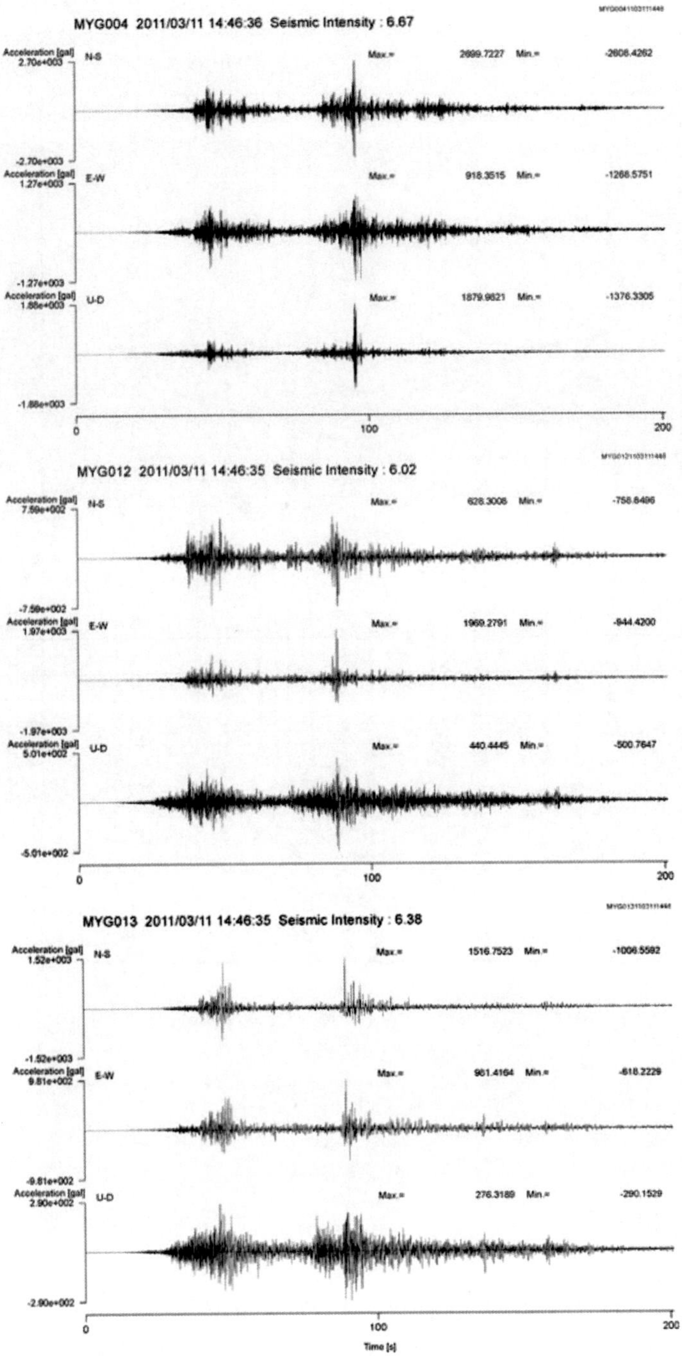

Fig. 2.11 Accelerograms at three K-NET sites in Miyagi Prefecture (NIED [10])

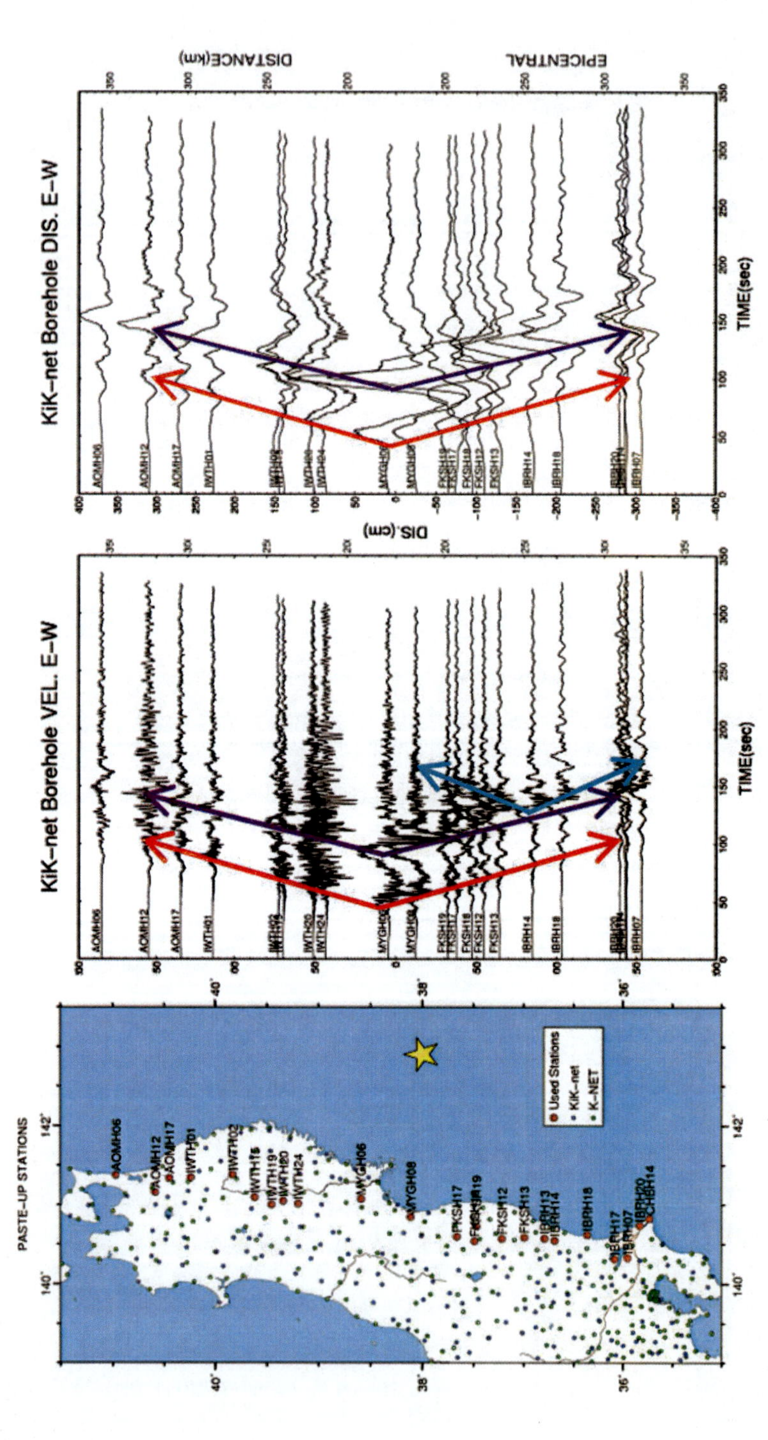

Fig. 2.12 Paste-up of the velocity and displacement seismograms calculated from the borehole accelerograms of selected KiK-net data shown in the *left panel*

as high as that of the first wave group. Unlike the velocity waveform, the displacement waveform does not show the wave group that appears to have been generated in Fukushima Prefecture.

Likewise, the velocity waveform shows distinctly the first and second wave groups in Miyagi, Fukushima and Iwate Prefectures, along with the third wave group propagating from the south of Fukushima Prefecture. In the next section it shall be seen that focusing on this propagating wave packets allowed us the area that had generated short-period strong motions to be identified by its inversion to reproduce strong motions, and that the subsequent image of rupture process is turned out to be greatly different from that determined by the long-period seismic motions, crustal movement, and tsunami.

Lastly for this section, and in relation to the fact mentioned above, the interpretation of the difference from the well-established attenuation formula shall be discussed. In the attenuation formula shown in Fig. 2.10 the shortest distance to the fault plane was used as a representative distance. However, if short-period motions were not uniformly generated on the fault plane, the distance would be underestimated (i.e., shorter than reality) for those points located closer to the areas of the fault plane where short-period motions were not actually generated. If the area that generated strong motions has been identified, the representative distance should be the shortest distance to that area. As the simplest test for understanding such effects, we made a comparison between the observed PGAs and the empirical formula, assuming the simple distance to the hypocenter by JMA as the representative distance. Figure 2.13 shows that with a sufficiently long (~200 km) hypocentral distance, PGAs are no longer underestimated and the results are generally more consistent in distant areas. This indicates that the main energy of short-period motions came from the area or the areas that generated strong motions, not the whole fault plane equivocally. However, PGAs in the south side were relatively higher than those in the north side, indicating the presence of another area that generated strong motions in the south side, as shown in the next section for strong motion inversions.

2.1.4 Strong Motion Generation Areas

Strong motion seismology, an academic field that advanced rapidly after the 1995 Hyogo-ken Nanbu (Kobe) earthquake, tells us that slip distribution on a fault plane is quite heterogeneous and that the area that accumulates a large amount of strain, the so-called asperity, only occupies a small part of a fault plane. Its proportion is almost always consistent regardless of the scale of the fault size; in other words, an asperity can be scaled from the scale of the fault. In inland earthquakes, in most cases so far, the asperity that generates long-period (~several seconds) motions and the place that generates short-period motions, the so-called strong motion generation area (SMGA), roughly coincide with each other, or the asperity contains the SMGA.

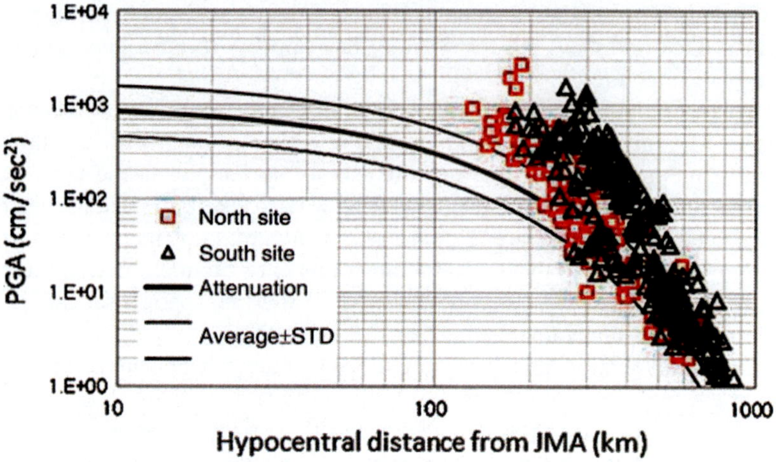

Fig. 2.13 Observed PGAs versus simple hypocentral distances from the JMA hypocenter and the attenuation formula of Si and Midorikawa [11] with standard deviation in *solid curves*

Following the 2003 Tokachi-Oki Earthquake, some researchers suggested that these two may not necessarily be the same in large-scale ocean-trench earthquakes, but this idea has not been widely accepted.

The latest earthquake was large in scale and the area that generated long-period motions located near the ocean trench, about 100 km wide and 200 km long, coincided with the area with a high final slip obtained from tsunami and crustal movement with a longer time constant, and differences of the gross pictures among researchers were small as shown above. However, if a significant amount of short-period motions were also generated near the ocean trench, it would be difficult to explain why such strong short-period motions were generated as seen in Fig. 2.11 by such a shallow part of the fault, from which short-period motions must propagate over a long distance.

Asano and Iwata [12] assumed that the first three wave groups in the observed waveform were emitted from different SMGAs and estimated the rupture initiation point for each SMGA using travel time analysis by identifying the first arrival of each wave group. As the result, the rupture initiation points (S1, S2, S3) of the SMGAs were determined to be at the locations indicated by stars in Fig. 2.14. The rupture starting times for S1–S3 were determined to be 25.0, 67.2, and 114.3 s, respectively.

They then performed waveform synthesis using empirical Green's function (0.1–10 Hz) to estimate the area, rise time, stress drop, rupture propagation velocity within SMGA and relative rupture initiation point within SMGA for SMGA1, 2 and 3 via grid search. Their targets were acceleration envelopes and displacement waveforms. The summation numbers of small earthquakes were determined by observed spectral ratio. EGF1, an aftershock that had occurred near the hypocenter, was used for SMGA1 and 2, and EGF2, an aftershock that had occurred near off Fukushima

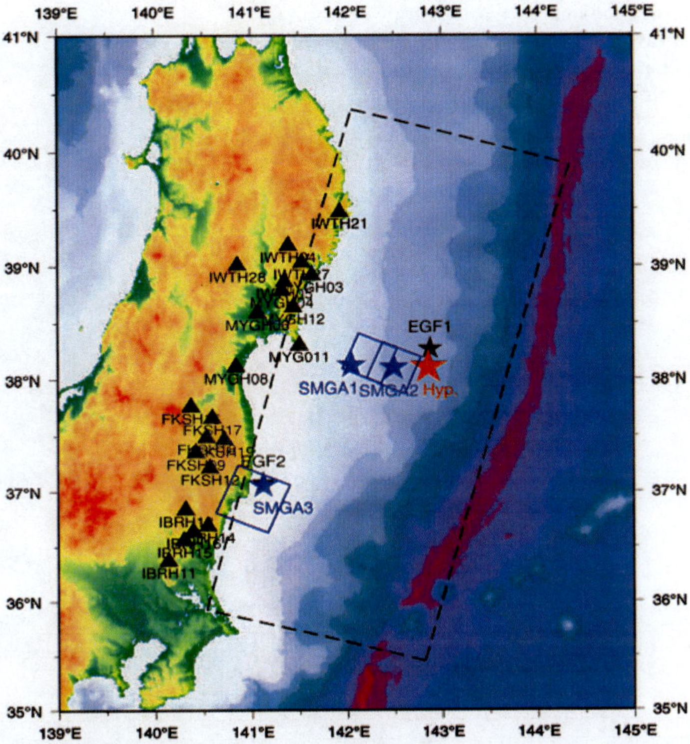

Fig. 2.14 Strong motion generation areas based on the inversion for acceleration envelop and low-cut displacement seismograms by using empirical Green function method [12]

Prefecture, was used for SMGA3. The characteristics of the observed waveform were explained by this model. Their model explains acceleration waveforms, including PGAs, very well.

Likewise, Kawabe et al. [13] identified SMGAs using empirical Green's function. They used the KiK-net (borehole) data for the Pacific coast to model SMGAs using forward modeling. The target frequency was 0.1–10 Hz, the same as that used by Asano and Iwata [12], and the fault plane was assumed to be a rectangular plane that runs through the hypocenter announced by the JMA (strike 195°, dip: 13°) based on the shape of the Pacific Plate. For the empirical Green's function, the data of the Mj6.3 earthquake that occurred off Ibaraki Prefecture at 20:44 on October 19, 2005, was used for Asp5 (SMGA) and the data of the Mj6.4 earthquake that occurred off Miyagi Prefecture at 3:16 on March 10, 2011 (an aftershock of the March 9 event) was used for all others.

First, they estimated the rupture process based on the propagation and arrival time of wave groups and other factors. They identified five wave groups and determined a SMGA for each. The results are shown in Fig. 2.15. Asp1 and Asp2 corresponded to SMGA1 and SMGA2 determined by Asano and Iwata [12] and Asp5

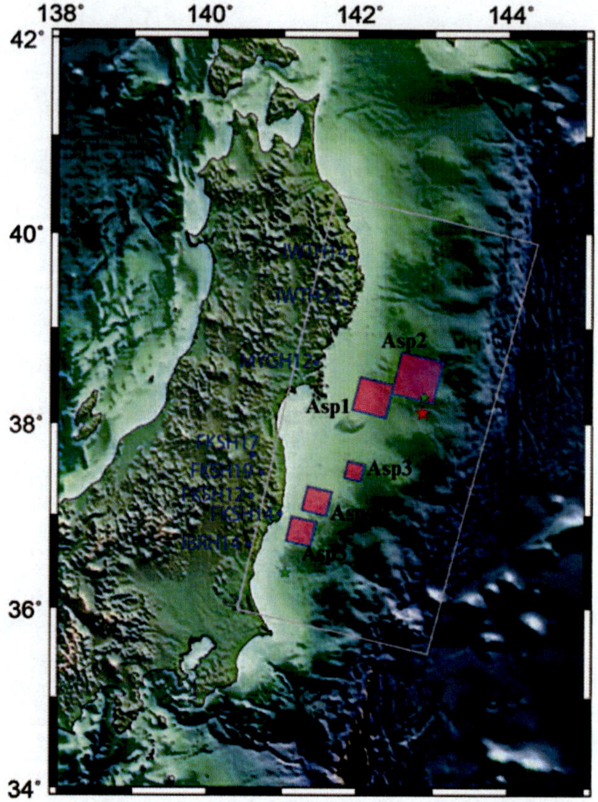

Fig. 2.15 Strong motion generation areas obtained by Kawabe et al. [13] by using the empirical Green function method

corresponded to SMGA3. For Asp3 and Asp4, while it depends on how you look at the wave groups between them, at least there seems to be a clear necessity for the existence of Asp4 from looking at the paste-up of waveforms. Note that Kawabe et al. [13] called them as asperities, but actually they obtained SMGAs as long as we are calling the area with long-period, large slip near the Japan Trench as the asperity of the main shock.

Kurahashi and Irikura [14] also found five SMGAs, although at slightly different locations. Figure 2.16 shows their final source model that was derived from the observed strong motion accelerations and velocities of 14 KiK-net stations shown by red triangles. In this model they put three SMGAs around the hypocenter of the main shock. SMGA4 is similar to the SMGA3 of Asano and Iwata [12].

We should note that the total seismic moments of these strong motion inversions are quite small; Asano and Iwata [12] give Mw = 8.0, Kawabe et al. [13] Mw = 8.3, and Kurahashi and Irikura [14] Mw = 8.5. This means these SMGA contributed only 3–18% of the total seismic moment of this Mw9.0 event.

These specific concentrations of the short-period (less than a couple of seconds) strong motions in the regions close to the shore line are confirmed by the back projection analysis of teleseismic measurement by dense arrays in the US and Europe.

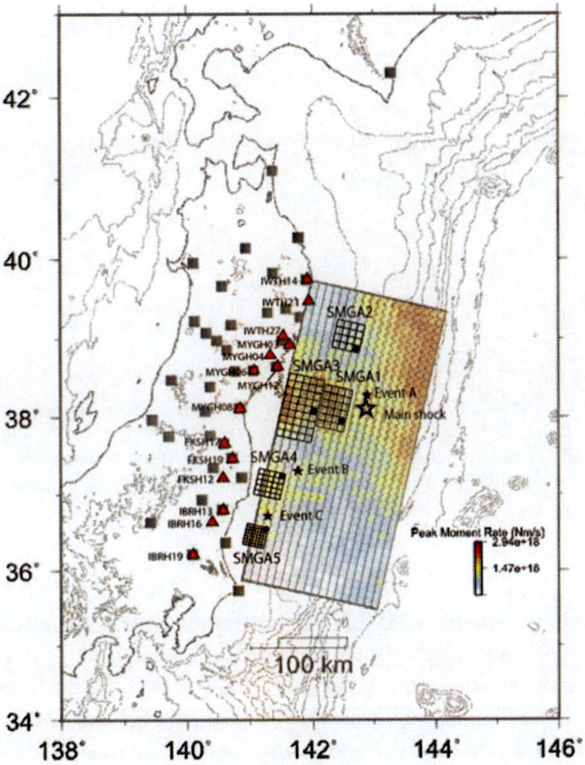

Fig. 2.16 Strong motion generation areas obtained by Kurahashi and Irikura [14] by using the empirical Green function method. Background contour is the distribution of peak moment-rate inverted using long-period strong motion data

For example, Wang and Mori [15] delineated the movement of the energy release center as the peak point of the back projected coherent amplitude by using the US array. They carried out the back-projection analyses using data low-cut filtered at 1 s, band-passed filtered between 1 and 5 s, and high-cut filtered at 5 s to find that short period motions are mainly generated from the deeper, western part of the fault plane, while long-period motions are from the middle of the fault plane.

2.1.5 *Site Effects in These Areas*

Site effect studies are intensively performed in the Tohoku region because of the subsequent moderate-size damaging earthquakes in the region after the advent of the K-NET and KiK-net as well as the JMA Shindokei network after the 1995 Kobe event. For example, back to 1998 Satoh et al. [16] found K-NET site effect in the Tohoku region which can vary strongly from site to site. They found that at MYG005 (Onikobe) there is quite a large (~50) site amplification at around 0.7 Hz and at MYG013 (Sendai, Oroshimachi) a moderate (~10) amplification at the same frequency.

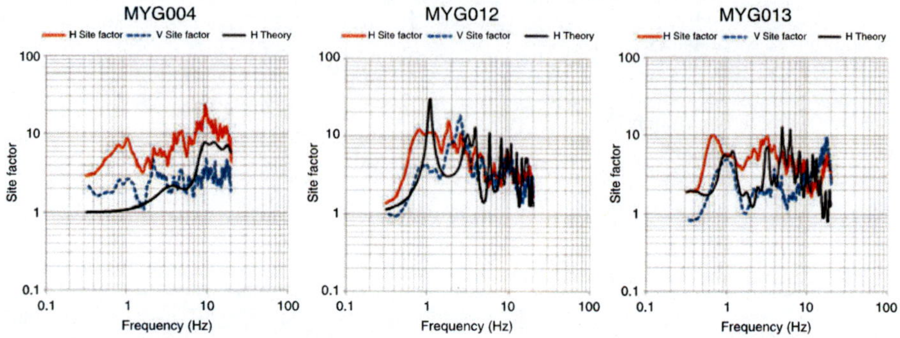

Fig. 2.17 Site amplification factors determined by the generalized spectral inversion by Kawase and Matsuo [17] and Kawase [18] for three sites in Miyagi Prefecture for *horizontal components* (two components' RMS value) with *red lines* and *vertical component* with *blue dotted lines*. One-dimensional theoretical amplification characteristics for S-wave velocity structures taken from the PS logging for top 10 or 20 m and inverted by using genetic algorithm down to the bedrock are shown with *black lines*

Later Kawase and Matsuo [17] and Kawase [18] used generalized spectral inversion by using YMGH01 as a reference to obtain site amplification factors from the seismological bedrock (outcrop). Although amplifications at all the sites were not shown in these references, they also found quite a large amplification at several sites in the Tohoku region, especially in Miyagi and Iwate Prefectures. In Fig. 2.17 horizontal and vertical site amplification factors at MYG004, MYG012 (Shiogama), and MYG013 relative to the seismological bedrock with the S wave velocity of 3.45 km/s. Note that these site amplifications were determined by small to moderate ground motions, whose PGAs are all less than 200 Gal (cm/s^2), observed from 1996 to 2002 or 2004. At MYG004 we can see a large amplification (~20) at around 10 Hz in the horizontal direction. As shown in the next section this peak were moved to lower frequency at around 5 Hz because of soil nonlinearity. There is a peak at around 1 Hz at MYG004, which does not have any corresponding peak in the theoretical prediction. We need to explore what kind of geological structure produces such a low-frequency peak. As for the sites MYG012 and MYG013 they also have peaks around 1 Hz or lower, however, they have corresponding peaks in the theoretical prediction, because of thick (0.5–1.5 km) sedimentary rock formation in Pleistocene and Pliocene as derived by Satoh et al. [19].

Here we would like to introduce specific information about the site conditions around MYG004 where the highest HPGA of 2.7 g was observed. Figure 2.18 shows surrounding environment near the site. As we can see the station is close to the sharp slope of a small hill with a height of about 3 m. Thus there should be apparent 2D topographic effect for the site. As for the bedrock structure based on the PS logging data by NIED, it has only 4 m of sediments on top of the rock formation. That rock formation is weathered so that top 6.25 m has relatively low S-wave velocity (550 m/s). Apparently the top three layers of sediments should be the source of 10 Hz peak seen in Fig. 2.17.

Fig. 2.18 A small hill on which MYG004 K-NET Tsukidate site is situated. The direction of photo is from north northwest to south southeast. The flat part in front of the site is the parking lot of the Kurihara Cultural Hall (Bunka-Kaikan)

To delineate the effect of the small hill where the station rests, we measured microtremors along the line perpendicular to the direction of the hill axis, which is in the north–south direction. Figure 2.19 shows the spectral ratios of EW component (i.e., the direction of the hill axis) microtremors relative to the EW component of microtremor at the point 30 m away from the foot of the hill. Z02 was placed in front of MYG004. As we can see from 10 to 20 Hz, we have large amplification at Z01 and Z02, both of which are on the hill.

2.1.6 Soil Nonlinearity

Because of the strong ground motions soil nonlinearity inevitably has emerged in wide areas with high PGAs and PGVs and soft sediments. Notable damage were caused by the severe liquefaction, especially around the Tokyo Bay area where we have been making artificial land for quite a long time as deep as 50 m. Apparently the unprecedentedly long duration of strong motions in the area contribute the severe and pervasive liquefaction there. Unfortunately the author is not a specialist of liquefaction and so the report of observations and subsequent analyses are provided by other experts in a different chapter. The authors only refer to the web pages of the Ministry of Land, Infrastructure, Transport related to liquefaction here [20].

We would like to show here some evidence of soil nonlinearity based on the horizontal-to-vertical (H/V) spectral ratios of strong motions during the Tohoku

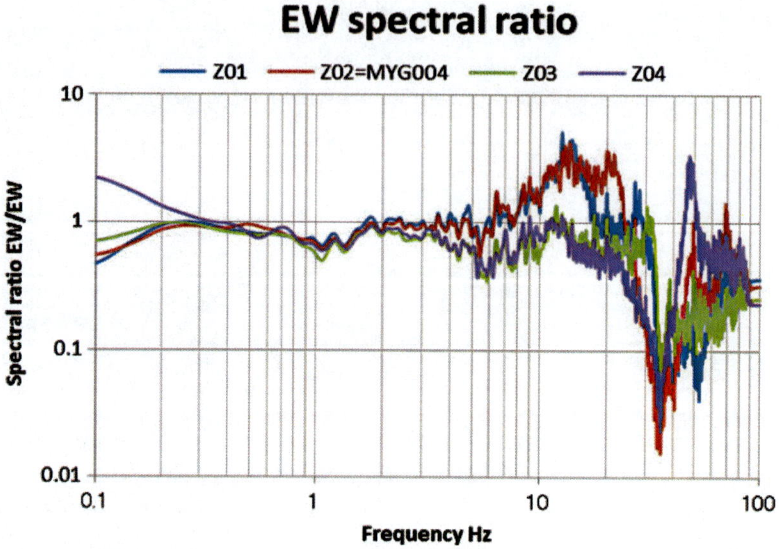

Fig. 2.19 Spectral ratios of microtremors at four sensors deployed near the MYG004 K-NET Tsukidate. The direction of measured line is along the north–south direction perpendicular to the hill axis. Z01 and Z02 are on the hill, while Z03 and Z04 are on the flat part. The reference sensor is placed on the flat part 30 m away from the hill

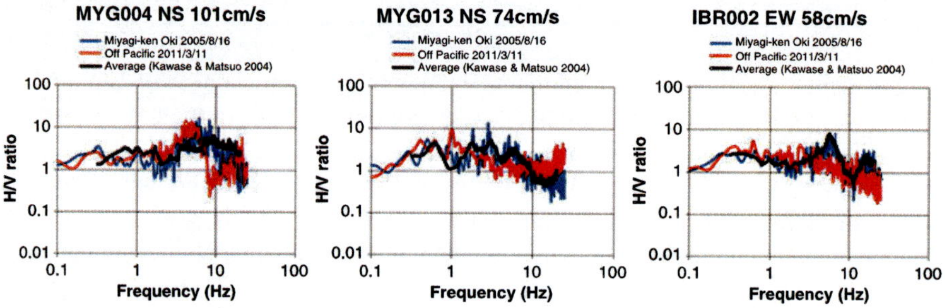

Fig. 2.20 Horizontal-to-Vertical (H/V) spectral ratios for observed strong motions during the main shock (*red lines*) and the 2005 Miyagi-ken Oki earthquake (*blue lines*), together with the H/V ratios of the site factors determined by the generalized spectral inversion by Kawase and Matsuo [17] for tens of weak to moderate ground motions (*black lines*)

event. Figure 2.20 shows H/V ratios of the observed strong motions during the main shock and the 2005 Miyagi-ken Oki earthquake at MYG004 (NS component), MYG013 (NS component), and IBR002 (EW component), together with those calculated from the horizontal versus vertical site factors separated by Kawase and Matsuo [17]. The number in the top of each panel is the PGVs at these components during the main shock. As we can see at these sites, the peak shifts of H/V ratios are quite apparent and the degrees of the shifts seem to be depending on their PGV values. If PGV would be less than 30 cm/s, there seems no strong peak frequency shift in the H/V ratios. Since the main shock durations were as long as or more than 300 s, we may also see non-stationary characteristics of H/V ratios. Since earthquake H/V ratios can be theoretically predicted by using the ratio of 1-D transfer function of vertically incident S-wave with respect to that of vertically incident P-wave, as long as the wave field is well diffused and consists mainly of body waves [21], we could delineate the reduction levels of S wave velocity directly from these H/V ratios during the main shock.

2.1.7 Long Period Motions

As expected long period motions were generated and propagating along the Japan's main island (Honshu), especially to the west because of the rupture directivity. However, based on the spectral shape of the observed strong motions in the Kanto Basin, there does not seem so much excitation of the basin-induced or basin-transduced surface waves in the long period range from 5 to 10 s as we have been observing in the previous earthquakes around the Kanto region. The durations of observed motions are quite long as expected but it is mainly a direct consequence of the large ruptured area and slow rupture propagation; that is, the source effect. These ground motions acted as input to high-rise buildings in Tokyo and Yokohama metropolitan areas and they produced quite large deformations, some of which can be seen as motion pictures on internet.

As for the long period motions we must mention about the conspicuously large ground motions at the ground floor and subsequent non-structural damage at one building in Osaka, Sakishima high-rise building now used as the Osaka Prefecture Government Office. Figure 2.21 show velocity seismograms and their Fourier spectra at the basement of the building on the reclaimed land in the Osaka Bay (by courtesy of Prof. Asano of DPRI, Kyoto University) observed and released by Building Research Institute [22]. It shows quite large amplitude at around 0.15 Hz. It is well known that in the coastal area of the Osaka Bay we can see large amplitude at around 4–6 s so that it is a serious question to engineers why they did not try hard to avoid resonance of the structure whose design period is said to be more than 5 s.

2.1.8 Damage Potential of Observed Strong Motions

Structural damages caused by the strong ground motions were rather small, considering that PGAs of the strong motions generated in this earthquake was very high. For example, around the K-NET Tsukidate station (MYG004), where the highest acceleration of 2.7 g was recorded, noticeable structural damages to buildings were difficult to confirm, except for the gymnasium of a nearby elementary school that had a minor damage in the non-structural elements. In this section, using the model developed by Nagato and Kawase [23] for estimating heavy damages to mid- to

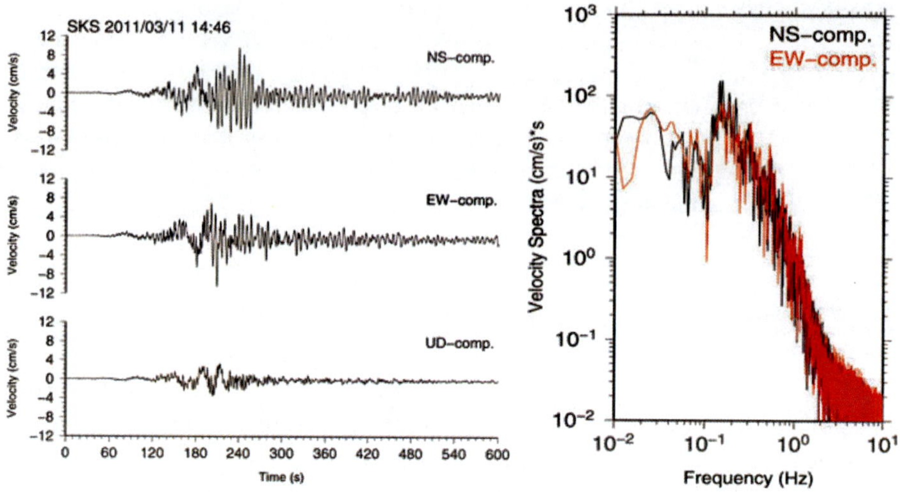

Fig. 2.21 Three component velocity waveforms and their Fourier spectra of two horizontal components observed at the basement floor of the Sakishima Osaka Prefecture Government Office. Building Research Institute installed the seismometers at the top, in the middle and at the basement [22]

low-rise reinforced concrete buildings, low-rise steel-frame buildings, and wooden houses to reproduce the damage ratios caused by the 1995 Hyogo-ken Nanbu earthquake, we shall estimate the structural damage potential of the observed strong motions and investigate the reason why the damages were relatively minor.

Figure 2.22 shows the estimated heavy-damage and collapse ratios calculated by inputting the observed strong motions from K-NET and KiK-net stations into the Nagato–Kawase Model. It can be seen that, with some exceptions, the damage ratio is lower than 10% at most points. While damage ratios of over 30% were estimated for several points with high acceleration, including K-NET Tsukidate, the estimated damage level was generally low.

The reason for this small damage potential of strong motions can be ascertained by approaching it from a different perspective. Figure 2.23 shows the diagram of PGA versus the equivalent predominant frequency, PGA/2πPGV, which has been used to examine the relationship between observed strong motions and structural damages. Equivalent predominant frequency is a simplified indicator of the dominant frequency of seismic motions assuming sinusoidal nature. In this diagram equi-PGV lines will be a slope from left-down side to right-up side. The symbols are the observed values from K-NET and KiK-net data. The purple lines indicate 800 and 2,700 Gal (cm/s^2) and the black lines indicate the uniform velocity line of 100 and 250 cm/s. These dotted lines are considered to be the danger lines based on the observation in Kobe, above which major damages are caused. The figure shows that the records with high PGAs all have a dominant frequency of 1 Hz or above and their PGVs are all lower than 100 cm/s (except for MYG004). Therefore, it is suggested that although the overall durations were much longer than those of the strong

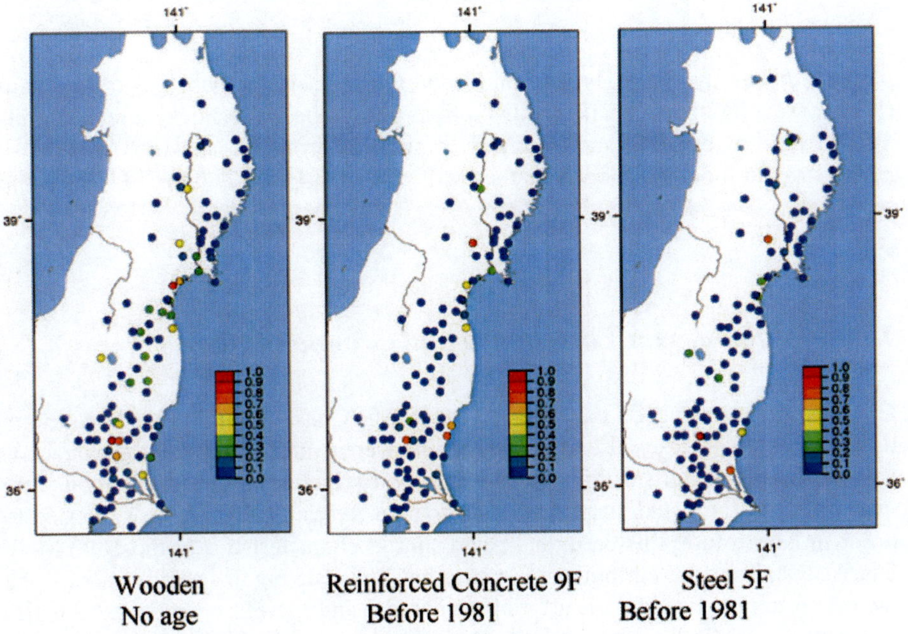

Fig. 2.22 Estimated heavy damage or collapse ratios for observed strong motions by using Nagato and Kawase's damage prediction models [23]

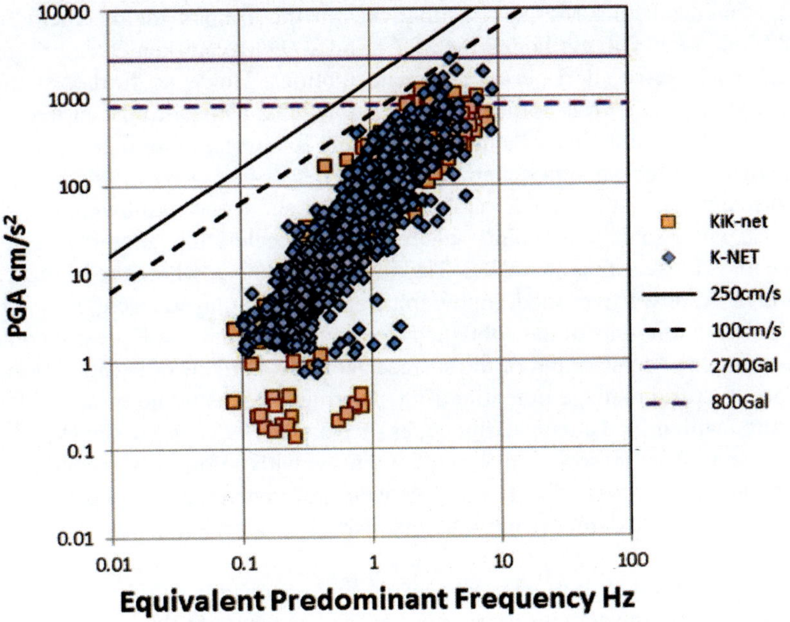

Fig. 2.23 PGA and equivalent predominant frequency relationship. *Symbols* are K-NET and KiK-net data. *Slant lines* are equi-PGV lines for 100 and 250 cm/s. *Red line* shows the PGA level of 800 cm/s² as the start line of heavy building damage

motions during the 1995 Hyogo-ken Nanbu Earthquake, the damages of the main shock were minor because the strong motions during the main shock were not dominant in the "moderately short-period" component around 1 s, which will cause heavy structural damages because they will give high PGAs and high PGVs at the same time.

2.1.9 Summary on Earthquake and Ground Motions

In this section, the data on the 2011 Off the Pacific Coast of Tohoku earthquake and its source characteristics as currently available were summarized, and the characteristics of the observed seismic motions and their structural damage potential were examined. As the Headquarters for Earthquake Research Promotion supposed to occur in near future, this earthquake had similar characteristics to the 1896 Meiji Sanriku-Oki Tsunami earthquake (i.e., a large slip occurring in a shallow area along the ocean trench, causing devastating tsunamis), and therefore generated a major slip causing a M9.0 earthquake. On the other hand, unlike usual tsunami earthquakes in this region that did not generate strong motions, this earthquake simultaneously generated rather strong, short-period dominant ground motions, which were generated in the areas deeper, closer to the land and does not coincide with the area that caused the M9.0 long-period slip. This came as no surprise, as it has always been pointed out that M7-class earthquakes in the Tohoku region, including the 1978 Miyagi-ken Oki earthquake of Mj7.4 and 2005 Miyagi-ken Oki earthquake of Mj7.2, would have short-period dominant motions. However, further studies are needed as to whether it is common or exceptional that a mega-thrust earthquake has always such a double face of a high stress event for strong motion generation and a slow slip event for tsunami generation.

With regards to site characteristics, the records show that significant nonlinearity was generated mainly at points where a high acceleration and/or velocity was recorded. It is clear that at K-NET Tsukidate (MYG004), where the highest acceleration of 2.7 g was recorded, topographic effects also played a role. It is expected that immense amounts of data obtained from this earthquake will greatly contribute to our further understanding of the surface geological effects of seismic motions.

The structural damage potential of this earthquake was by no means high when compared with inland crustal earthquakes such as the 1995 Hyogo-ken Nanbu earthquake. One of the reasons is that velocity pulses with a dominant frequency in the "moderately short-period" range (~1 s) were not generated, as these would have caused the severest damages to buildings. However, a large number of records with extremely high PGAs were obtained, and it is still an issue why major damage was not created at such high acceleration levels, an examination of which must include the evaluation of the actual seismic resistance of Japanese buildings.

2.2 Ground Motion Records in Tohoku District

2.2.1 Ground Motion Records

2.2.1.1 Strong-Motion Characteristics in Tohoku District

Large accelerations were observed over wide area of Tohoku district during the 3/11 Off the Pacific Coast of Tohoku earthquake. Also, large accelerations were observed during the 4/7 Mj7.1 intermediate-depth intra-slab earthquake occurred off Miyagi Prefecture, and the 4/11 Mj7.0 shallow inland earthquake occurred at the coastal area of Fukushima Prefecture. In this section, strong motion characteristics observed in the Tohoku district are briefly described.

Strong-Motion Distribution by the Main Shock

Figure 2.24 shows distributions of Peak Ground Acceleration (PGA) and Peak Ground Velocity (PGV) of the main shock records obtained at K-NET and KiK-net of the National Research Institute for Earth Science and Disaster Prevention (NIED), Japan Meteorological Agency (JMA), the Port and Airport Research Institute (PARI), Disaster Control Research Center of Tohoku University (DCRC), and Small

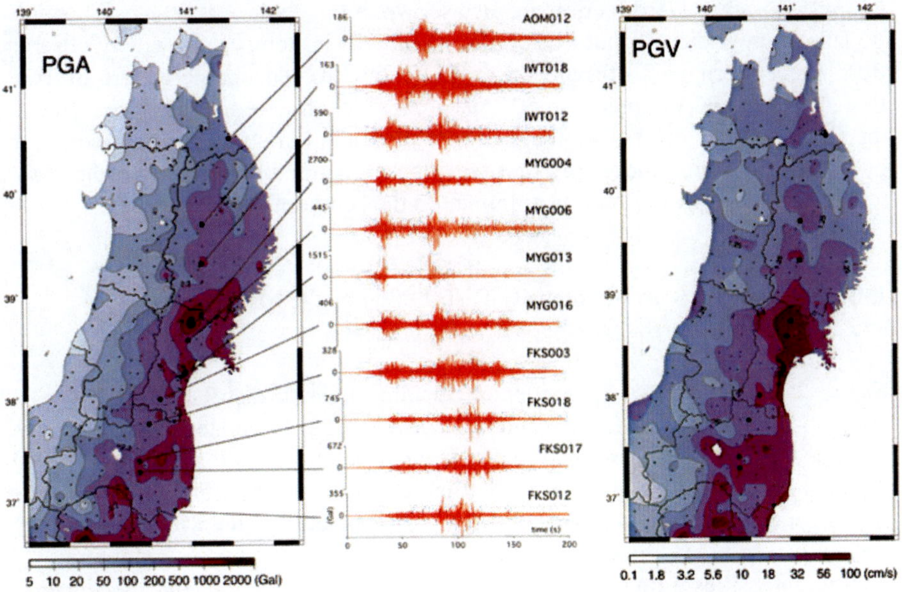

Fig. 2.24 Distribution of PGA and PGV calculated from the 3/11/2011 earthquake records at ground surface

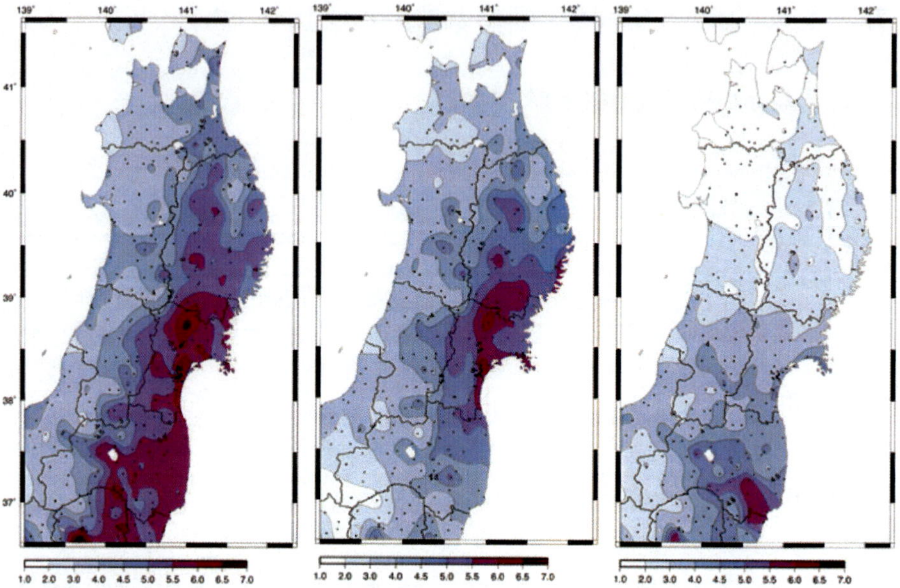

Fig. 2.25 Distribution of JMA intensities for the 3/11/2011 (M9.0), 4/7(Mj7.1), and 4/11(Mj7.0) earthquakes

Titan Network of Tohoku Institute of Technology (TOHTECH). PGV was estimated from integration of acceleration data with cut off frequency of 50 s. Major acceleration waveforms in NS direction are also shown in this figure.

This figure indicates that PGA larger than 500 Gal (cm/s^2) are widely observed from the eastern part of Fukushima Prefecture to the southern part of Iwate Prefecture, and PGV larger than 50 cm/s are mainly observed in the lowland area of the central part of Miyagi Prefecture. Also, acceleration waveforms are composed of two major wave groups; the first wave group is dominant in the northern part, while the second wave group is dominant in the southern part.

JMA Instrumental Seismic Intensity Distribution by the Main Shock and Two Major Aftershocks

Figure 2.25 shows JMA instrumental seismic intensity distributions of the 3/11 main shock, as well as the 4/7 and 4/11 aftershocks, calculated form the above-mentioned strong-motion records (Note: Local government stations are not included). The main shock intensity has aerial distribution similar to PGA distribution: intensities larger than or equal to 6 upper (partly 7) distributed over the wide areas from the eastern part of Fukushima Prefecture to the southern part of Iwate Prefecture.

Still, intensities from 6 lower to 6 upper distributed in the lowland areas from the central part of Miyagi Prefecture to the southern part of Iwate Prefecture during the 4/7 aftershock. During the 4/11 shallow inland aftershock, 6 lower was observed at the coastal area of Fukushima Prefecture.

Table 2.2 Large amplitude records observed during the 2011 Off the Pacific Coast of Tohoku earthquake (seismic intensity larger than or equal to 6.1)

No.	Organization	Station	Address	JMA Int.	PGA (cm/s/s)	PGV (cm/s)[a]
1	NIED	MYG004	Tsukidate, Kurihara	6.6	2,700	94
2	TOHTECH	smt.CCHG	Arai, Sendai	6.5	1,074	–
3	TOHTECH	smt.IWAK	Iwakiri, Sendai	6.4	859	–
4	DCRC	dcr.009	Matsumori, Sendai	6.4	821	88
5	NIED	MYG013	Nigatake, Sendai	6.3	1,517	74
6	JMA	4B9	Mikka-machi, Ohsaki	6.2	550	85
7	NIED	MYG006	Kita-machi, Ohsaki	6.1	572	98
8	DCRC	dcr.018	Okino, Sendai	6.1	512	79
9	NIED	FKSO16	Shin-shirakawa, Fukushima	6.1	1,295	59
10	TOHTECH	smt.NAKI	Nanakita, Sendai	6.1	1,853	–
11	DCRC	dcr.023	Oroshimachi, Sendai	6.1	613	77

[a]Cut-off period of 50s

Table 2.3 Large amplitude records observed during the 4/7 earthquake (seismic intensity larger than or equal to 5.8)

No.	Organization	Station	Address	JMA Int.	PGA (cm/s/s)	PGV (cm/s)[a]
1	NIED	MYG013	Nigatake, Sendai	6.2	1,002	63
2	DCRC	dcr.009	Matsumori, Sendai	6.2	767	76
3	NIED	MYG004	Tsukidate, Kurihara	6.1	1,242	44
4	NIED	IWT010	Ichinoseki, Iwate	5.9	801	52
5	NIED	MYG006	Kita-machi, Ohsaki	5.9	478	62
6	NIED	MYG012	Asahicho, Shiogama	5.8	1,447	37
7	NIED	IWT012	Kitakami, Iwate	5.8	708	45

[a]Cut-off period of 10s

Table 2.4 Large amplitude records observed during the 4/11 earthquake (seismic intensity larger than or equal to 5.6)

No.	Organization	Station	Address	JMA Int.	PGA (cm/s/s)	PGV (cm/s)[a]
1	NIED	FKSO13	Furudono, Fukushima	5.8	434	40
2	NIED	FKSH12	Hirata, Fukushima	5.6	510	30
3	NIED	FKSO12	Iwaki, Fukushima	5.6	490	49

[a]Cut-off period of 10s

Large Amplitude Records of the Main Shock and Aftershocks

Tables 2.2, 2.3, and 2.4 show the list of observed records at the main shock and the two aftershocks. These lists are sorted in the descending order of intensity. In the main shock, stations observed large intensities are concentrated in Miyagi Prefecture, especially in the cities of Kurihara, Sendai, and Ohsaki. As for the records in Sendai, detailed information will be described in Sect. 2.2.1.2.

Figure 2.26 shows 5% damped pseudo velocity response spectra for the major stations in Table 2.3. Spectra of Takatori record at the 1/17/1995 Kobe earthquake

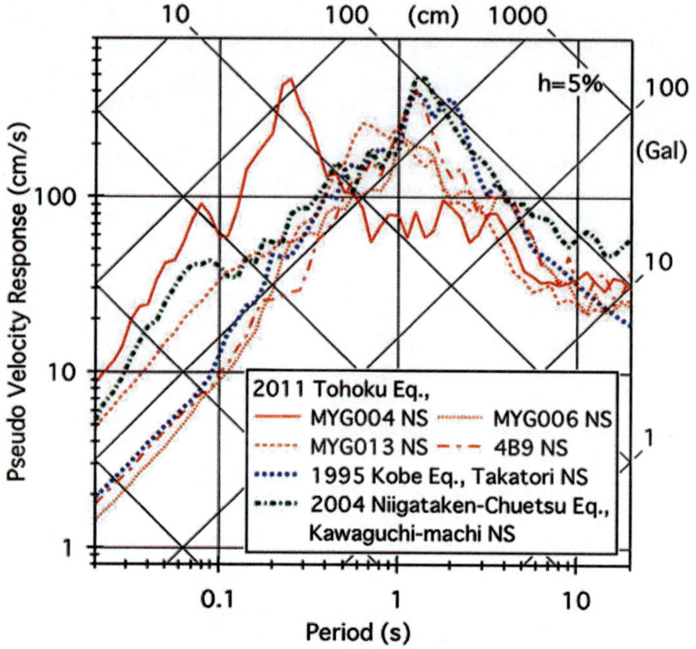

Fig. 2.26 Pseudo velocity response spectra of the 3/11 main shock large-amplitude records in comparison with those of the past disastrous earthquakes in Japan

and Kawaguchi-machi record at the 10/23/2004 Mid Niigata Prefecture earthquake are also plotted as representatives of the earthquake records in the heavily damaged zone by the past disastrous earthquakes.

Spectra of Takatori and Kawaguchi-machi records are predominant around 1–2 s, while spectrum amplitude at MYG004 (Terrace) is large at the shorter periods but the 1–2 s amplitude is smaller. On the other hand, MYG006 (Ohsaki plain) and MYG013 (Sendai plain) have larger 1–2 s amplitude compared with MYG004, and 4B9 at the center of Furukawa (Ohsaki plain) have similar amplitude level to that of Takatori. Spectral characteristics of MYG004 and MYG006 will be discussed in detail in Sect. 2.2.3.

During the 4/7 aftershock, large amplitudes were observed at the same stations where large amplitude had been also observed during the main shock (Table 2.3). Figure 2.27 shows K-NET Sendai (MYG013) acceleration waveforms, where the largest intensity was observed. At this place boiled sand was observed near the station, and acceleration spikes, probably due to soil liquefaction, were observed in the horizontal components. Such spikes were also observed during the main shock at this station.

Figure 2.28 shows acceleration waveforms at K-NET Nakoso (FKS012), the station closest to the 4/11 epicenter. Response spectra at this station of the main shock and the 4/11 aftershocks are plotted in Fig. 2.29. The 4/11 amplitude was larger than that of the main shock at short periods.

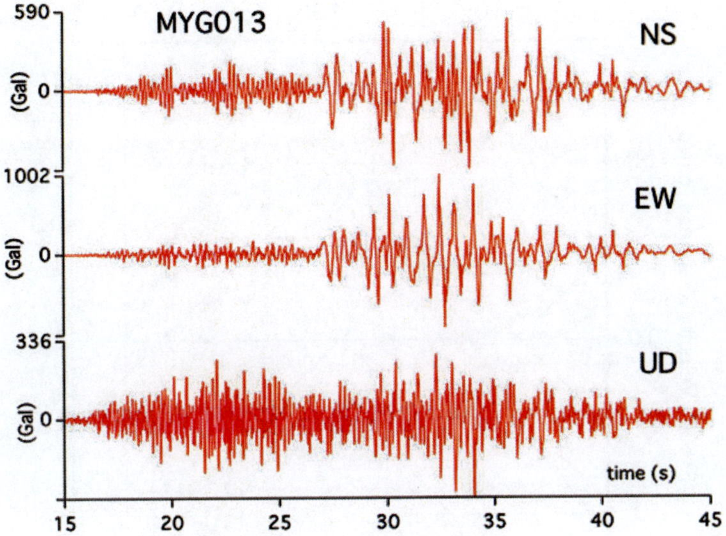

Fig. 2.27 Seismograms at MYG013 during the 4/7 earthquake

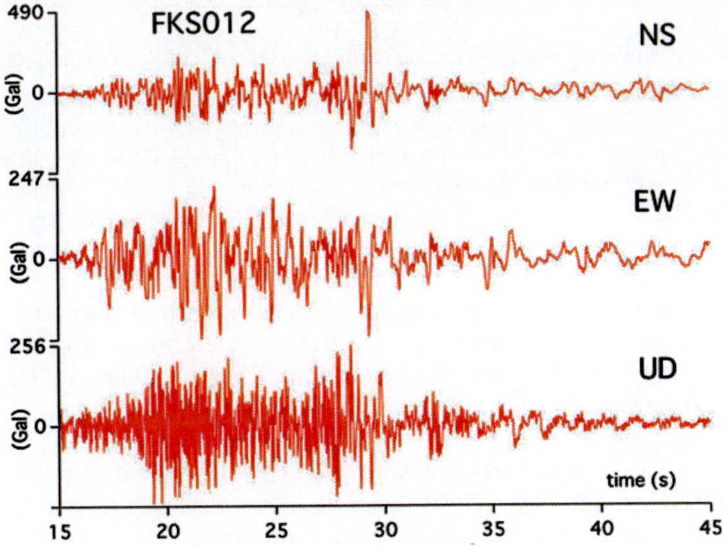

Fig. 2.28 Seismograms at FKS012 during the 4/11 earthquake

2.2.1.2 Strong-Motion Characteristics in Sendai

In Sendai, there are two major strong ground motion networks: Small Titan [24] by Tohoku Institute of Technology and DCRC network [25] by Disaster Control Research Center of Tohoku University. Small Titan is a free-field ground motion network which station shelter is the same as that of K-NET, while almost all the

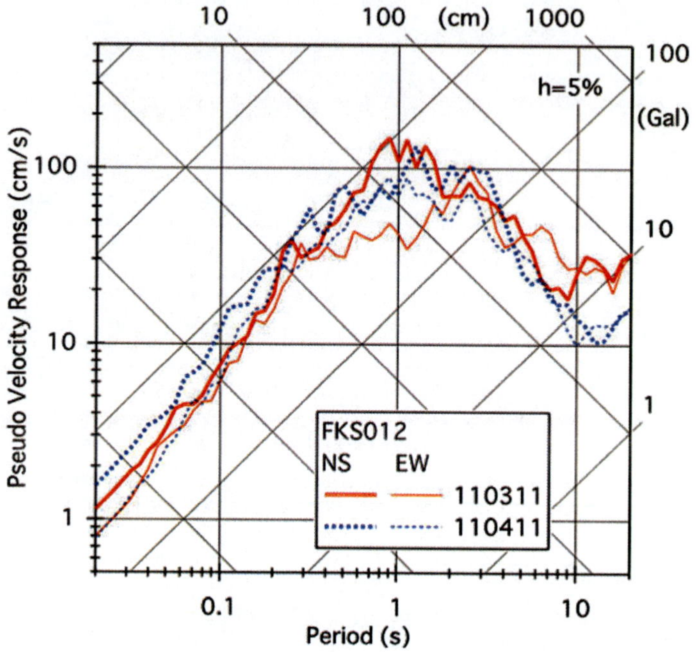

Fig. 2.29 Pseudo velocity response spectra of observed records at FKS012 during the 3/11 and 4/11 earthquakes

DCRC stations are located on the first floor of low-rise building (simultaneous observation with top floor at some places).

During the 3/11 main shock, records at 17 out of 20 stations were obtained in the Small Titan network as shown in Table 2.5 [26]. Also, records at 14 out of 21 stations were obtained in the DCRC network as shown in Table 2.6 [27].

Figures 2.30 and 2.31 show station locations and acceleration waveforms of the Small Titan and the DCRC network, respectively. Waveforms at JMA E06, K-NET MYG013, and ground station of Izumi electric power building (IZU), Tohoku Electric Power Co., are also plotted in Fig. 2.31.

The largest acceleration observed in the Small Titan network was 1,853 Gal (cm/s^2) at Nanakita Jr. High School (NAKI), the largest seismic intensity was 6.5 at Shichigo Jr. High School (CCHG). Figure 2.32 shows velocity response spectra of these records. On the other hand, the largest acceleration (822 Gal) and seismic intensity (6.2) in the DCRC network was observed at Matsumori elementary school. By the strong motion network, K-NET, operated by NIED PGA of 1,517 Gal was observed at MYG013 (Nigatake, Sendai), where boil sand and acceleration spikes probably due to soil liquefaction were observed, as described in the previous section.

Figure 2.33 compares pseudo velocity response spectra at stations located in the east and west sides of the Nagamachi–Rifu fault. The spectrum at No. 27, locating near the Sendai railway station, is commonly plotted in each side as a reference. Spectra in the west side are equal to or relatively larger at short period (less than 1 s)

Table 2.5 Outline of earthquake records by Small Titan network, operated by Tohoku Institute of Technology

No.		Station	Code	Latitude	Longitude	PGA (cm/s²)	JMA Int.	JMA Int. Scale
S1	(AKA001)	Shokei-Gakuin University	SHOK	38.192	140.832	215	5.1	5 upper
S2	(AKA002)	Yanagiu Jr. High School	YAGI	38.186	140.876	764	5.9	6 lower
S3	(AKA003)	Higashi Shiromanu Elementary School	HSHR	38.191	140.927	521	5.9	6 lower
S4	(AKA004)	Arahama Elementary School	ARAH	38.22	140.983	542	5.9	6 lower
S5	(AKA005)	East-Sendai Senior High School	SENH	38.217	140.936		No record	
S6	(AKA006)	Higashi Nagamachi Elementary School	HNAG	38.217	140.894	720	6	6 upper
S7	(AKA007)	Taihaku Elementary School	TAH	38.228	140.821	709	5.6	6 lower
S8	(AKA007)	Kuryu Elementary School	KURI	38.264	140.789	564	5.5	6 lower
S9	(AKA008)	Kasumicho Campus, TOHTECH	TITK	38.243	140.854	471	5.8	6 lower
S10	(AKA008)	Futatsusawa Campus, TOHTECH	TITF	38.231	140.874	429	5.6	6 lower
S11	(AKA009)	Shichigo Jr. High School	CCHG	38.234	140.949	1,074	6.5	7
S12	(AKA009)	Sendai Technical High School	SIKO	38.256	140.921	542	5.8	6 lower
S13	(AKA010)	Renbo Elementary School	RENB	38.249	140.891	521	5.7	6 lower
S14	(AKA010)	Sakuragaoka Elementary School	SAKR	38.3	140.854	681	5.8	6 lower
S15	(AKA011)	School for the Visually Impaired	MOGA	38.275	140.882	700	5.6	6 lower
S16	(AKA011)	Nankodai-Higashi Elementary School	NANK	38.296	140.916	699	5.8	6 lower
S17	(AKA012)	Tago elementary School	TAKA	38.271	140.961		No record	
S18	(AKA012)	Tagajo Daini Jr. High School	TGNI	38.298	140.983		No record	
S19	(AKA013)	Iwakiri Jr. High School	IWAK	38.299	140.949	859	6.4	6 upper
S20	(AKA013)	Nanakita Jr. High School	NAKI	38.321	140.896	1,853	6.1	6 upper

Table 2.6 Outline of earthquake records by DCRC strong-motion network, operated by Tohoku University

No.	Sensor	Station	4/7/2011 PGA (cm/s^2)	4/7/2011 PGV* (cm/s)	4/7/2011 JMA Int.	3/11/2011 PGA (cm/s^2)	3/11/2011 PGV** (cm/s)	3/11/2011 JMA Int.	3/9/2011 PGA (cm/s^2)	3/9/2011 PGV* (cm/s)	3/9/2011 JMA Int.
2	ETNA	Rokugo Elementary School	311	42.1	5.7		No record			No record	
3	ETNA	Furujiro Elementary School	251	22.4	5.1	320	59.5	5.7	24	3.1	3.3
4	ETNA	Higashi Rokugo Elementary School		Removed		613	74.2	6.0	29	3.4	3.4
5	QDR	Daiichi Jr. High School	230	19.3	5.1	383	39.4	5.6	28	2.9	3.5
8	QDR	Shogen-Chuoh Elementary School	534	25.3	5.5	840	60.4	6.0	30	2.2	3.2
9	QDR	Matsumori Elementary School	767	75.5	6.2	822	85.7	6.4	46	4.2	3.7
10	QDR	Miyagi Prefecture Library 1F	279	18.0	5.0	407	62.7	5.6	20	2.4	3.2
11	QDR	Miyagi Prefecture Library 3F		No record			No record		34	3.1	3.5
12	QDR	Seiryo Secondary School 1F		No record			No record		19	3.5	3.3
14	QDR	Tsurugaya Elementary School 1F	432	30.6	5.7		No record		20	1.9	3.1
16	QDR	Nakano Jr. High School 1F		No record			No record		40	3.2	3.6
18	QDR	Okino Elementary School 1F	360	31.8	5.6	512	77.6	6.2	37	3.5	3.5
20	QDR	Minami Koizumi Elementary School	220	25.7	5.3	381	63.0	5.6	19	2.4	3.1
21	QDR	Nishitaga Jr. High School	186	16.4	5.0	400	45.1	5.5	23	3.0	3.4
22	QDR	Tomizawa Jr. High School	232	21.1	5.2	416	54.6	5.7	29	3.2	3.4
23	QDR	East Water Supply Center	472	37.3	5.8	613	75.4	6.1	30	2.6	3.3
24	QDR	Ryutaku-Ji		Removed			No record			No record	
25	QDR	Nagamachi Minami Community Center	264	29.5	5.5	494	69.3	6.0	59	6.0	4.0
26	QDR	Aoba Ward Office	318	21.9	5.2		No record		24	3.2	3.3
27	SSA-1	Sumitomo Seimei Bldg.	167	14.0	4.9	318	29.2	5.3	15	2.2	3.1
28	SMAC-MD	Tohoku University 1F		No record		333	53.7	5.6	35	4.4	3.7

*cut-off period of 10s, **50s

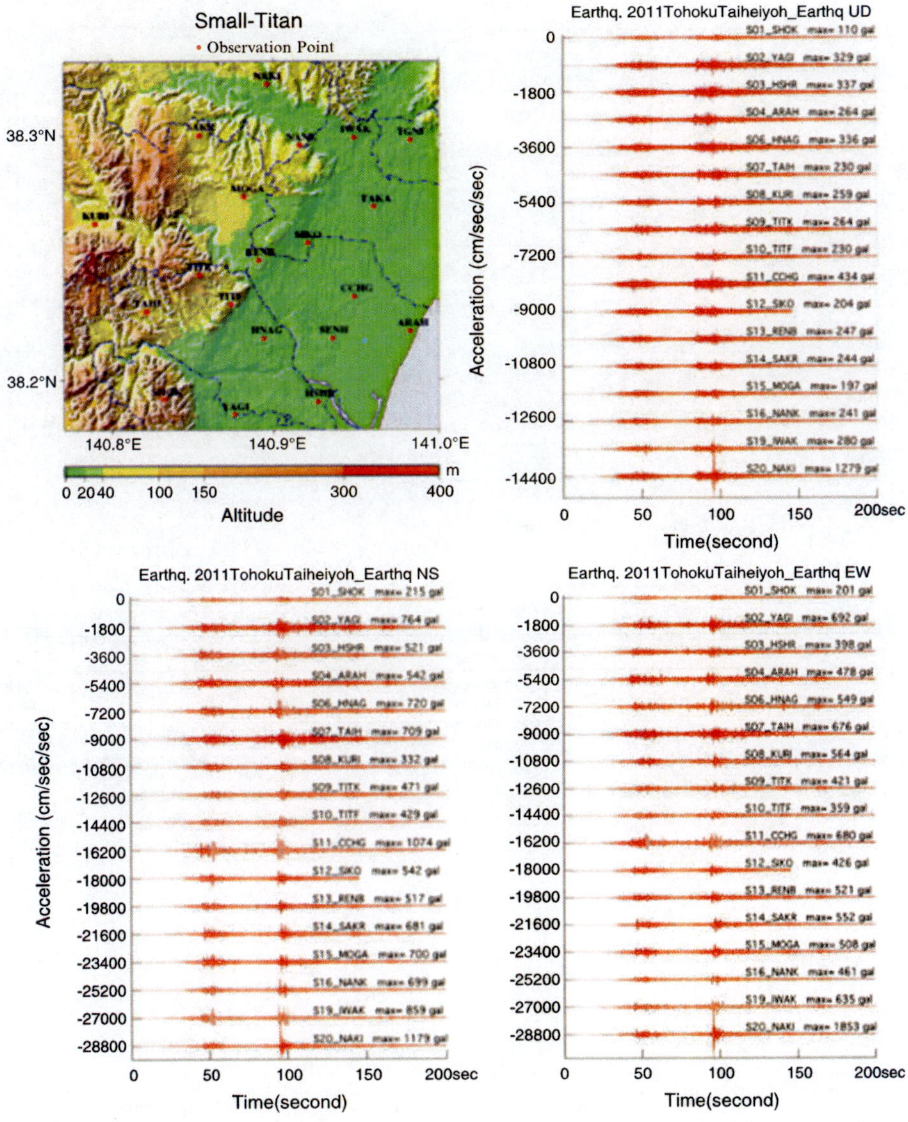

Fig. 2.30 Earthquake records observed by the Small-Titan strong-motion network operated by Tohoku Institute of Technology [26]

than the No. 27 spectrum, while the spectra in the east side are significantly larger than the No. 27 spectrum, especially around 1 and 3 s.

Such spatial difference may be due to the difference of subsurface structure; the west side of the fault is terrace and the east side is lowland (alluvial deposits) in Sendai, as shown in Fig. 2.30 [28]. Figure 2.34 shows distribution of instrumental seismic intensities calculated by spatial interpolation of Small Titan and DCRC

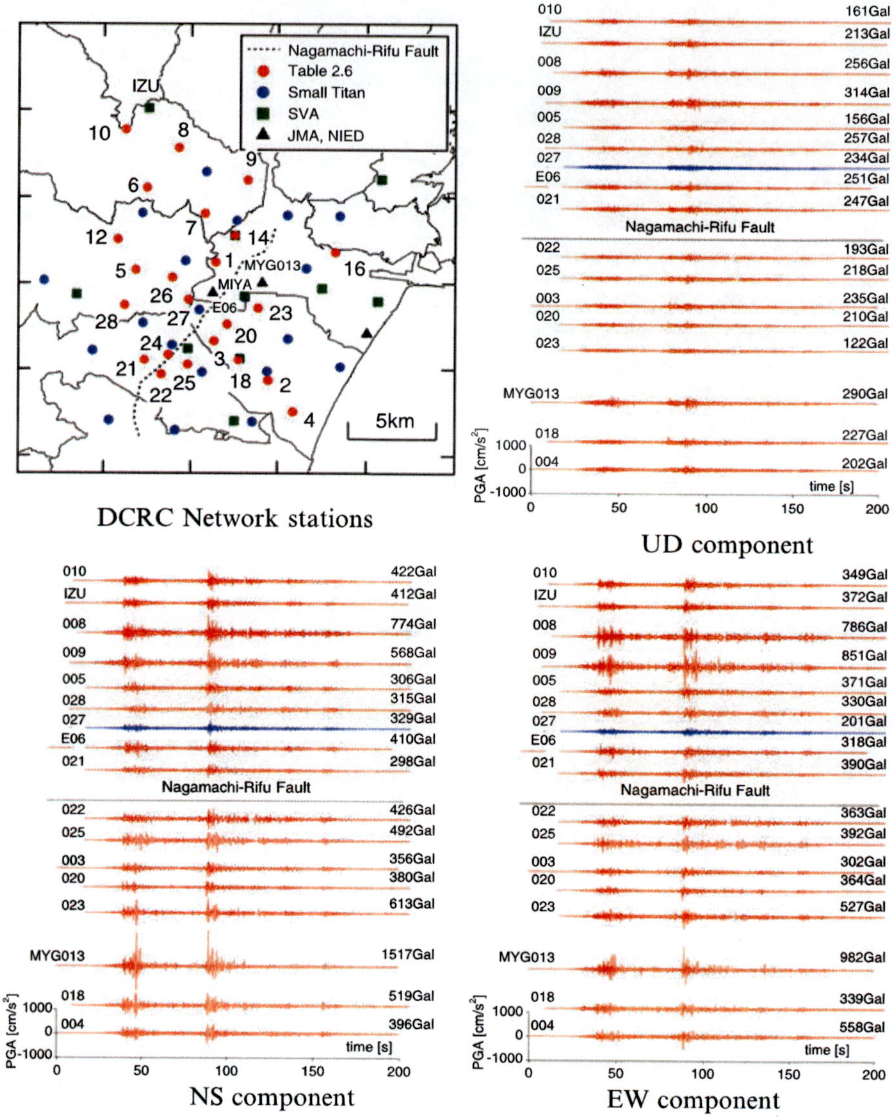

Fig. 2.31 Earthquake records observed by the DCRC strong-motion network operated by Tohoku University

network data. Large Intensities are obtained at east side of the fault and along the Nanakita River at northern area of Sendai.

Figure 2.35 compares pseudo velocity response spectra in EW component for the 3/11 main shock and the 4/7 aftershock at No. 27 (near Sendai station) and No. 23 (Oroshi-machi, east side of the fault) stations. The main shock amplitude is larger than the aftershock amplitude at No. 27, but almost equals at No. 23. As shown in this example, maximum amplitude is similar between the main shock and the aftershock at some sites.

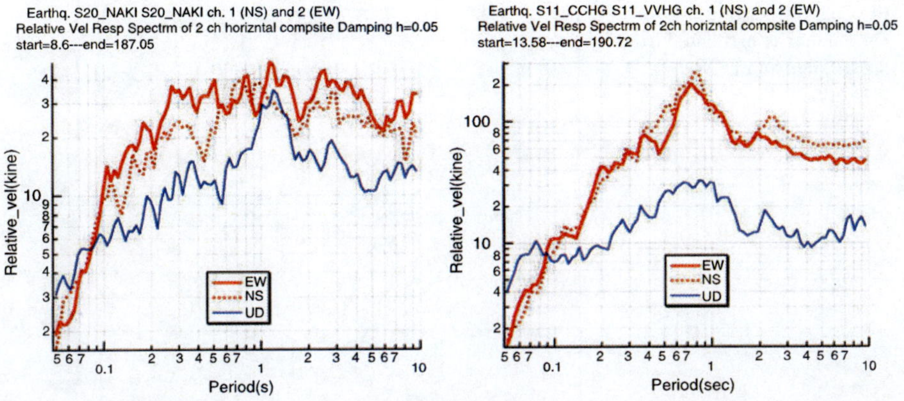

Fig. 2.32 Velocity response spectra observed at NAKI and CCHG stations of Small-Titan strong-motion network [26]

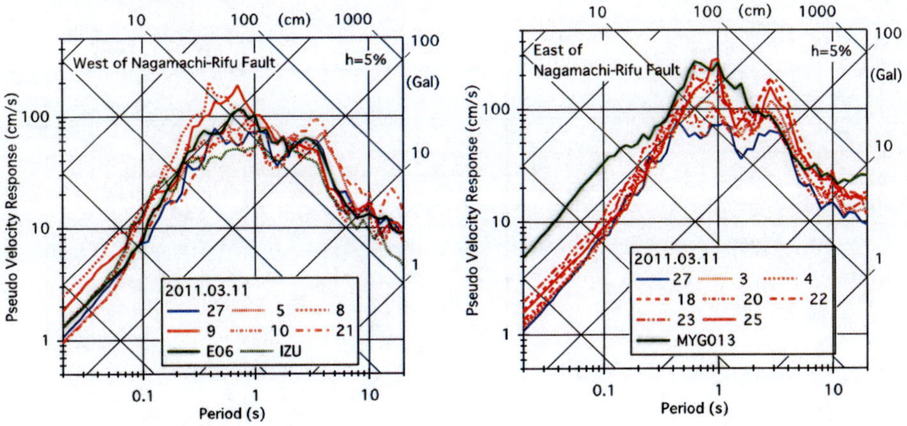

Fig. 2.33 Comparison of pseudo velocity response spectra of the DCRC strong-motion network records

2.2.2 Earthquake Records in Buildings

In this sub section earthquake records observed in buildings (including soil-structure simultaneous observation) for the 3/11/2011 Off the Pacific Coast of Tohoku earthquake are reported.

Tables 2.7, 2.8, and 2.9 show outlines of observation records in base-isolated, structural control, and ordinary buildings, respectively. Simultaneous observations of base-isolated and ordinary buildings with ground are included in Table 2.7. These tables indicate peak accelerations not only at the 3/11 main shock but also at the 4/7 aftershock (only where PGAs are provided), because the damages became severe by the aftershock at some buildings.

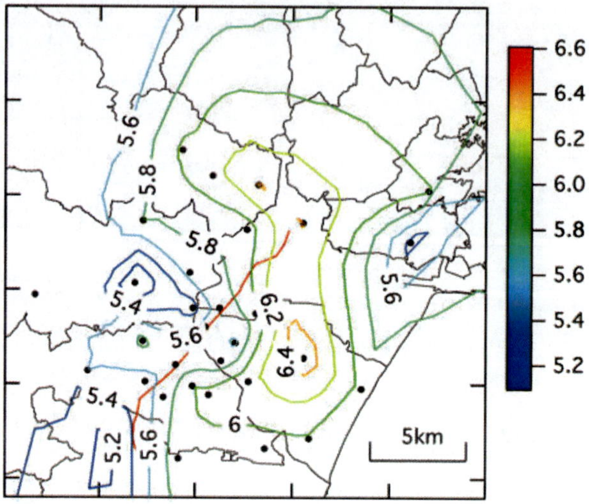

Fig. 2.34 Distribution of JMA intensity estimated from Small Titan and DCRC networks

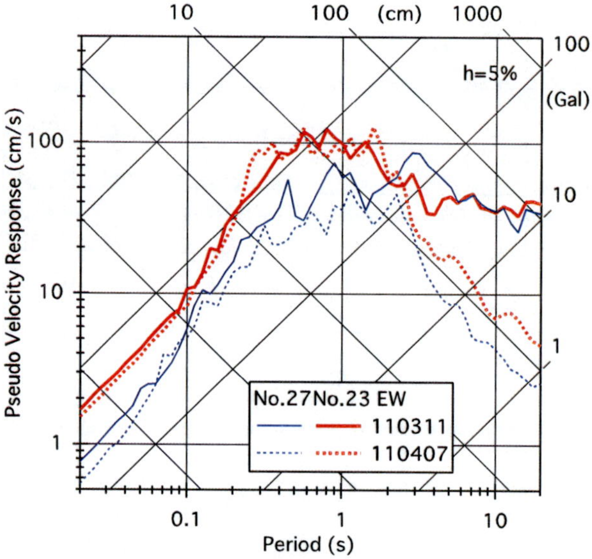

Fig. 2.35 Comparison of pseudo velocity response spectra of observed records at DCRC No .023 & No. 027 stations during the main shock and the 4/7 aftershock

Detailed observation information is shown in the references of the tables above. Figure 2.36 shows the building locations in the tables. Many of the buildings are located in Sendai.

Horizontal peak accelerations of 300 Gal (cm/s^2) or around were observed at building basements in Sendai. In simultaneous observation of base-isolated and ordinary buildings, 700 Gal to 1 g were observed at the top floors of ordinary buildings (maximum acceleration was 1,054 Gal at BI2), while peak acceleration at the top floors of base-isolated buildings are almost equal to or smaller than those at the basements.

Table 2.7 Outline of earthquake records at base-isolated buildings

No.	Building name (data-providing organization)	Building	Outline of structure	Completion year	References	Sensor location	3/11/2011 Peak acceleration (cm/s^2) H1	H2	UD	4/7/2011 Peak acceleration (cm/s^2) H1	H2	UD
BI1	Tohoku University Base-isolation Test Building (Shimizu Co.)	Base-isolation	RC, 3-stories, spread foundation, HD-LRB	1986	[29]	Base-ment Center	301	241	243			
						IF Center	361	219	280			
						3F	282	180	–			
						RF	344	244	–			
		Ordinary	RC, 3-stories, spread foundation	1986	[29]	IF	327	258	249			
						3F	702	824	–			
BI2	Izumi Electric Power Building (Tohoku Electric Power Co., Higashi Nihon Kougyo Co.)	Ground base-isolation				GL	417	378	228	331	331	172
			RC, 5-stories, spread foundation, HD-LRB	1990	[30]	B1F	327	345	218	207	217	181
						1F	183	177	240	119	118	186
						3F	139	174	489	106	116	413
						RF	199	224	768	149	143	682
		Ordinary	RC, 5-stories spread foundation			RF	1054	1043	718	473	743	509
BI3[a]	City Building Sendai–Daiichi (Tohoku Electric Power Co., Obayashi Co.)	Base-isolation	6 stories above ground and 2 below, braced frame (column: SRC, beam: S), HD-LRB	1995	[30]	GL	454	569	303	392	527	295
						Base-ment[b]	278	381	213	184	175	201
						B2F[c]	214	200	254	137	127	190
						5F	241	209	424	208	154	368

(continued)

Table 2.7 (continued)

No.	Building name (data-providing organization)	Building	Outline of structure	Completion year	References	Sensor location	3/11/2011 Peak acceleration (cm/s^2) H1	H2	UD	4/7/2011 Peak acceleration (cm/s^2) H1	H2	UD
BI4	Hachinohe City Hall (BRI)	Ground			[31]	GL-105 m	36	46	32	20	19	15
						GL-30 m	86	89	49	45	48	26
						GL	286	210	61	97	114	47
		New Bldg. (base-isolation)	10 stories above ground and 1 below, SRC, pile foundation, LRI	1998	[31]	B1F	100	104	58	73	56	51
						01F	91	122	73			
						10F	120	123	205			
		Main Bldg. (ordinary)	5 stories above ground and 1 below, SRC	1980	[31]	B1F	97	110	55	71	55	24
						06F	348	335	78	175	194	46

Note: *H1*, *H2* two horizontal components, H1 is near N–S component

[a]The 3/11 records at BI3 is only the latter part of two wave groups

[b]Lower of Isolation Device

[c]Upper of Isolation Device

Table 2.8 Outline of earthquake records at structural control buildings

No.	Building name (data-providing organization)	Outline of structure	Completion year	References	Sensor location	3/11/2011 Peak acceleration (cm/s²) H1	3/11/2011 H2	3/11/2011 UD	4/7/2011 Peak acceleration (cm/s²) H1	4/7/2011 H2	4/7/2011 UD
SC1	Tohoku Electric Power Co. Head Office (Tohoku Electric Power Co., Higashi Nihon Kougyo Co.)	2-stories below GL(RC,SRC), 28 stories above(S), rooftop 2-stories, low yield point steel wall	2002	[30]	B2F	206	175	162	159	163	146
					1F	289	240	209	182	183	147
					14F	384	284	324	291	253	180
					28F	503	375	317	281	349	299
SC2	DOCOMO Tohoku Bldg. (NTT Facilities Inc.)	2-stories below GL (SRC), 21-stories above(S), steel tower spread foundation, RDT+LED	2004	[32]	B2F	257	158	147			
				[33]	1F	312	211	153			
					RF	444	338	400			
SC3	Sendai City Hall (Sendai City, Tohoku Branch of Yamashita Sekkei, TOHTECH, Tobishima Co.)	2-stories below GL, 6-stories above, rooftop 3-stories. SRC, spread foundation, retrofit by brace damper	1965, 2008 retrofit	[34]	1F	413	284	207	230	274	173
					RF	853	808	461	418	600	368

Note: *H1, H2* two horizontal components, H1 is near N–S component

Table 2.9 Outline of earthquake records at ordinary buildings

No.	Building name (data-providing organization)	Building	Outline of structure	Completion year	References	Sensor location	3/11/2011 Peak acceleration (cm/s^2) H1	H2	UD	4/7/2011 Peak acceleration (cm/s^2) H1	H2	UD
OR1	Sendari Government Office Bldg. #2 (BRI)		Steel, 15-stories	1973	[35]	B2F	259	163	147	199	161	130
						15F	346	361	543	270	309	380
OR2	Tohoku University (BRI)		SRC, 9-stories, pile foundation	1969 (2000 retrofit)	[35–37]	01F	333	330	257			
						09F	908	728	640			
OR3	Miyako City Hall (BRI)		RC, 6-stories, pile foundation	1972	[31, 35]	GL	174	174	240			
						01F	138	122	277			
						07F	246	197	359			
OR4	Iwaki City Hall (BRI)		RC, 8-stories pile foundation	1973	[35]	B1F	175	176	147			
						09F	578	449	260			
OR5	Tsuruoka Government Office Bldg.		RC, 4-stories	1987	[35]	01F	34	36	14	28	23	10
						04F	37	39	15	32	25	11
OR6	Akita Prefectural Office (BRI)		RC, 6-stories	1959	[35]	B1F	47	50	24	36	33	24
						08F	192	175	44	194	140	44

OR7	Kasumicho Campus, TOHTECH (TOHTECH)	No. 5	3-stories below GL, 5-stories above, rooftop 2-stories, RC (partly SRC) pile foundation	1968 (1978 damaged then retrofit)	[38]	B3F	–	–	–			
						5F	681	820	368			
		Library	4-stories, rooftop 1-story, RC, pile foundation	1968 (2002 retrofit)	[38]	1F	363	295	257			
						RF	933	585	303			
		No. 3	6-stories, rooftop 2-stories, RC, pile foundation	1969	[38]	1F	230	336	259			
						6F	847	1420	693			
		No. 6	4-stories, SRC, pile foundation	1969	[38]	1F	376	351	304			
						4F	566	833	347			
		No. 1	1-story below GL, 4-stories above, rooftop 1 story, S (partly SRC) pile foundation	2006	[38]	1F	280	354	147			
						4F	785	1194	228			
OR8	High-Rise Bldg. A (TOHTECH)		2-stories below GL, 31-stories above, rooftop 3-stories, RC, spread foundation	1995	[38]	B1F	262	202	188			
						1F	301	209	193			
						RF	545	527	393			
OR9	High-Rise Bldg. B (TOHTECH)		2-stories below GL, 21-stories above, rooftop 1 stories, RC, pile foundation	1993	[38]	1F	216	161	171			
						RF	600	452	334			
OR10	Natori Campus, Sendai-NCT (Sendai-NCT)	Department of Architecture	RC, 3-stories, rooftop 1-story, spread foundation	1964 (2002 retrofit)	[30]	3F	720	766	534	639	613	452
		Advanced Course	RC, 4-stories, rooftop 1-story, spread foundation	1999	[30]	1F	451	451	397	331	355	347
						4F	654	718	412	608	787	424
OR 11	Oshika Branch Office (Tohoku University)		RC, 3-stories with 2-stories extension, spread foundation	1975	[39–41]	1F	555	626	565			
						2F	865	757	499			
						3F	1,056	853	1,157			

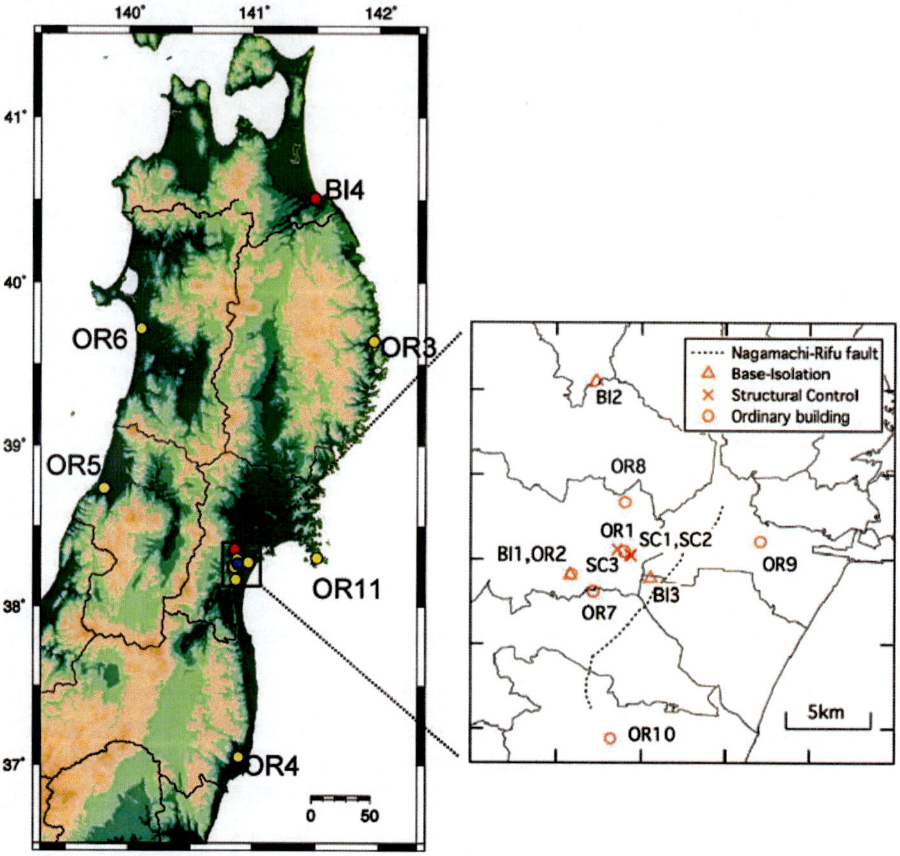

Fig. 2.36 Location of the buildings where earthquake records were observed. BI with *red symbol* means a base isolation building, SC with *blue symbol* means a structural controlled building, and OR with *yellow symbol* means ordinary buildings

As for structural control buildings, 500 Gal or lower were observed at the top of high-rise buildings (SC1, SC2), and 853 Gal was observed at the 8-story building (SC3). As for ordinary buildings, peak accelerations of 900 Gal or larger were observed at five stations (OR2, 3 buildings at OR7, and OR11).

Hereafter, examples of the observed records in structural controlled buildings and ordinary buildings will be shown consecutively in Figs. 2.37, 2.38, 2.39, 2.40, 2.41, 2.42, and 2.43. In the last part of this section, comparison of 2011 records with the past disastrous earthquake records is shown for OR2, where many earthquakes including the 1978 Miyagi-ken Oki earthquake have been observed.

Here comparison of the observation records with those due to past damage earthquakes will be presented. Among the examples described above, in the building of Departments of Civil Engineering and Architecture, Faculty of Engineering, Tohoku University (THU Building) shown in Fig. 2.40, long-term monitoring of dynamic

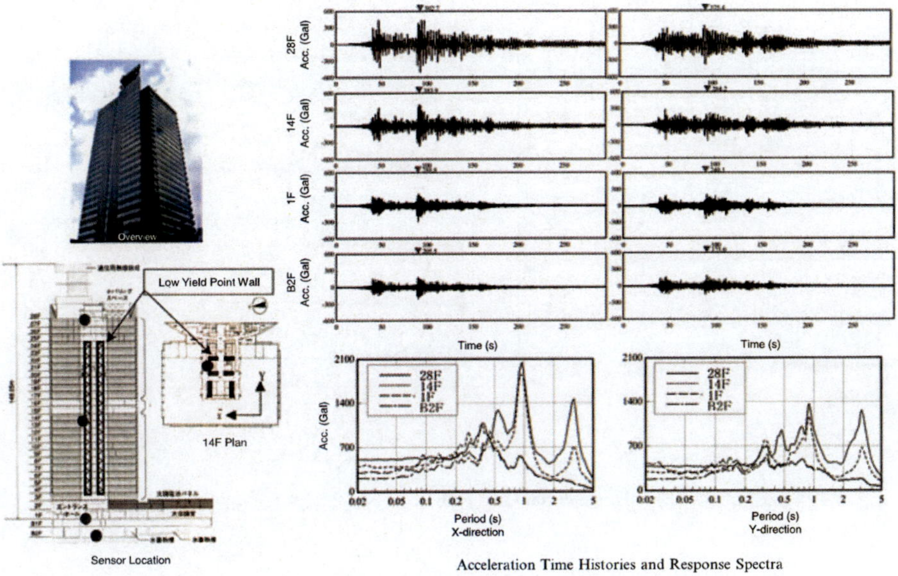

Fig. 2.37 Earthquake records at the head office building, Tohoku Electric Power Co.

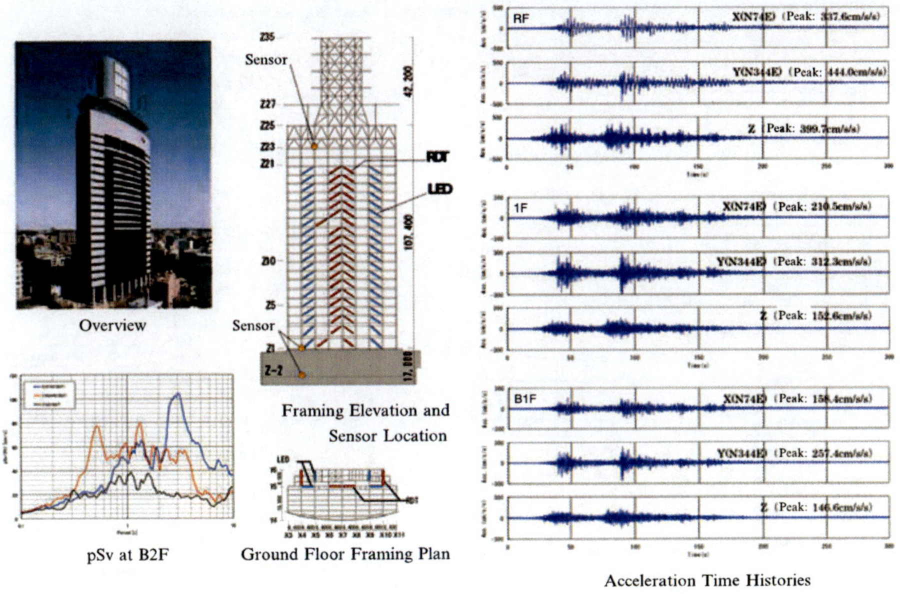

Fig. 2.38 Earthquake records at DOCOMO Tohoku Building, NTT DOCOMO, INC.

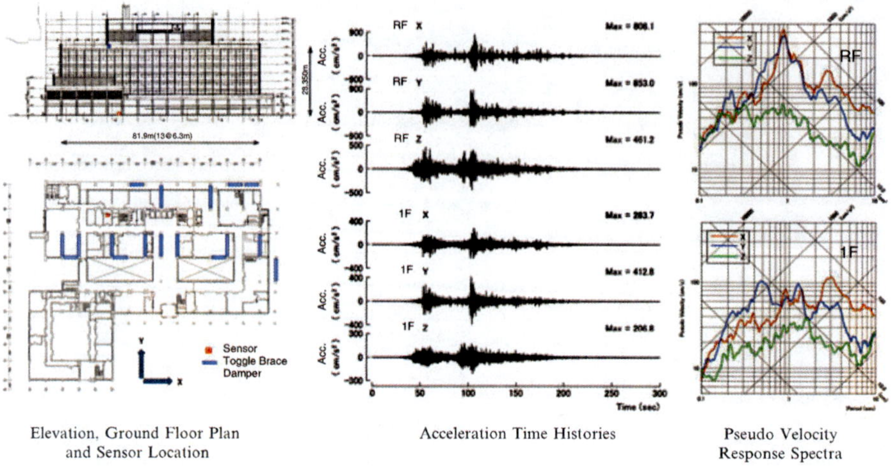

Fig. 2.39 Earthquake records at Sendai City Hall

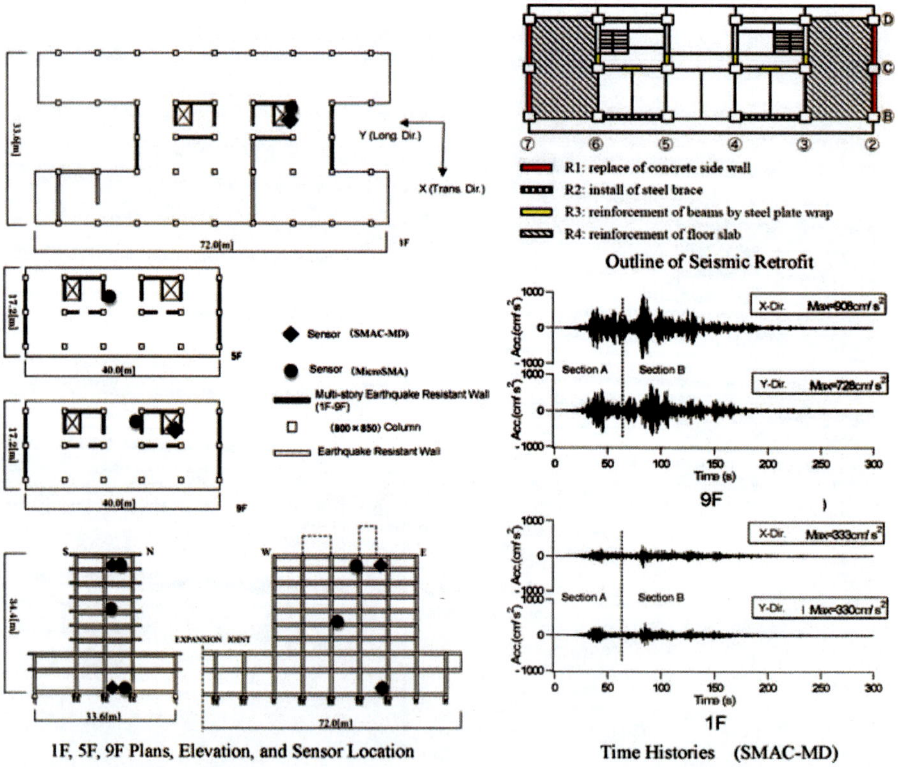

Fig. 2.40 Earthquake records at the research building of Department of Civil Engineering and Architecture, Tohoku University

Overview

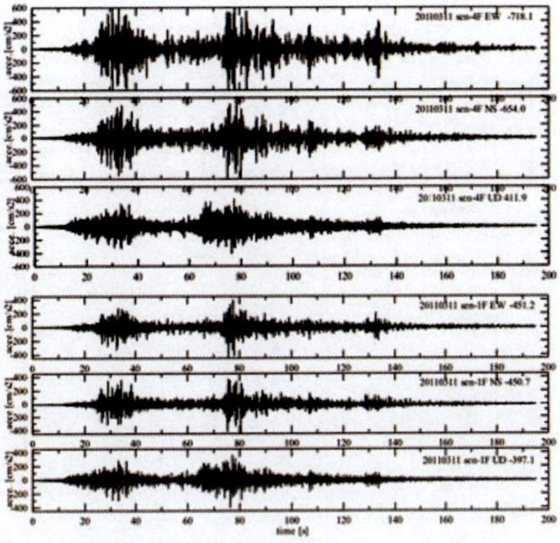

Time Histories (Upper: 4th Floor, Lower: 1st Floor)

Fig. 2.41 Earthquake records at the building of Advanced Engineering Course, Sendai National College of Technology

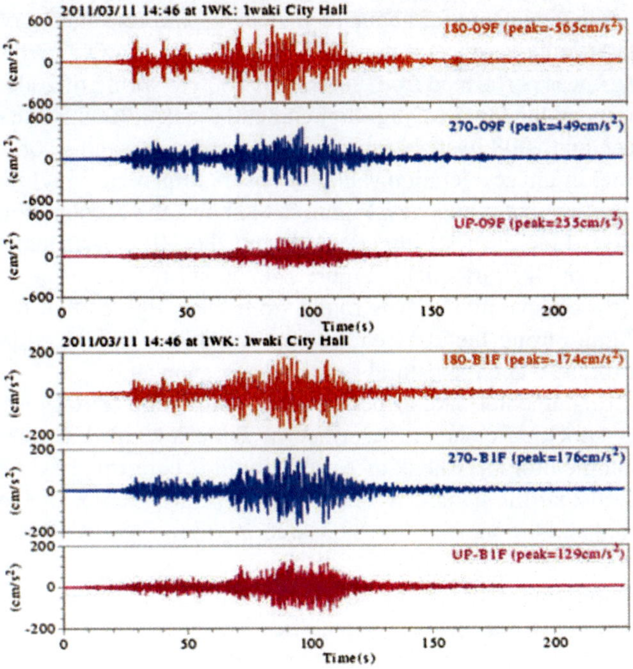

Time Histories (Upper: 9th Floor, Lower: Underground Floor)

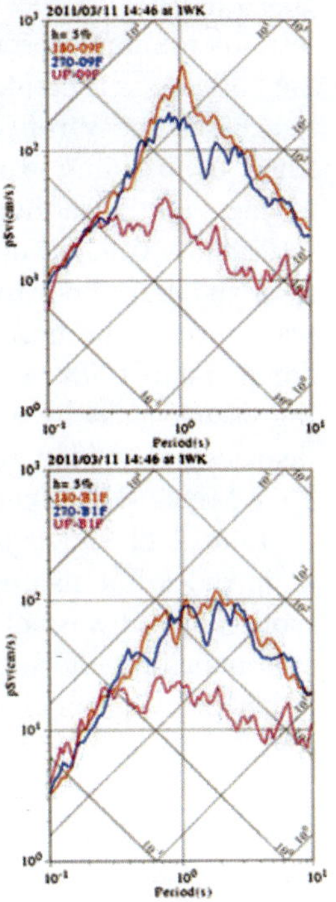

Pseudo Velocity Response Spectra

Fig. 2.42 Earthquake records at Iwaki City Hall

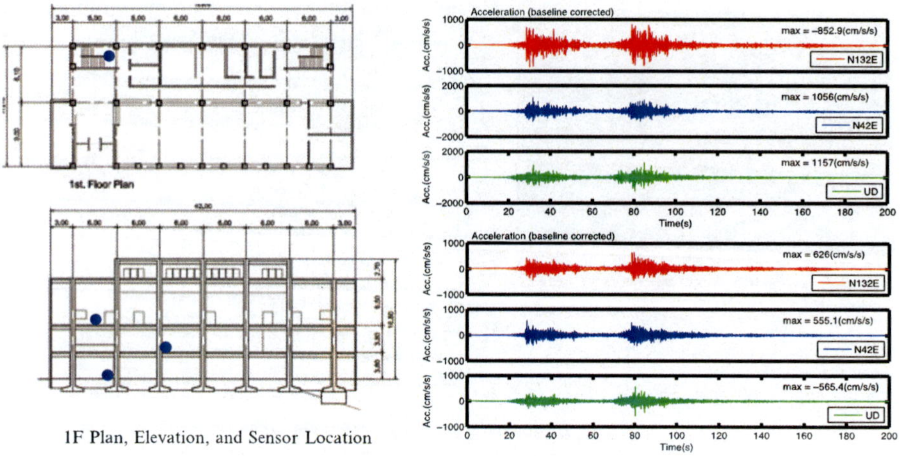

1F Plan, Elevation, and Sensor Location

Time Histories (Upper: 3rd Floor, Lower: 1st Floor)

Fig. 2.43 Earthquake records at the Oshika branch office building, Ishinomaki City

characteristics has been performed by strong motion observation, forced vibration test, microtremor observation, and so on for about 40 years since the completion of the building in1969 [42]. After experience of the 1978/6/12 Miyagi-ken Oki earthquake, seismic retrofit work was performed from autumn of 2000 to spring of 2001. Then, the building experienced the 2005 Miyagi-ken Oki earthquake, 2008 Iwate–Miyagi Nairiku earthquake, and 2008 Iwate Northern Coast earthquake, and so on.

Table 2.10 shows the maximum acceleration values of observation records at first floor and ninth floor for past main earthquakes. Figure 2.44 shows the acceleration waveform in the transverse direction (NS direction) during the 2011 earthquake compared to 1978 Miyagi-ken Oki earthquake. Figure 2.45 shows the corresponding pseudo velocity spectra comparatively. It is found from these figures that the displacement of the building during the 2011 earthquake was larger compared to 1978 Miyagi-ken Oki earthquake and the natural period became longer.

Table 2.11 shows the fundamental natural period of the building for not only main shock but also, fore-shock, aftershocks described in Table 2.12 [43]. Strong motion record was not obtained for aftershock of April 7, but it is confirmed by the continuous monitoring [40] that the natural frequency due to microtremor of the aftershock was not changed and the same as that on March 19 (transverse: 1.17 Hz, longitudinal: 1.37 Hz).

Figure 2.46 shows the relation between the amplitude dependent fundamental natural period and deflection angle in the transverse direction, based on the long-term monitoring data. It is found from this figure that the deflection due to the first

Table 2.10 Peak accelerations observed at Tohoku University for the 2011 and past disastrous earthquakes

		1F Peak Acc. (cm/s/s)		9F Peak Acc. (cm/s/s)	
Date	M	NS	EW	NS	EW
2011/3/11	9.0	333	330	908	728
1978/6/12	7.4	258	203	1,040	524
2005/8/16	7.2	87	81	329	287
2008/6/14	7.2	88	70	392	293

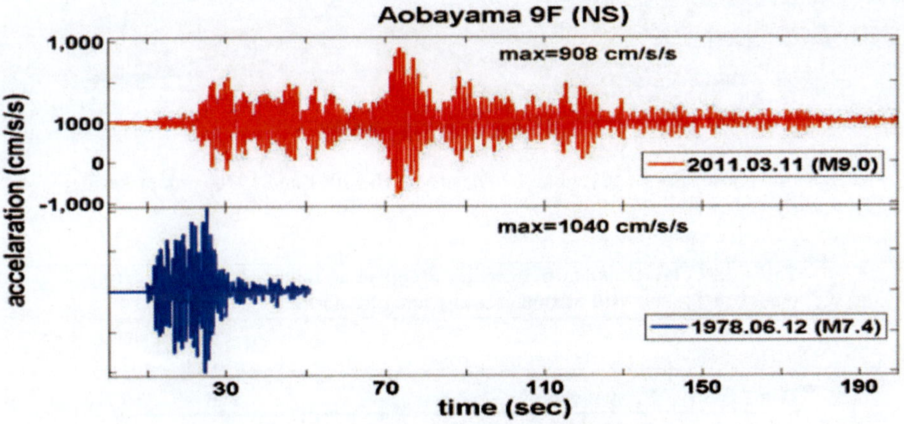

9F Acceleration Time Histories

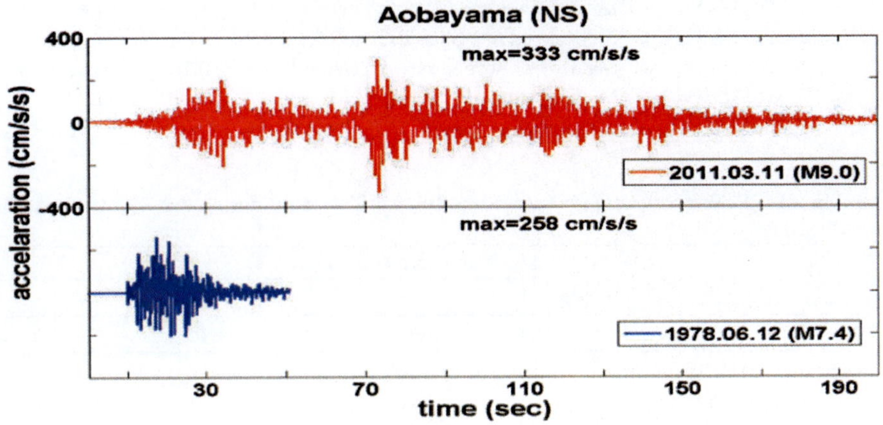

1F Acceleration Time Histories

Fig. 2.44 Acceleration waveforms at Tohoku University during the 2011 and 1978 earthquakes (transverse direction)

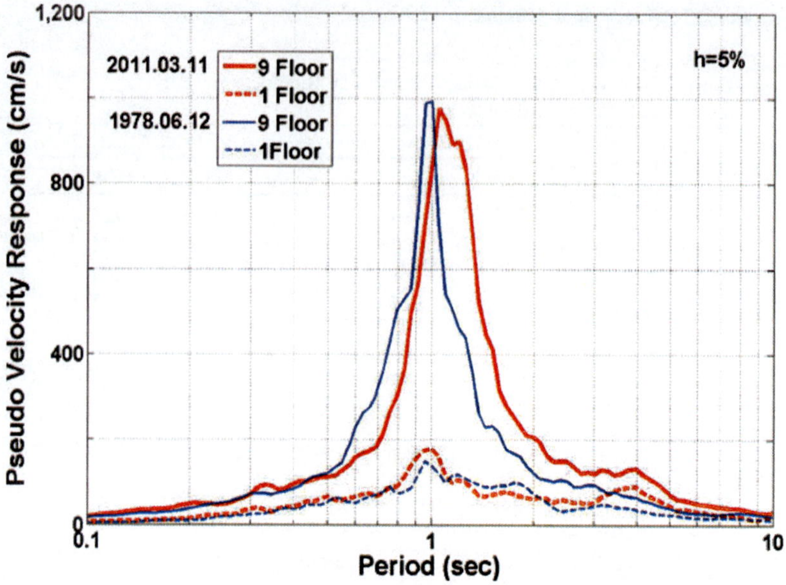

Fig. 2.45 Response spectra at Tohoku University for the 2011 and 1978 earthquakes (transverse direction)

Table 2.11 Predominant frequencies observed at Tohoku University during the 3/9, 3/11, and 3/19 earthquakes and microtremors

Date	Event name	Predominant frequency (Hz)	
		TR	LN
2011/3/9	Microtremors	1.61	1.61
2011/3/9	Foreshock Sanriku	1.26	1.26
2011/3/11	Microtremors	1.61	1.61
2011/3/11	Main shock first phase (section A)	1.05	1.05
2011/3/11	Main shock second phase (section B)	0.78	0.88
2011/3/19	Microtremors	1.17	1.37
2011/3/19	Aftershock	0.93	1.16

Table 2.12 Peak accelerations observed at Tohoku University during the 3/9, 3/11, and 3/19 earthquakes

Event name	Maximum acceleration (cm/s/s)					
	1F			9F		
	TR	LN	UD	TR	LN	UD
Microtremors (3/9)	–	–	–	–	–	–
Foreshock Sanriku (3/9)	37	34	23	171	89	51
Microtremors (3/11)	–	–	–	–	–	–
Tohoku earthquake first phase (section A) (3/11)	207	216	128	594	617	377
Tohoku earthquake second phase (section B) (3/11)	333	330	257	908	728	640
Microtremors (3/19)	–	–	–	–	–	–
Aftershock (3/19)	15	18	10	34	56	29

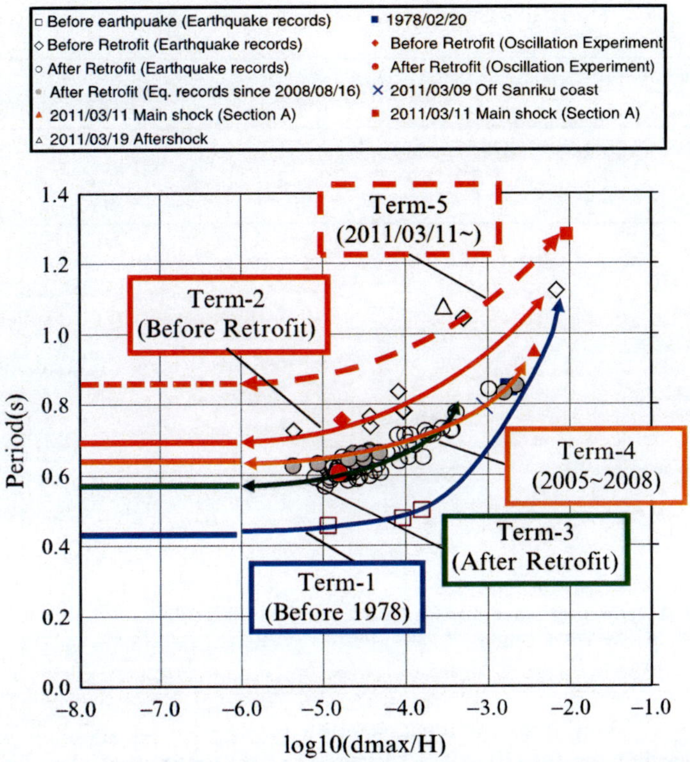

Fig. 2.46 Relation between predominant period and deflection angle of the Tohoku University building based on 40-years structural monitoring data after its completion (transverse direction)

phase of the 2011 earthquake is smaller compared to 1978 Miyagi-ken Oki earthquake, but became lager in the second phase, and that the stiffness of the building became smaller compared to that after 1978 earthquake.

2.2.3 Comparison of Strong-Motions Between the 2011 and Other Earthquakes Records

2.2.3.1 Comparison of the Observation Records During the 2011 and the 1978 Earthquakes at Basement Floor of Sumitomo Building

The observed records at Sumitomo building near Sendai station by DCRC, Tohoku University, are very valuable to compare the records during the 1978 Miyagi-ken Oki earthquake at the same observation point. It is noted that the observed record was situated as engineering bedrock motion. The observed ground motions at this observation point during the 2011 earthquake are compared with those due to 1978 Miyagi-ken Oki earthquake and also 2005 Miyagi-ken Oki earthquake.

Table 2.13 PGA observed at Sumitomo-seimei building during the 2011 and past disastrous earthquakes

Earthquake					
Date	M	NS (S25E)	EW (S65W)	UD	Seismometer
2011/3/11	9.0	317.7	234.1	180.3	SSA-1
1978/6/12	7.4	250.9	240.9	90.8	SMAC-Q
2005/8/16	7.2	131.2	90	64.6	SSA-1

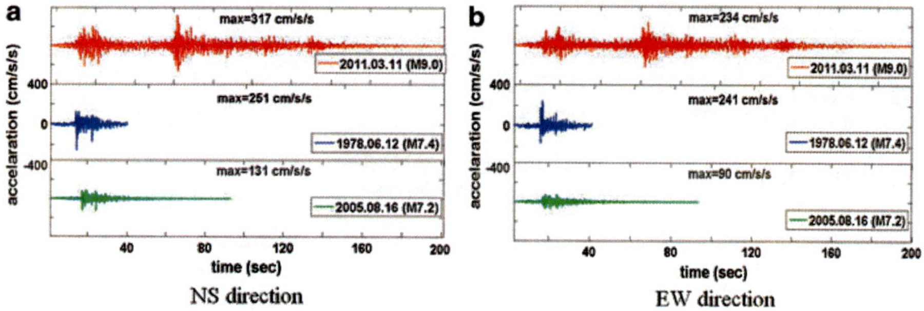

Fig. 2.47 Acceleration waveforms at Sumitomo-seimei building for the 2011 and 1978 earthquakes (transverse direction)

Table 2.13 shows maximum acceleration values of three components for 2011 Tohoku earthquake (M9.0), 1978 Miyagi-ken Oki earthquake (Mj7.4), and 2005 Miyagi-ken Oki earthquake (Mj7.2). Figure 2.47 shows the acceleration waveforms of two horizontal directions for the three earthquakes comparatively. Figure 2.48 shows the corresponding pseudo velocity spectra. Findings focused on the comparison of the 1978 earthquake from these figures are as follows:

1. The envelope shape of the 1978 earthquake is almost same as the first phase of the 2011 earthquake.
2. Period contents shorter than 1.5 s of the 2011 earthquake is larger than those of 1978 earthquake, about 20% larger at 1 s, and twice as large at 0.5 s.
3. The period contents around 3 s of the 2011 earthquake is twice as large compared to those of 1978 earthquake.

2.2.3.2 Ground Motion Amplification Characteristics in Aobayama Hill

It is important to investigate the ground motion characteristics at Aobayama campus of Tohoku University where some 8- and 9-story buildings were severely damaged. Ground motion amplification compared to Sumitomo building was investigated for the 2011 earthquake and 1978 Miyagi-ken Oki earthquake.

Figure 2.49 shows acceleration waveforms in the NS direction (transverse direction) at THU building during the 2011 earthquake and the 1978 Miyagi-ken

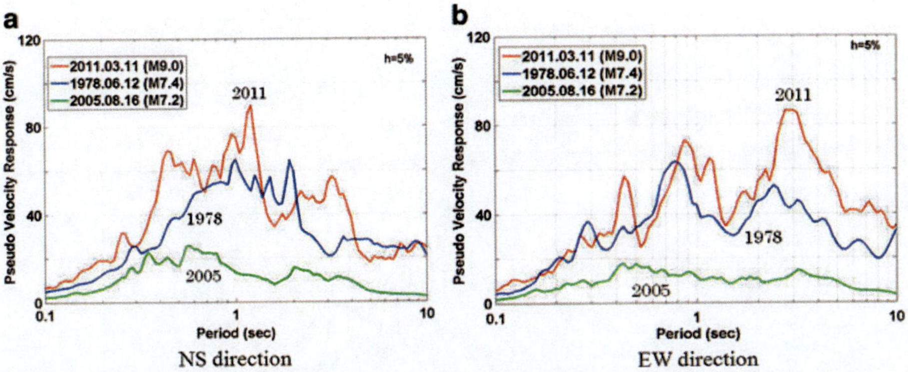

Fig. 2.48 Pseudo velocity response spectra at Sumitomo-seimei building for the 2011 and past disastrous earthquakes

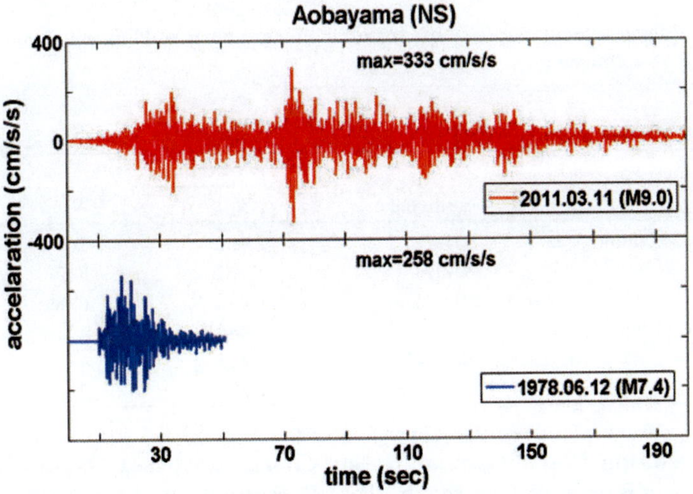

Fig. 2.49 Acceleration waveforms of the THU building during the 2011 and past disastrous earthquakes

Oki earthquake. Figure 2.50 shows the comparison of pseudo velocity spectra for the two earthquakes. It is found from these figures that the ground motion content at around 1 s period was amplified by more than two times at Aobayama campus compared to Sumitomo building near Sendai station not only for the 2011 earthquake but also for the 1978 Miyagi-ken Oki earthquake. It would be necessary to consider the geological amplification in seismic design and seismic retrofit of building structures.

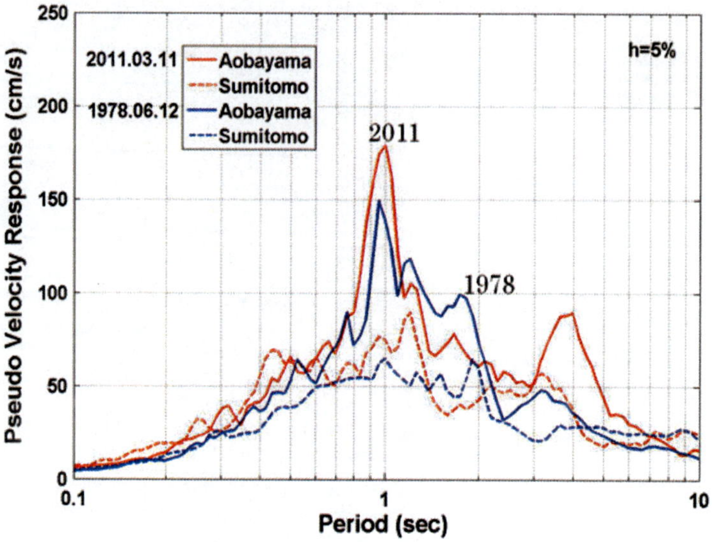

Fig. 2.50 Pseudo velocity response spectra at Sumitomo-seimei building and Aobayama for the 2011 and 1978 earthquakes

Table 2.14 PGA observed at MYG004 and MYG006

		PGA (cm/s/s)		
Station	Earthquake	NS	EW	UD
MYG004 (Tsukidate)	2011/3/11	2,700	1,268	1,880
	2011/4/7	1,242	886	476
	2008/6/14	740	678	224
MYG006 (Furukawa)	2011/3/11	444	571	239
	2011/4/7	415	478	233
	2008/6/14	238	233	104

2.2.3.3 Comparison of Ground Motion Characteristics in Osaki City and Kurihara City for the 2011 Earthquake and the 2008 Iwate–Miyagi Nairiku Earthquake

Severe ground motions were observed in the northern part of Miyagi Prefecture during the 2011 earthquake. The area was damaged during the 2008 Iwate–Miyagi Nairiku earthquake. The observation records at K-NET Tsukidate with JMA seismic intensity 7 and K-NET Furukawa were compared to those during the 2008 Iwate–Miyagi Nairiku earthquake.

Table 2.14 shows PGA values at K-NET Tsukidate (MYG004) and K-NET Furukawa (MYG006) for the two earthquakes. Figures 2.51 and 2.52 show acceleration–displacement response spectra (Sa–Sd spectra) in the two horizontal directions at K-NET Tsukidate and K-NET Furukawa for the two earthquakes comparatively.

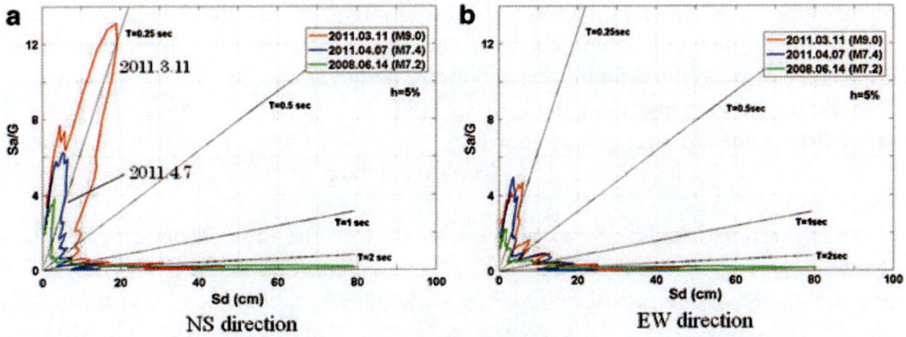

Fig. 2.51 Sa–Sd spectra for the earthquake records at MYG004

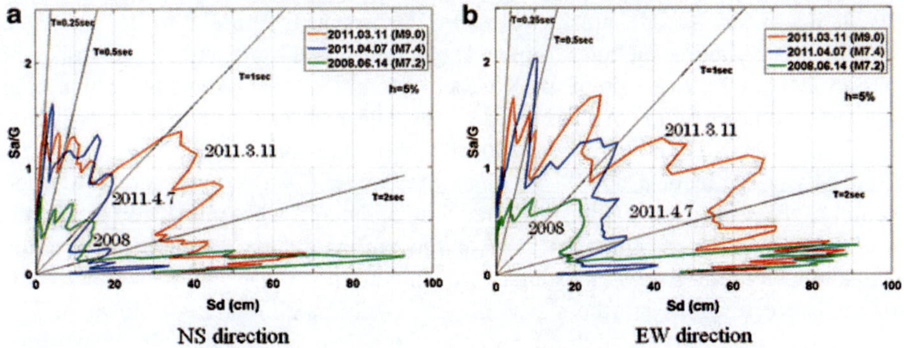

Fig. 2.52 Sa–Sd spectra for the earthquake records at MYG006

It is found from these figures that the observed motion at Tsukidate with JMA seismic intensity 7 is consisted of shorter period contents than 0.3 s and the displacement at around 1 s is as small as 10 cm although the amount is larger than that for the 2008 earthquake. In the case of K-NET Furukawa, it is found that the displacement during the 2011 earthquake was larger than that of the 2008 earthquake by comparing the displacement spectral values for the two earthquake and that the response displacement in the period range shorter than 1 s by the aftershock on April 7 is almost the same as that due to the main shock. These ground motion characteristics at Tsukidate of Kurihara City and Furukawa of Osaki City are consistent with damage features in the two areas.

2.3 Topography and Geology

The sequence of Tohoku earthquake caused extensive damage over much of eastern Japan. Damage was particularly heavy in Iwate, Miyagi, and Fukushima Prefectures, all in the Tohoku region; Ibaraki and Chiba Prefectures, both in the Kanto region;

Sakaemura, in Nagano Prefecture; and Fujinomiya, in Shizuoka Prefecture. The geological and geo-morphological characteristics of these areas are diverse; included are basins, hills, alluvial plains, coastal areas, and even landfill.

Topographical map of the Tohoku district [44] is shown in Fig. 2.53. Topographical map of the Kanto district [44] is shown in Fig. 2.54. Geological map of the Tohoku district [45] is shown in Fig. 2.55. Geological map of the Kanto district [46] is shown in Fig. 2.56.

Trending north–south along the western edge of the three Tohoku Prefectures (Iwate, Miyagi, and Fukushima) is the Ou Range, composed primarily of Cretaceous granitic and Neogene rock. The Ou Range includes Mount Iwate (2,038 m, Iwate), Mount Kurikoma (1,627 m, Miyagi and Iwate), Mount Zao (1,825 m, Miyagi), Mount Azuma (2,035 m, Fukushima), and numerous other mountains.

In Iwate Prefecture, the Ou Range, together with the Kitakami Mountains, which trend north–south along the eastern edge of the prefecture, forms a basin where cities such as Morioka, Hanamaki, Kitakami, Oshu and Ichinoseki are located. The Kitakami Mountains includes Mount Hayachine (1,917 m) and Mount Yakushi (1,645 m). On the east side of the Kitakami Mountains, Sanriku coast, which is in the eastern coastal region of the prefecture, is positioned. The cities of Rikuzentakata, Ofunato, Kamaishi, and Miyako are located along the coast.

In Miyagi Prefecture, an alluvial plain (the Sendai Plain) extends throughout the center of the prefecture. On it are located Sendai, Natori, Ishinomaki, and numerous other cities. In the northeastern part of the prefecture, the southern part of the Kitakami Mountains and the Oshika Peninsula are positioned. In the southern part of the prefecture, the northern part of the Abukuma Mountains is located. The Abukuma River flows through the mountain, and then empties into the Sendai Bay.

In Fukushima Prefecture, the Ou Range, together with the Abukuma Mountains running roughly parallel to the east, forms the Naka-dori, or central valley, within which the cities of Fukushima, Koriyama, and Sukagawa, among others, are located. The coastline runs to the east of the Abukuma Mountains, forming between the two the Hama-dori, or coastal region, on which sit the cities of Soma and Iwaki. The highest point in the Abukuma Mountains is Mount Otakine (1,193 m). In the Aizu Basin, which is a basin on the west of the Abukuma Mountains, Aizuwakamatsu City and Kitakata City are located.

The northern part of Ibaraki Prefecture contains several mountainous regions, most notably the Abukuma and Yamizo mountains, while the southern part, comprising the Hitachi Upland and the downstream basin of the Tonegawa River, is notable for a considerable degree of urban development, including the cities of Mito and Tsukuba. The Hitachi Up-land itself consists of numerous plateaus, the terraces, which tend to be covered with a thick layer of Kanto loam over relatively soft sand and gravel strata.

Chiba Prefecture likewise can be generally classified into a northern part, specifically the Shimousa Upland and its peripheral low-lands, and a southern part where hills predominate. The Shimousa Upland itself comprises the Narita Group (marine strata formed during the Pleistocene epoch) and, on top of that, layers of Kanto loam. Numerous cities are located on the upland, including Noda, Funabashi,

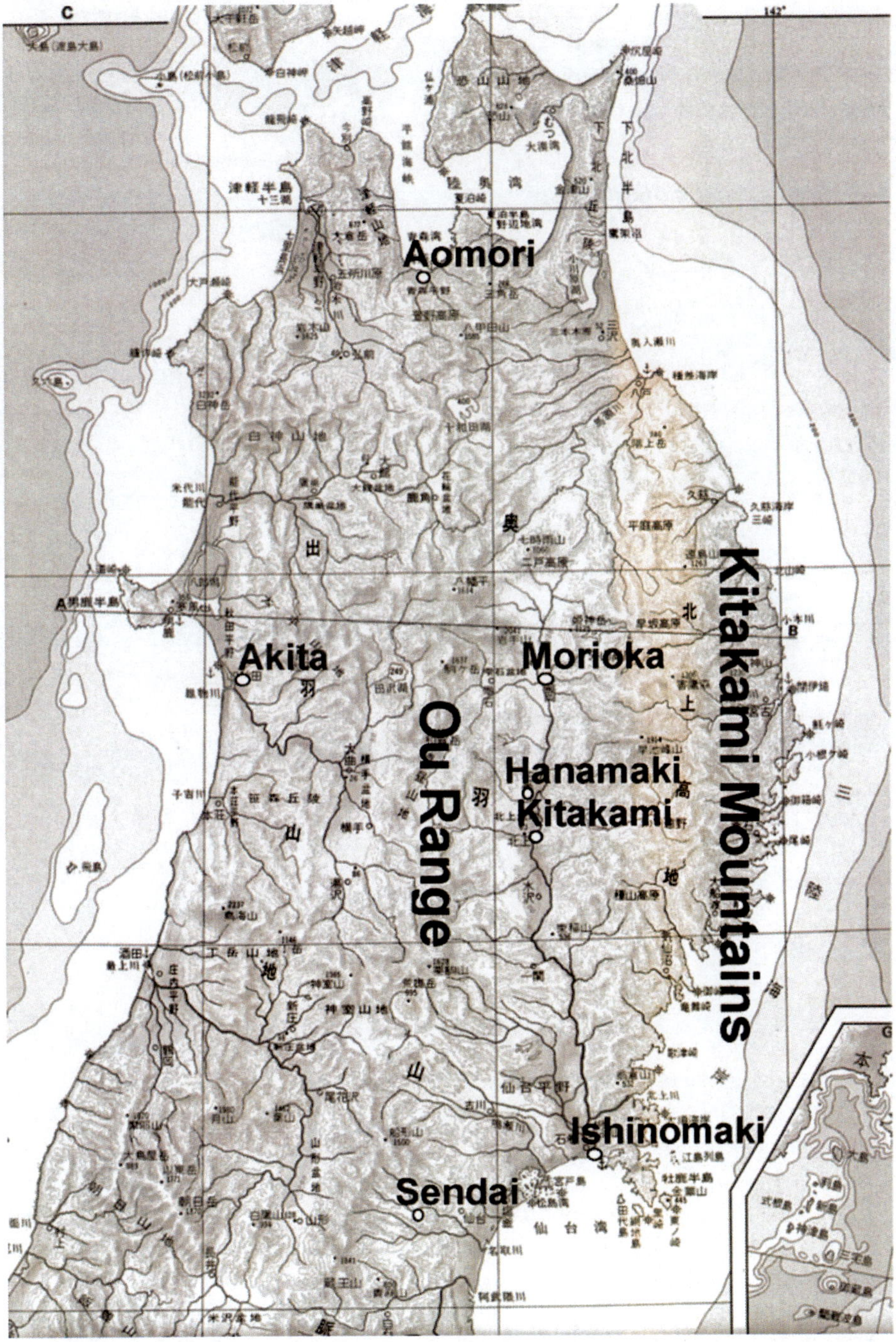

Fig. 2.53 Topographical map of the Tohoku district [44]

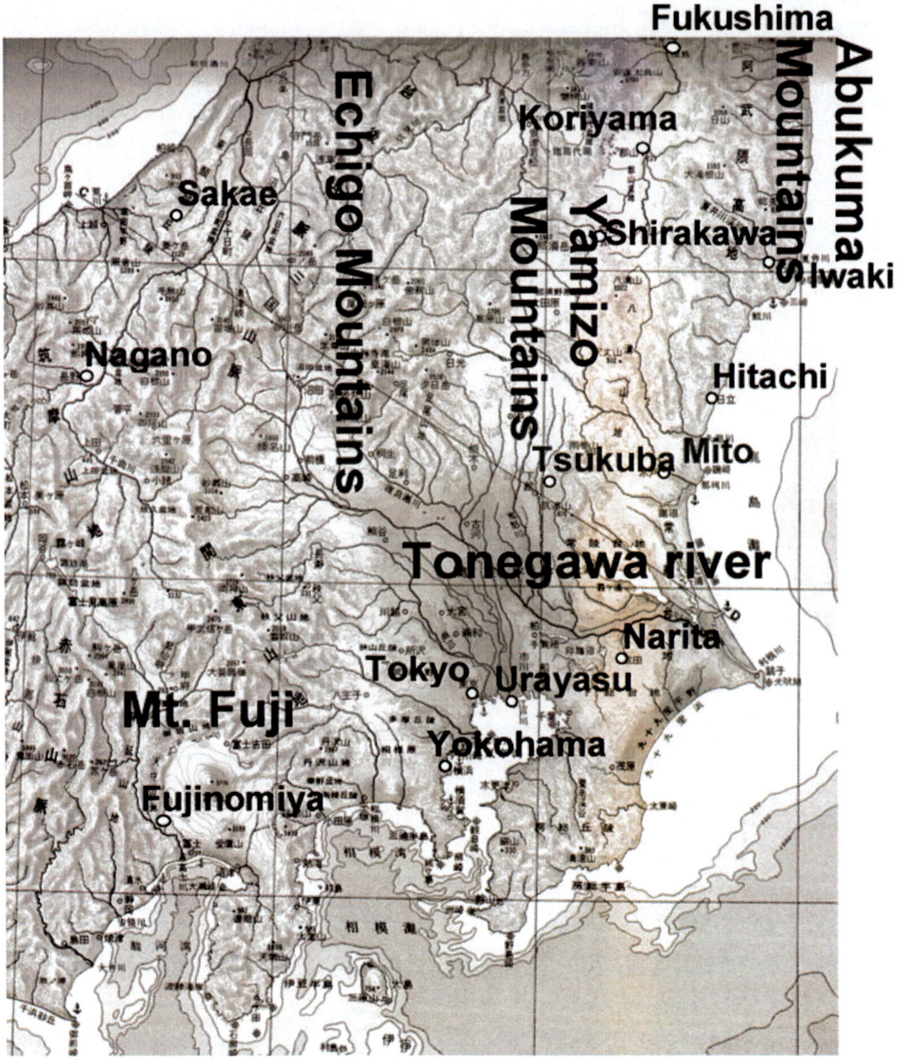

Fig. 2.54 Topographical map of the Kanto district [44]

and Narita. The northwest section of the Chiba Prefecture runs along the north side of Tokyo Bay, where the ground is naturally low and, in many cases, presently covered by landfill. Indeed, more than 70% of the City of Urayasu is built on landfill.

Sakae-mura, a village in the mountainous area, is located on the Chikuma River basin within the interior of northern Nagano Prefecture. Fujinomiya, a small rural city, is located to the west of Mount Fuji in the interior of eastern Shizuoka Prefecture. The center of the city lies within the Fujinomiya Lowlands, a fault angle basin.

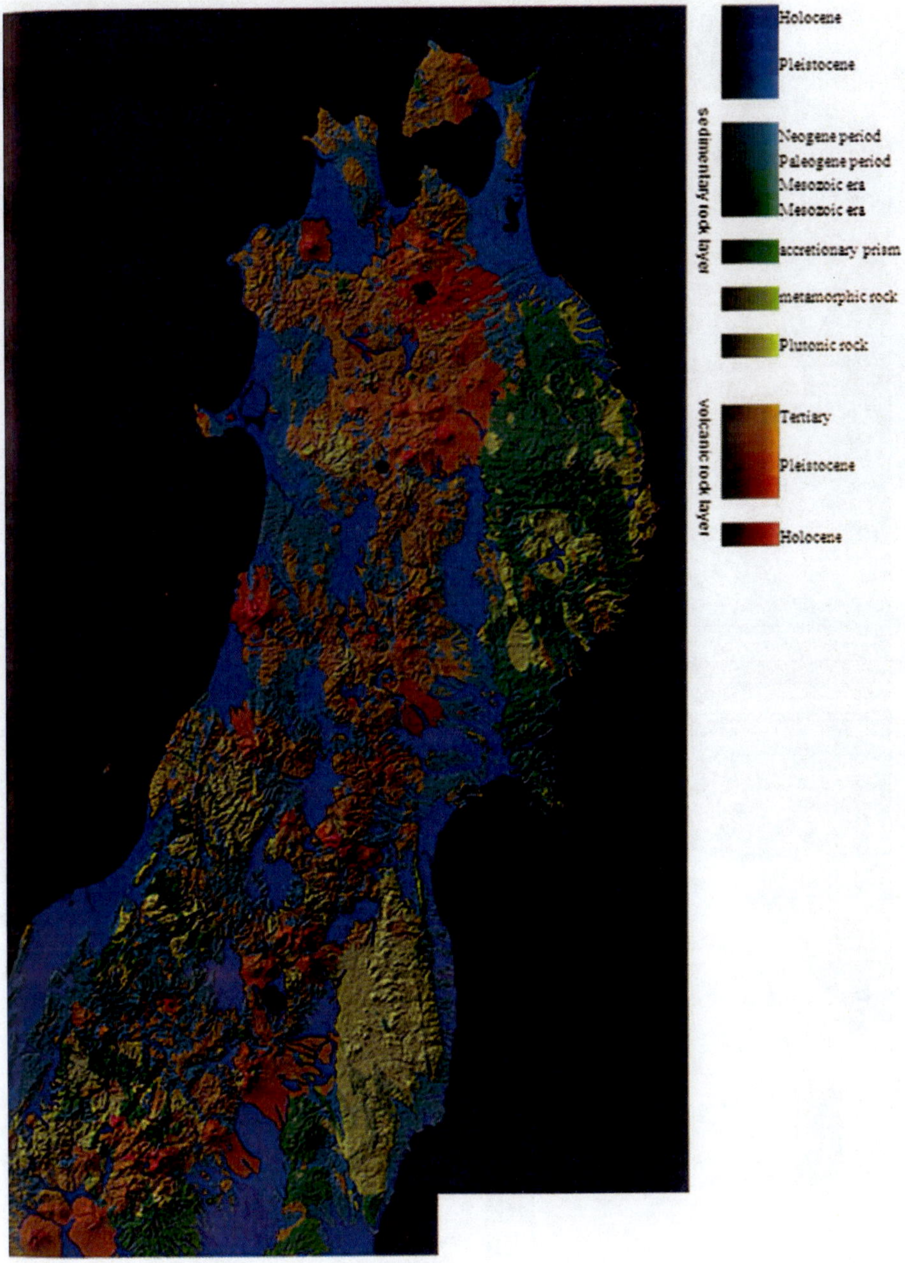

Fig. 2.55 Geological map of the Tohoku district [45]

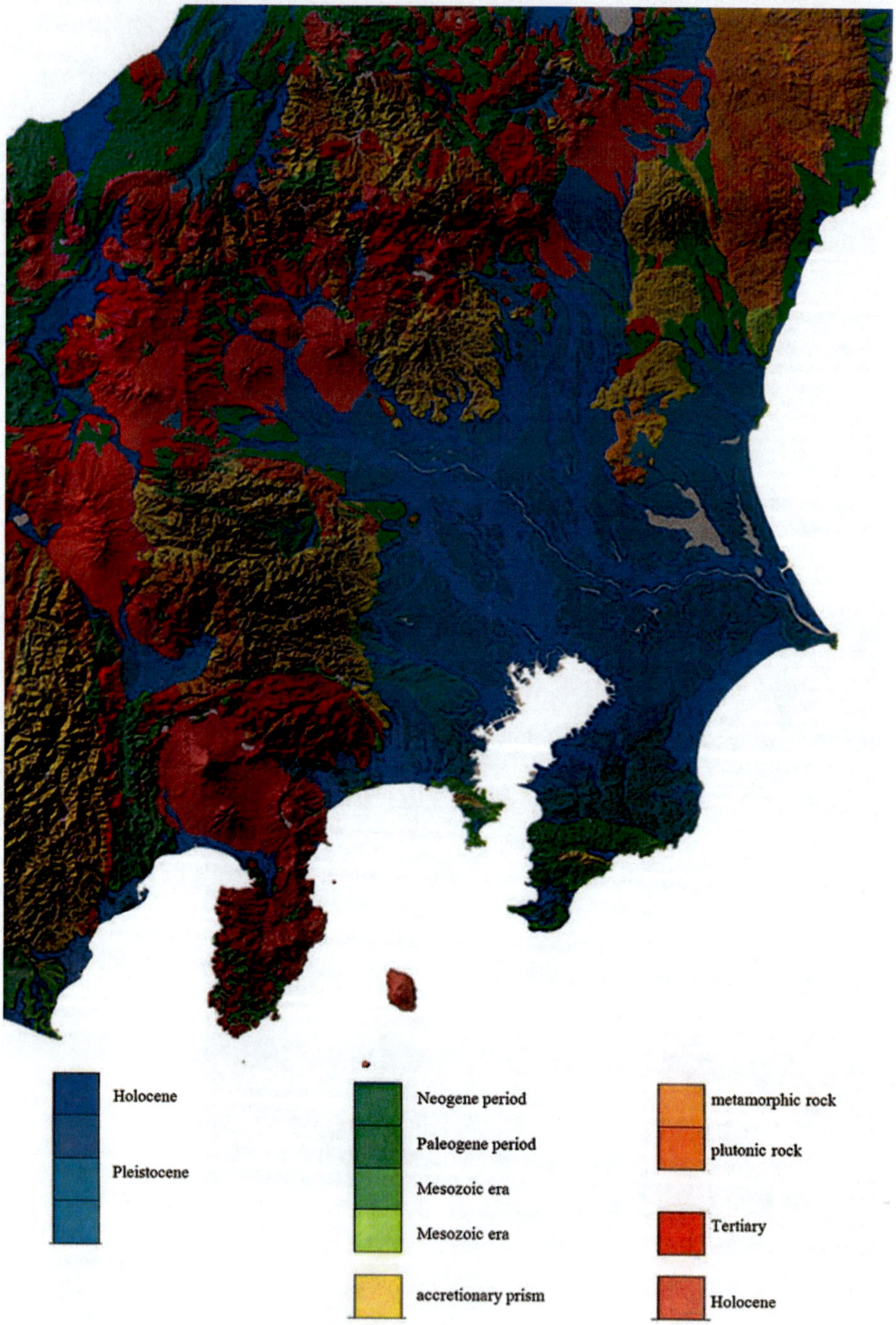

Fig. 2.56 Geological map of the Kanto district [46]

2.4 Tsunami

2.4.1 Mechanism of the Tsunami Generation

The Meteorological Research Institute analyzed the data observed on March 11, and examined the source region of the tsunami waves. The results are summarized in a report entitled "Analytical results on 2011 Tohoku-Chiho Taiheiyo-Oki earthquake" [47]. The results are shown in Fig. 2.57. The tsunami waves were recorded at 19 stations located along the Pacific Ocean coast through the Hokkaido region to the Kanto region. Tsunami source area has been estimated by inverse calculation using the tsunami arrival times. The area of tsunami source was found to be 550 km long and 200 km wide which stretched from offshore of Ibaraki Prefecture to Iwate Prefecture.

To validate the tsunami source model, the estimated area has been compared with the aftershock distribution, which has already shown in Sect. 2.1. The number of aftershocks with intensity larger than 4 by JMA scale is shown in Fig. 2.58 [48]. Aftershocks have occurred in the focal region of 500 km long and 200 km wide as well as in the eastern side of the trench axis. Thirty-three aftershocks occurred on

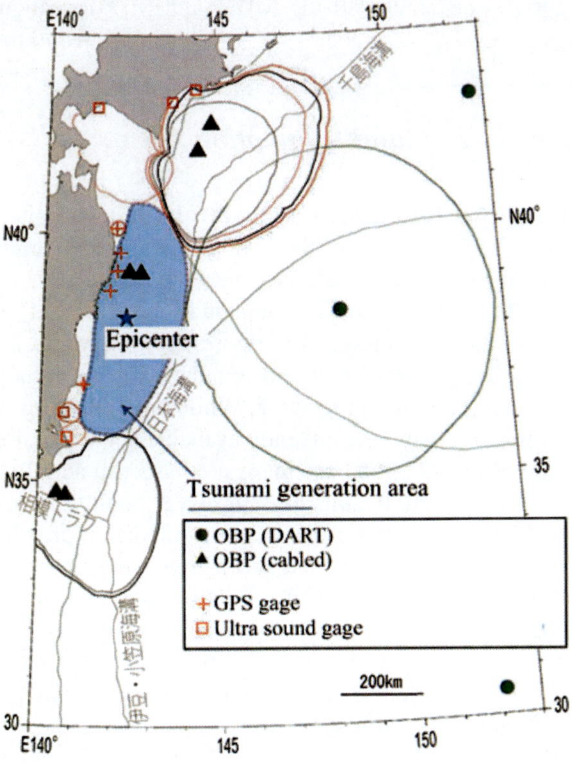

Fig. 2.57 Estimated tsunami source [47]

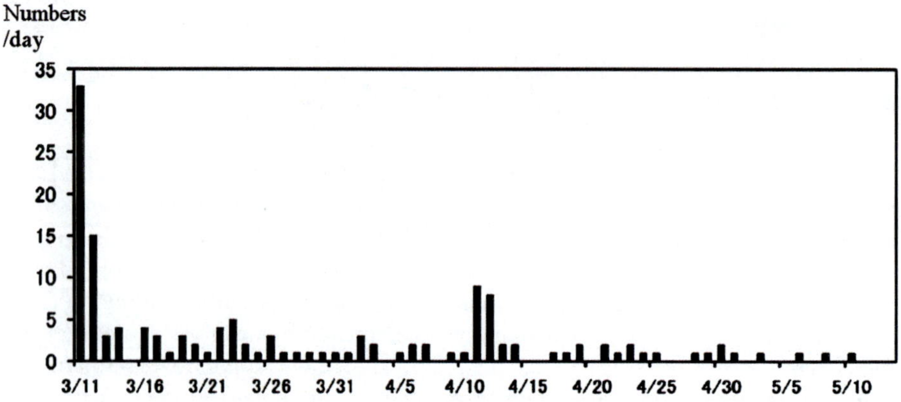

Fig. 2.58 Occurrence of aftershocks per day that generate ground motions with JMA seismic intensity higher than 4 [2]

March 11 and their number decreased gradually over time in spite of a certain increase in mid-April. Several earthquakes, which occurred outside of the focal region such as in southern Ibaraki, in northern Nagano, in eastern Shizuoka, and in northern Akita, generated ground motions with seismic intensity higher than 5 upper. They included the Mj7.1 and Mj7.0 earthquake of April 11 and April 7, respectively, and tsunami warnings were issued in their coastal areas.

2.4.2 Tsunami Inundation Area

Geospatial Information Authority Institute had taken aerial photos immediately after the occurrence of the earthquake and had published them on its website [49]. Aerial photos are useful to map the inundation area. An example is shown in Fig. 2.59. The photo was taken on March 13, 2011 at Motoyoshi, Kesen-numa City in Miyagi Prefecture. This figure shows that the low lying areas were inundated along Tsuyagawa River and inundation is bordered by the road embankment surrounding the low-lying areas. While only one inundation maps are shown here, 21 inundation maps are currently available at the website of the Geospatial Information Authority [50]. As these maps are based on aerial photos, they are not necessarily precise enough. Results of field survey could improve the preciseness in future.

Figure 2.60 shows that map of inundation area in Aomori Prefecture where the height of the tsunami height was relatively lower than that in Iwate Prefecture or Miyagi Prefecture. Therefore, inundation is generally seen in the narrow area of coastal plains. However, Hachinohe harbor and its vicinity was inundated extensively, especially along Mabuchi-gawa River, at which the tsunami was confirmed to run-up and reach beyond Mabuchi Dam which locating at 2.5 km upstream from the coast.

Figure 2.61 shows the map of the Sanriku coast of Iwate Prefecture. There are little broad plains in this area. Hence the inundated area is formed as a narrow

Fig. 2.59 Inundation area in Motoyoshi, Kesen-numa City [49]

band near the back of the bay or along river flowing into the bay. For example, the inundation observed in Ofunato City is along Sakari-gawa River which flows through the center of the city. Meanwhile, Rikuzentakata City, experienced unusual high waves and most of the urban area was inundated, because the tsunami arrived via both stretching of sandy beach 2 km along coast and Kesen-gawa River flowing into the bay and penetrating the urban area. The sandy beaches near Takata pine forest had entirely disappeared, and the beach was extensively eroded whereas there was heavy sand deposition in the urban area. In particular, inundated area extensively developed along the Yahagi River and Kesen-gawa River as well as along Hamada-gawa River which is located on the eastern side of the urban area. In Kesennuma City, development of inundation area was observed along Okawa River and Shikaori-gawa River similarly.

Figure 2.61 also shows the Motoyoshi district of Kesennuma City, Miyagi Prefecture. This area is located along Tsuya-gawa River. The tsunami run up to this district, which is 4.5 km upstream from the coast, also propagated along Makago-gawa River which is the right tributary of the Tsuya-gawa River. There is no record of tsunami run-up into the region so far and the resident did not feel the necessity of preparation for tsunami disaster. In the downstream of Tsuya-gawa River, the Koizumi Large Bridge, spanning the national highway No. 45 (180 m long) and JR Bridge (200 m long) over JR Kesen-numa Line was washed away by the tsunami.

Figure 2.62 shows the inundation area on Sendai Plain in Miyagi Prefecture. Unlike the previous Sanriku coast, this area is a plain of 40 km long stretching from

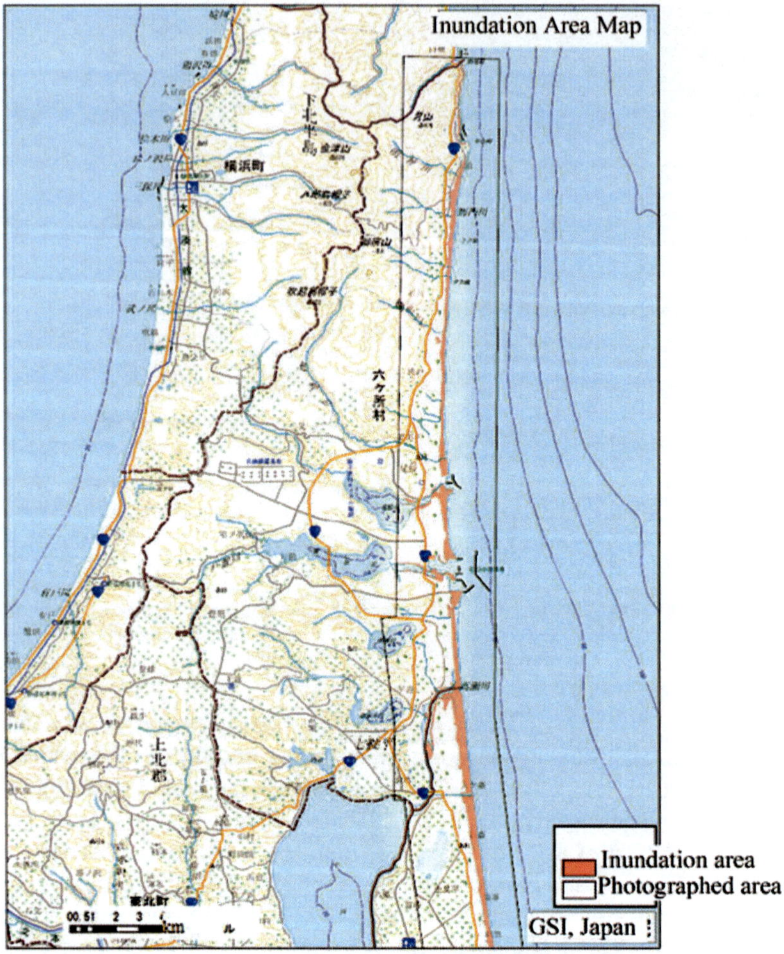

Fig. 2.60 Inundation area in Aomori Prefecture [50]

south (Yamamoto Town) to north (Sendai Port) and then inundation was uniform along the coast. The maximum inundation distance from the coastline was 5 km. The bank of Tobu-Sendai Road constructed parallel to the coast in this region probably contributed to the prevention of tsunami inundation. In the area near Abukuma-gawa River or Natori-gawa River, the tsunami run-up of longer distances was due to less resistance of the flow compared to the inland. A similar run-up in these rivers was observed when tsunami was transmitted after the 2010 Chile earthquake [51].

Sugawara et al. [52] have studied the tsunami inundation area due to the Jyogan Tsunami of 869 on the Sendai Plain [52]. By comparison, it is revealed that the inundation area in 869 was smaller than that in 2011. Figure 2.63 shows the inundation area in southern part of Fukushima Prefecture and northern part of Ibaraki Prefecture. In some limited places such as Onahama Port, the estuary of Same-gawa

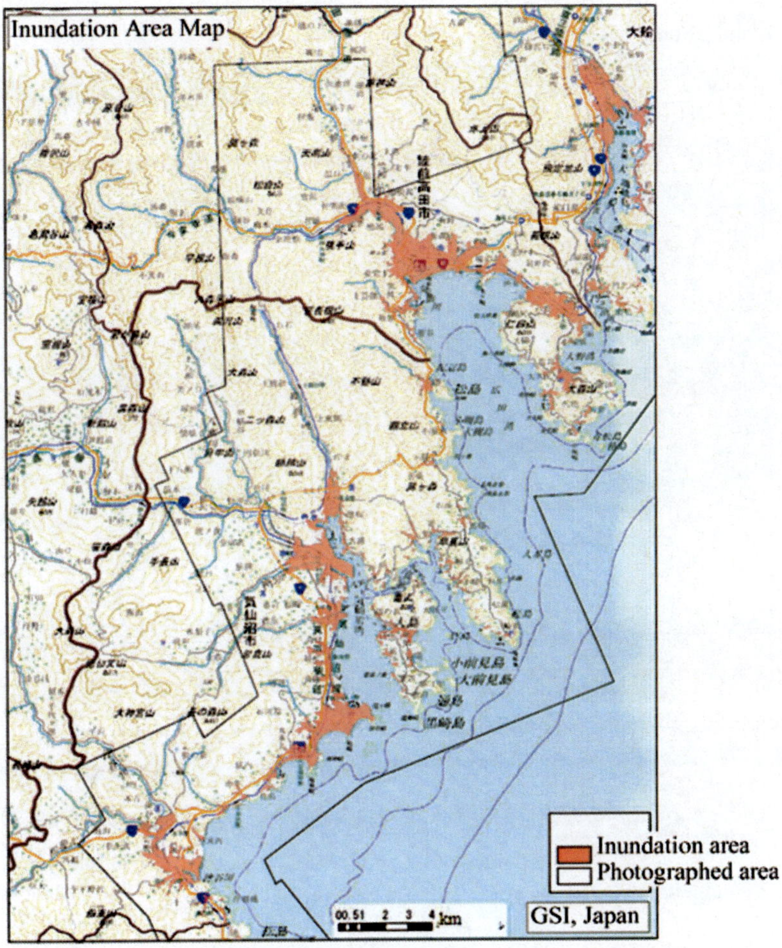

Fig. 2.61 Inundation area on Sanriku coast [50]

River, and Kitaibaraki City, were inundated. The large scale tsunami run-up had impacts on the sandy beach, leading to major changes of topography. It should be specially noticed that concentration of tsunami pulling flow, which is intensified depending on the effect of local topography around a structure, resulted in destruction of many structures [53].

Table 2.15 shows the number of families around the inundation area and tsunami affected areas [54, 55].

The area of inundation in Aomori Prefecture is 24 km^2; 2.8% of the total area of Aomori Prefecture, as seen from Table 2.15. The total people suffering the tsunami is about 15,000 people consisting in 5,300 households, and the ratio to the total population and households are 4.7% and of 4.1%, respectively.

The area of inundation in Iwate Prefecture is the 58k km^2; 1.2% of the total area of Iwate Prefecture. The population and the number 取る households in inundation

Fig. 2.62 Inundation area in Sendai Plain [50]

area is 107,000 and 40,000, respectively. They are 39.2% and 38.9% of the total population and total households in Iwate Prefecture, respectively.

In Miyagi Prefecture, inundate area is 327 km^2 and much larger than the other prefectures. The population and the number of households in the inundation area is 332,000 and 11,700, respectively, while the ratio to the total population and number of households in Miyagi Prefecture is 27.5% and 25.0%, respectively.

2.4.3 Time History of Observed Tsunami

The data related to sea level changes are open to public in real time on the website for coastal disaster prevention and other purposes in Japan. Such systems enabled

Fig. 2.63 Inundation area in southern Fukushima and northern Ibaragi Prefectures [50]

us to obtain also tsunami time histories during past tsunamis. However, at this time the tsunami heights were much larger than anticipated, and the measuring systems were destroyed at most of coastal stations. Therefore, detailed data such as time history of tsunami were obtained only at several limited places.

The observation station of JMA in Hachinohe was swept away by the tsunami. Fortunately, the monitoring instrumentation was retrieved from the sea floor in Hachinohe port near the station on May 11, 2011 [56]. Then the records were collected and analyzed and valid observation data were found to be recorded immediately after the spill, except for first 6 min. From the results, it was found that the observed tsunami had reached a height of 4.2 m above the sea level. Figure 2.64 shows the time history of the sea level.

Table 2.15 Inundated area and population in cities and towns in four prefectures in the Tohoku district [54, 55]

	Inundation area (km^2)	Whole area (km^2)	Ratio (%)	Estimated ratio of inundation (%)	
				In population	In household
Aomori Prefecture	24	844	2.8	4.7	4.1
Hachinohe	9	305	3.0	2.2	1.9
Misawa	6	120	5.0	4.7	3.6
Rokkasho Village	5	253	2.0	31.1	28.4
Oirase Town	3	72	4.2	15.8	14.4
Higashidori Village				3.1	3.0
Hashikami Town	0.5	94	0.5	8.1	6.9
Iwate Prefecture	58	4,946	1.2	39.2	38.9
Miyako	10	1,260	0.8	30.9	32.0
Ofunato	8	323	2.5	46.8	47.0
Kuji	4	623	0.6	19.4	18.2
Rikuzentakata	13	232	5.6	71.4	71.7
Kamaishi	7	441	1.6	33.3	32.5
Ohtsuchi Town	4	201	2.0	78.0	81.3
Yamada Town	5	263	1.9	61.3	63.2
Iwaizumi Town	1	993	0.1	10.5	9.9
Tanohata Village	1	156	0.6	41.2	40.2
Fudai Village	1	70	1.4	36.1	36.5
Noda Village	2	81	2.5	68.6	67.8
Hirono Town	1	303	0.3	15.3	15.2
Miyagi Prefecture	327	2,003	16.3	27.5	25.0
Sendai Miyagino	20	58	34.5	9.1	7.6
Sendai Wakabayashi	29	48	60.4	7.1	4.6
Sendai Taihaku	3	228	1.3	1.5	1.2
Ishinomaki	73	556	13.1	69.9	72.9
Shiogama	6	18	33.3	33.1	34.3
Kesen-numa	18	333	5.4	54.9	54.9
Natori	27	100	27.0	16.6	15.8
Tagajo	6	20	30.0	27.2	27.6
Iwanuma	29	61	47.5	18.2	15.0
Higashi-Matsushima	37	102	36.3	79.3	80.4
Watari Town	35	73	47.9	40.4	38.5
Yamamoto Town	24	64	37.5	53.8	55.7
Matsushima Town	2	54	3.7	26.9	28.7
Shichirigahama Town	5	13	38.5	44.8	42.9
Rifu Town	0.5	45	1.1	1.6	1.8
Onagawa Town	3	66	4.5	80.1	79.5
Minami-Sanriku Town	10	164	6.1	82.5	82.6
Fukushima Prefecture	112	2,456	4.6	13.5	11.9
Iwaki	15	1,231	1.2	9.5	8.8
Soma	29	198	14.6	27.6	23.2
Minami-Soma	39	399	9.8	18.9	15.7

(continued)

Table 2.15 (continued)

Hirono Town	2	58	3.4	25.6	24.5
Naraha Town	3	103	2.9	22.7	21.1
Tomioka Town	1	68	1.5	8.8	9.0
Ohkuma Town	2	79	2.5	9.8	9.1
Futaba Town	3	51	5.9	18.4	16.8
Namie Town	6	223	2.7	16.1	14.0
Shinchi Town	11	46	23.9	56.8	56.9

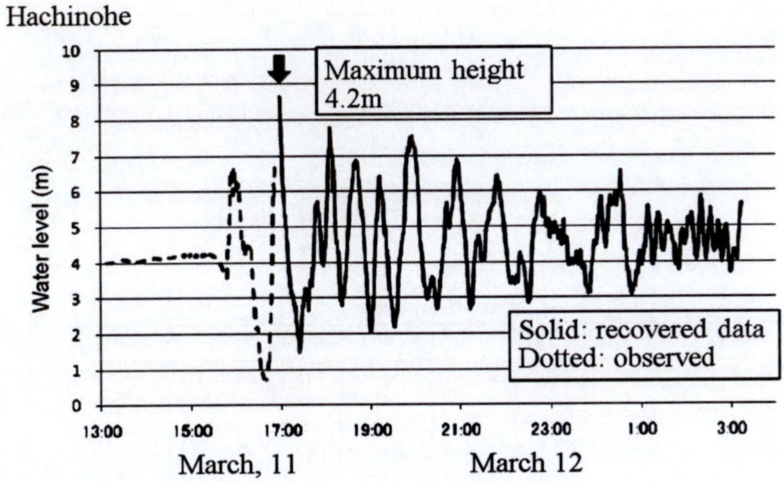

Fig. 2.64 Tsunami water level variation in Hachinohe [56]

Unlike the monitoring stations at coastal locations that were swept away, GPS sea level recorders floating on the sea at offshore stations left valuable records. The variation of sea level due to the tsunami at offshore location is usually smaller than that observed at coast. But it is reasonable to conclude that a GPS sea level recorder has enough sensitivity to detect moderate tsunamis necessary to issue tsunami warnings.

Figure 2.65 shows the time history of water level variation measured by a GPS recorded offshore of Iwate Prefecture [57]. The first wave is prominently large and following waves are gradually decreased from the second wave to the seventh wave. There was no clear periodicity from the first wave to the third wave, while the following waves starting from the fourth wave seemed to be repeated with a period of about 50 min.

Figure 2.66 shows the details of the early part of the waveform shown in Fig. 2.65. This plot shows gradual increase of the water level began on 15:00 and it took 6 min for the 2 m increase. Then sharp rise of water level of 4 m in 4 min followed.

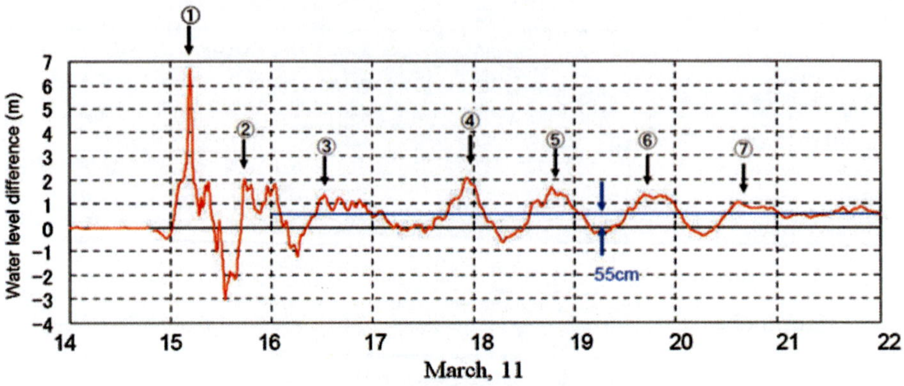

Fig. 2.65 Sea level variation measured by the GPS wave gauge at Iwate-ken Nanbu Oki station [57]

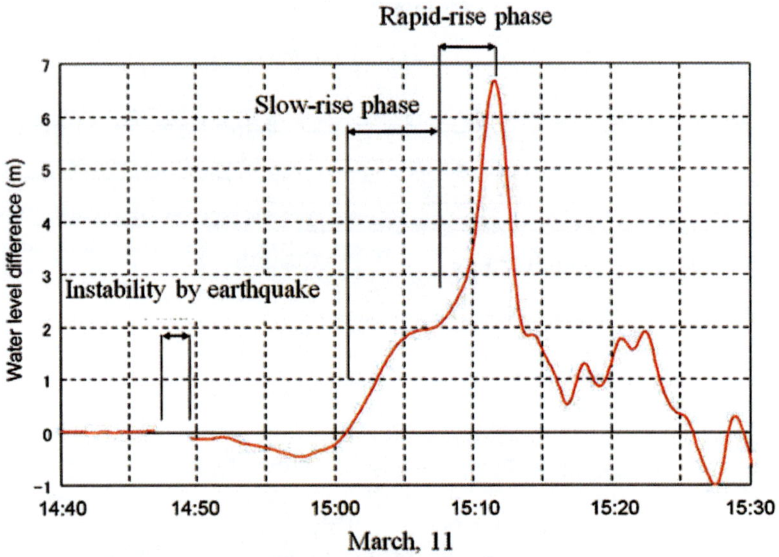

Fig. 2.66 Exaggerated sea water level variation measured by the GPS wave gauge shown in Fig. 2.65 [57]

Figure 2.67 compares the wave form after the 2011 Off the Pacific Ocean of Tohoku earthquake with that after the 2010 Great Chilean Tsunami. It is observed that the tsunami height of the 2011 tsunami is much larger that of the 2010 tsunami. Even the seventh peak of the 2011 tsunami (*a* in the figure) is much larger than the largest peak of the 2010 tsunami.

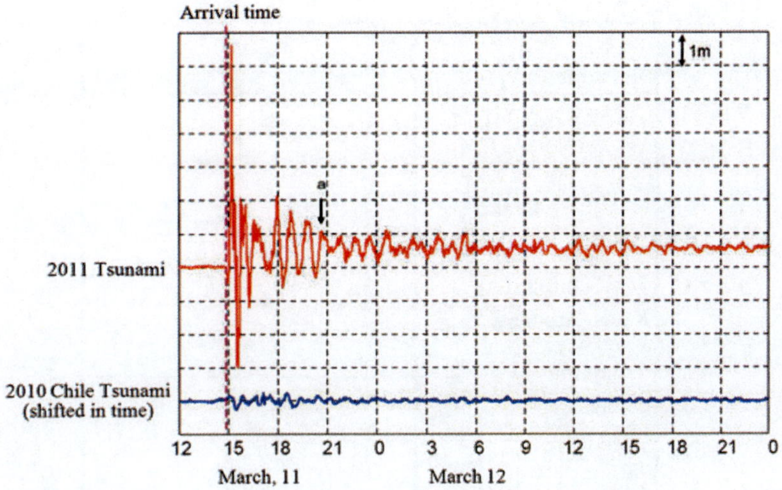

Fig. 2.67 Comparison between the 2011 Tsunami and the 2010 Chile Tsunami [57]

2.4.4 Maximum Inundation Heights and Maximum Run-Up Heights

A survey on the maximum inundation heights and the maximum run-up heights of tsunami water was conducted by the Port and Airport Research Institute. Figure 2.68 shows the data obtained at 105 points along the pacific coast from Kashima Port, Ibaraki Prefecture to Hachinohe Port, Aomori Prefecture. The definitions of inundation heights and run-up heights are as follows in this report:

- Inundation height: the altitude of the traces of tsunami water remaining on walls of the building from the level of the astronomical tide at the time of tsunami arrival
- Run-up height: the altitude of run-up traces of tsunami water from the level of the astronomical tide at the time of tsunami arrival

Table 2.16 shows the tsunami heights and inundation heights at several locations along the coast from the north end of Misawa City in Aomori Prefecture, through the south end of Soma City in Fukushima Prefecture. The dataset are compiled by the JMA. The dataset consists of data by JMA, the data of joint tasks by numerous institutions, as well as the data by a group "Joint Task Force for Reconnaissance of 2011 Tohoku-Chiho Taiheiyo-Oki Earthquake" established in the coastal engineering committee of Japan Societies of Civil Engineers (JSCE), where the definition of tsunami height and inundation height is explained as follows:

- Tsunami height: the altitude of sea water at tidal station or GPS sea level recorder from the level of the astronomical tide at the time of tsunami arrival

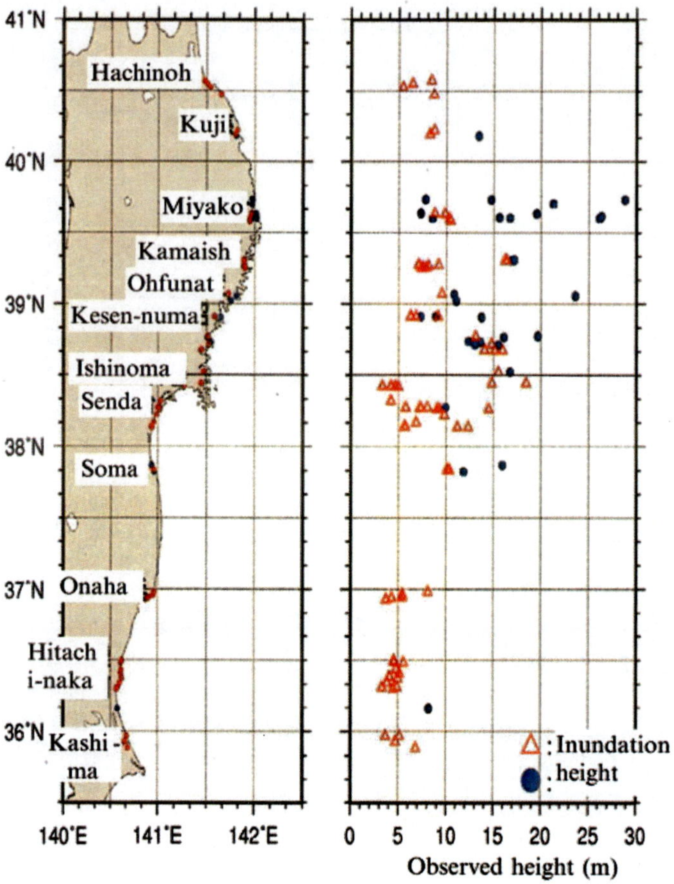

Fig. 2.68 Observed tsunami height [58]

- Inundation height: the altitude of the traces of tsunami water remaining on walls from the level of the astronomical tide at the time of tsunami arrival

It should be noted that the altitude of the traces of tsunami water from the ground is called "inundation depth." The terminology describing tsunami related quantity requires distinguishing "height" from "depth".

As shown in Fig. 2.69, the tsunami heights along the coast from Miyako City to Soma City were approximately from 8 to 9 m in general. As shown in Figs. 2.68 and 2.69, inundation height exceeds 10 m at the cities in the south of Kuji City. The coastal locations from the northern Iwate Prefecture to Oshika peninsula in Miyagi Prefecture have inundation heights from 10 to 15 m. Whereas the inundation heights were around 5 m and relatively low in Matsushima Bay, the inundation heights from Sendai Bay to Souma City increased to 10 m. It is noted that some inundation heights of adjacent cities is quite different for Kamaishi City and Ishinomaki City.

Table 2.16 Corresponding table for the observed tsunami height [59]

Prefecture	Name of the area	Inundation height (m)	Run-up height (m)
Aomori	Hachinohe Port, Hachinohe	5.4–8.4	
	Daija, Hashikami Town	8.6	
Iwate	Kuji Port, Kuji	8.2–8.7	13.4
	Taoi, Miyako		7.8–28.8
	Miyako Port, Miyako	8.7–10.4	7.3–16.7
	Outside of Miyako Port, Miyako		19.5–26.3
	Ryoishi, Kamaishi	16.2–16.4	16.9–17.1
	Kamaishi Port, Kamaishi	6.6–9.1	
	Ayarishirahama/Nagasaki, Ofunato		11.0–23.6
	Ofunato Port, Ofunato	9.5	10.8
Miyagi	Karakuwa, Kesen-numa		13.7
	Kesen-numa Port Kesen-numa	6.3–9.1	7.3–8.8
	Motoyoshi, Kesen-numa	13.1	16.1–19.7
	Utatsu, Minami-Sanriku Town	14.8	12.3–15.6
	Shizugawa, Minami-Sanriku Town	14.0–15.9	
	Ogatus, Ishinomaki	15.5	16.7
	Ishinomaki Port, Ishinomaki	3.3–5.0	
	Onagawa Port, Onagawa	14.8–18.4	
	Shiogama Port, Shiogama	4.2	
	Sendai Port, Sendai	5.7–14.5	9.9
	Arahama, Wakabayashi-ku Sendai	9.8	
	Sendai Airport, Natori	5.6–12.3	
Fukushima	Tsurishihama, Shinchi Town		15.9
	Soma Port, Soma	10.1–10.4	11.8
	Toyoma Todoro/Nakanosaku, Iwaki	5.4–8.1	
	Onahama Port, Iwaki	3.7–5.4	
Ibaraki	Hitachi Port, Hitachi	4.5–5.6	
	Hitachi-naka Port, Hitachi-naka	4.8–5.1	
	Nakaminato/Ajigaura/Isozaki Hitachi-naka	3.8–5.0	
	Oharai Port, Oharai	3.3–4.9	
	Ohtake, Hokota		8.2
	Kashima Port, Kashima	3.7–6.8	

These cases may be attributed to the tsunami shielding effect of the breakwater for Kamaishi City and Oshika peninsula for Ishinomaki City [59].

2.4.5 *An Example of Tsunami Damage: Taro-Cho Case in Miyako City*

Figure 2.70 shows the distribution of tsunami inundation in Taro-Cho in Miyako City [50]. Taro-Cho, a town that belongs to Miyako City, is small but has been recognized as one of the notorious places devastated by tsunamis many times in its

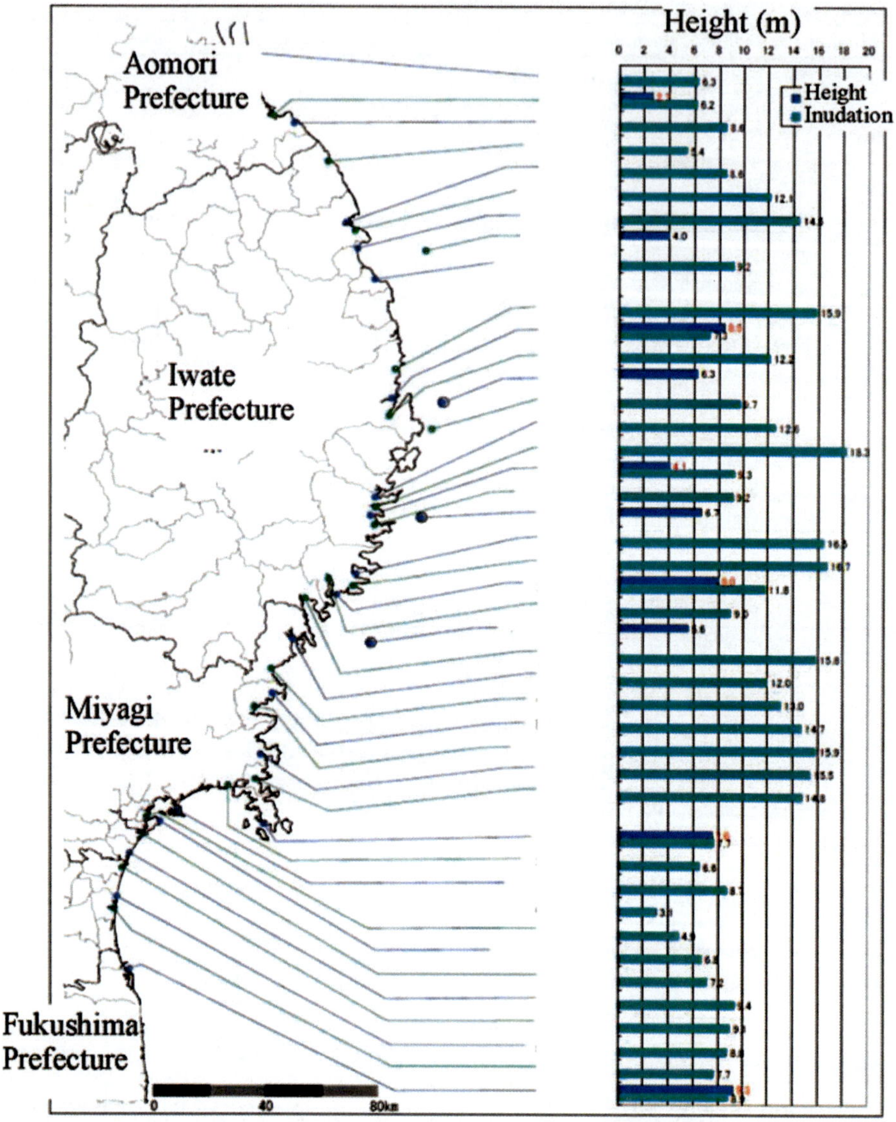

Fig. 2.69 Observed tsunami height (*dark blue*) and inundation height [60]

history. Figure 2.71 shows the devastation of the central area seen from the top of the seawall in the sea. The Town Hall, a 3-story RC building, was not damaged by the tsunami because it is constructed on a hill, the ground level of which is 5 m higher than the road as shown in Fig. 2.72.

Fig. 2.70 Inundation area in Taro-Cho [4]

Fig. 2.71 Overview of Taro-Cho central area from the seawall top

Fig. 2.72 Taro-Cho town office (3-storied RC building on high ground, not affected)

The seawall is designed to protect the entire Town of Taro. The south and the north side views from seawall top are shown in Figs. 2.73 and 2.74. The seawall is spanning from the northern end to the southern end of the town continuously. While most of the wooden houses were swept away as shown in the Figs. 2.73 and 2.74, several reinforced concrete and steel buildings remained structurally stable.

Figure 2.75 shows seawall with height of 10 m from the sea level. All the buildings have some damages except those buildings on the hillside. Figure 2.76 shows several remaining reinforced concrete and steel buildings after tsunami. Figure 2.77 shows a 5-storied building of steel construction in which the inundation depth reached 7 m. Figure 2.78 shows a 2-storied building of steel construction in the central area of Taro-Cho. The exterior cladding of the first story was totally destroyed. This building survival may be attributable to the destruction of the cladding, which resulted in reduction of tsunami pressure.

Figure 2.79 shows a 2-storied and a 3-storied reinforced concrete buildings in west end area of Taro-Cho. They remained after tsunami. Located in the western edge of Taro-Cho, the elevation of the ground level is slightly higher than the sea level and so the inundation height was below the second floor level of these buildings.

Figure 2.80 shows a hotel near the southern end of the Taro-Cho. It is a 5-storied steel structure. It remained structurally intact while tsunami water washed away the non-structural elements below the second floor.

Fig. 2.73 South-side view from the seawall top

Fig. 2.74 North-side view from the seawall top

Fig. 2.75 A huge seawall in Taro-Cho, 10 m high from the sea level

Fig. 2.76 Reinforced concrete buildings and steel-made buildings that remained after tsunami

Fig. 2.77 A steel-made 5-storied building in the central area of Taro-Cho that remained after tsunami

Fig. 2.78 A steel-made 2-storied building in the central area of Taro-Cho that remained after tsunami

Fig. 2.79 2-storied and 3-storied RC buildings in the western end area of Taro-Cho

Fig. 2.80 A steel-made 5-storied hotel in the southern end area of Taro-Cho

Fig. 2.81 A wooden house destroyed by tsunami with 3.2 m inundation height

Figure 2.81 shows a collapsed wooden house located at a lower site near the Town Hall of Taro-Cho. The inundation depth of the house is estimated to be 3.2 m.

2.4.6 History of the Tsunamis in Japan

The coastal regions along Pacific Ocean from Kanto district through Tohoku district have been hit by several tsunamis and earthquakes in the past. Similarities of the historic earthquakes such as the 869 Jyogan earthquake, 1611 Keicho Sanriku earthquake, 1896 Meiji Sanriku earthquake, and 1933 Showa Sanriku earthquakes have been noted in recent years. Other earthquakes including 1952 Tokachi-oki Earthquake and 1960 Chile earthquake caused similar devastation. Meanwhile, no critical tsunami was generated and consequently no serious damage occurred after the 2010 Chile earthquake. Some evacuated residents returned homes in spite of the evacuation alarm based on the individual decision without sufficient information. It is our serious subject how to transfer knowledge for seriously needed immediate evacuation in case that the official tsunami alarm is issued to those people who live in areas where historically suffered from devastating tsunamis repeatedly.

Acknowledgement Microtremor observations at MYG004 were performed by the joint investigation team of DPRI (H. Kawase, S. Matsushima, Baoyintu, F. Nagashima, K. Nakano) and Shimizu Corp. (T. Satoh, T. Hayakawa, M. Ohshima). Data from K-NET and KiK-net distributed promptly by NIED are highly appreciated. Thanks are also given to Profs. K. Irikura, H. Kawabe, and K. Asano for their kind and prompt supply of their unpublished materials and to all the institutions who provided us their nice scientific findings on the web.

References

1. Yoshida Y, Ueno H, Muto D, Aoki S (2011) Source process of the 2011 off the Pacific Coast of Tohoku earthquake with the combination of teleseismic and strong motion data, Earth Planets Space, Special Issue, 63, 565–569, also at http://www.mri-jma.go.jp/Dep/sv/2011tohokutaiheiyo/index.html. Accessed Feb. 15, 2012
2. Japan Meteorological Agency (2011) Aftershock area of the 2011 off the Pacific coast of Tohoku earthquake, http://www.jma.go.jp/jma/menu/ yoshin_area.pdf. Accessed Feb. 15, 2012 (Aftershock data in Fig. 2.2 was taken from the daily earthquake location map at http://www.seisvol.kishou.go.jp/eq/daily_map/sendai/20110311.shtml. Accessed March 31, 2011)
3. The Headquarters for Earthquake Research Promotion (2009) Long-term evaluation of seismic activity from Sanrik-Oki to Ibaraki-Oki, Fig.1, http://www. jishin.go.jp/main/chousa/09mar_sanriku/f01.htm. Accessed Feb. 15, 2012
4. Japan Meteorological Agency (2011) Daily earthquake location map, http://www.seisvol.kishou.go.jp/eq/daily_map/sendai/20110309.shtml. Accessed March 31, 2011
5. Ozawa S, Nishimura T, Suito H, Kobayashi T, Tobita M, Imakiire T (2011), Coseismic and postseismic slip of the 2011 Magnitude-9 Tohoku-Oki earthquake, Nature, 475, 373–376, doi:10.1038/nature10227, also The Geospatial Information Authority of Japan, at http://www.gsi.go.jp/cais/topic110314.2-index.html. Accessed Feb. 15, 2012
6. The Geospatial Information Authority of Japan (2011) Crustal deformation on land and off shore regions and slip distribution model of the 2011 off the Pacific coast of Tohoku Earthquake, http://www.gsi.go.jp/cais/topic110520-index.html. Accessed Feb. 15, 2012
7. Fujii Y, Satake K , Sakai S , Shinohara M, Kanazawa T (2011) Tsunami source of the 2011 off the Pacific coast of Tohoku Earthquake, Earth Planets Space, Special Issue, 63, 815–820, also Building Research Institute, at http://iisee.kenken.go.jp/staff/fujii/OffTohokuPacific2011/tsunami_ja.html. Accessed Feb. 15, 2012
8. United States Geological Survey (2011) Finite fault model, Updated result of the March 11, 2011 Mw 9.0 earthquake offshore Honshu, Japan, http://earthquake.usgs.gov/earthquakes/eqinthenews/2011/usc0001xgp/finite_fault.php. Accessed Feb. 15, 2012
9. Earthquake Research Institute (2011) Source process and strong ground motion for the 2011 Off the Pacific Coast of Tohoku earthquake, http://outreach.eri.u-tokyo.ac.jp/eqvolc/201103_tohoku/danwakaishiryou/. Accessed Feb. 15, 2012
10. National Inst. Earth Science and Disaster Prevention (2011) Strong ground motions during the 2011 Off the Pacific Coast of Tohoku earthquake, http://www.kyoshin.bosai.go.jp/kyoshin/topics/TohokuTaiheiyo_20110311/nied_kyoshin2j.pdf. Accessed Feb. 15, 2012
11. Si S, Midorikawa S (1999) New attenuation relationships for peak ground acceleration and velocity considering effects of fault type and site condition. J Struct Constr Eng Architect Inst Jpn 523:63–70
12. Asano K, Iwata T (2011) Strong ground motion generation during the 2011 Tohoku-Oki Earthquake. AGU 2011 fall meeting, U42A-03, December 2011. Also at http://sms.dpri.kyoto-u.ac.jp/k-asano/pdf/jpgu2011.pdf. Accessed Feb. 15, 2012
13. Kawabe H, Kamae K, Uebayashi H (2011) Source model of the 2011 Tohoku-Chiho Taiheiyo-Oki earthquake. Seismological Society of Japan Fall Meeting, B22-05, 2011. Also at http://www.rri.kyoto-u.ac.jp/jishin/. Accessed Feb. 15, 2012

14. Kurihashi S, Irikura K (2011) Source model for generating strong ground motions during the 2011 Off the Pacific Coast of Tohoku earthquake. Earth Planets Space 63(Special Issue):571–576
15. Wang D, Mori J (2011) Rupture process of the 2011 Off the Pacific Coast of Tohoku earthquake (Mw 9.0) as imaged with back-projection of teleseismic P-waves. Earth Planets Space 63(Special Issue):603–607
16. Satoh T, Kawase H, Matsushima S (1998) Source spectra, attenuation function, and site amplification factors estimated from the K-NET records for the earthquakes in the Border of Akita and Miyagi Prefectures in August, 1996, Zisin. J Seismol Soc Jpn Ser-2 50(4):415–429
17. Kawase H, Matsuo H (2004) Amplification characteristics of K-NET, KiK-net, and JMA shindokei network sites based on the spectral inversion technique. In: 13th world conference on earthquake engineering, Vancouver, Paper No. 454
18. Kawase H (2006) Site effects derived from spectral inversion method for K-NET, KiK-net, and JMA strong-motion network with special reference to soil nonlinearity in high PGA records. Bull Earthquake Res Inst Univ Tokyo 81:309–315
19. Satoh T, Kawase H, Matsushima S (2001) Estimation of S-wave velocity structures in and around the Sendai Basin, Japan, using array records of microtremors. Bull Seismol Soc Am 91:206–218
20. Ministry of Land, Infrastructure, Transport and Tourism, Kanto Regional Development Bureau and The Japanese Geotechnical Society (2011) Soil liquefaction survey in Kanto district during the Off the Pacific Coast Tohoku earthquake. http://www.ktr.mlit.go.jp/ktr_content/content/000043569.pdf. Accessed 15 Feb 2012 (in Japanese)
21. Kawase H, Sánchez-Sesma FJ, Matsushima S (2011) The optimal use of horizontal-to-vertical spectral ratios of earthquake motions for velocity inversions based on diffuse field theory for plane waves. Bull Seismol Soc Am 101(5):2001–2014
22. Building Research Institute (2011) http://www.kenken.go.jp/japanese/contents/topics/20110311/pdf/quickreport/0311quickreport_40.pdf. Accessed 15 Feb 2012
23. Nagato K, Kawase H (2004) Damage evaluation models of reinforced concrete buildings based on the damage statistics and simulated strong motions during the 1995 Hyogo-ken Nanbu earthquake. Earthquake Eng Struct Dynam 33(6):755–774
24. Shoji Y, Kamiyama M (2000) A High-Density Array Observation System Small-Titan: An Introduction and It's Strong-Motion Records, Proc. 12th World Conf. Earthq. Eng., 1–8. Also at http://smweb.tohtech.ac.jp/smalltitan/english/index.html. Accessed Feb. 15, 2012
25. Ohno S, Motosaka M, Sato T, Yamamoto Y (2004) Strong-motion observation network for seismic capacity evaluation of buildings in Sendai, Japan. Summaries of technical papers of annual meeting. Architectural Institute of Japan, Structures II, pp 1075–1076 (in Japanese)
26. Kamiyama M, Matsuzawa T, Anazawa M (2011) Strong-Motion Records Obtained by Small-Titan of Tohoku Institute of Technology During the 2011 Great East Japan Earthquake, Quick Report Version 1 Issued for the Tohoku Branch of the Japanese Geotechnical Society. http://smweb.tohtech.ac.jp/smalltitan/english/Reference_paper/Strong-motion%20records%20furing%203.11Earthq_Eng.pdf. Accessed Feb. 15, 2012
27. Ohno S, Motosaka M, Shibayama A (2011) Strong-motion characteristics in Sendai during the 2011 Off the Pacific Coast of Tohoku earthquake. Summaries of technical papers of annual meeting. Architectural Institute of Japan, Structures II, p 21095 (in Japanese)
28. Sendai City (2002) Earthquake Damage Estimation Report 2002, http://www.city.sendai.jp/kurashi/shobo/shiryo/__icsFiles/afieldfile/2010/09/01/jisin.pdf (in Japanese). Accessed Feb. 15, 2012
29. Saruta M, Watanabe H, Izumi M (1989) Proof test of base-isolated building using high damping rubber bearing. In: Transactions of the 10th international conference on SMiRT, vol K2, pp 631–636
30. Tohoku University (2008) Strong motion records inside buildings during the Iwate-Miyagi Inland earthquake of June, 14, 2008, http://www.disaster.archi.tohoku.ac.jp/Saigai/tohoku/20080614_BuildingRecords_r1.pdf (in Japanese). Accessed Feb. 15, 2012
31. Tohoku University (2008) Strong motion records inside buildings during the Iwate-ken Hokubu earthquake of July, 24, 2008, http://www.disaster.archi.tohoku.ac.jp/Saigai/wiki/→2008/7/24→20080724_EqRecords.pdf (in Japanese). Accessed Feb. 15, 2012

32. Tsukuda S et al. (2006) Seismic Response of the Buildings during the Miyagi-ken-Oki Earthquake: Part 1 to Part 3. Summaries of technical papers of annual meeting. Architectural Institute of Japan, Structures II, pp 7–12 (in Japanese)
33. Goto W et al. (2009) Analysis of strong-motion records during the Iwate-Miyagi: Nairiku earthquake and Iwate-ken-Engan-Hokubu earthquake. Summaries of technical papers of annual meeting. Architectural Institute of Japan, Structures II, pp 549–550 (in Japanese)
34. Ikeda T, Abe Y, Mori K, Takase Y (2010) Earthquake observation of the retrofitted Sendai City Hall using Toggle system. Summaries of technical papers of annual meeting. Architectural Institute of Japan, Structures II, pp 13–14 (in Japanese)
35. National Institute for Land and Infrastructure Management and Building Research Institute (2011) Quick report of the field survey and research on "The 2011 Off the Pacific coast of Tohoku Earthquake" (the Great East Japan Earthquake), Technical Note of National Institute for Land and Infrastructure Management, vol 363, and Building Research Data, vol 132. http://www.kenken.go.jp/japanese/contents/topics/20110311/0311quickreport.html. Accessed 15 Feb 2012 (in Japanese)
36. Shiga T, Shibata A, Shibuya J, Takahashi J (1981) Observations of strong earthquake motions and nonlinear response analyses of the Building of Architectural and Civil Engineering Department, Tohoku University. Trans Architect Inst Jpn 301:119–129 (in Japanese with English abstracts)
37. Motosaka M, Suzuki H, Sato K (2002) Evaluation of seismic damage and reinforcement effects of an existing building based on forced vibration tests before and after the reinforcement work. In: Proceedings of the 11th earthquake engineering symposium, pp 2015–2020 (in Japanese with English abstracts)
38. Architectural Institute of Japan (2004) Report on the damage investigation of the May 26, 2003 off Miyagi Prefecture Earthquake, 3.4 earthquake records in buildings, pp 80–97 (in Japanese)
39. Takahashi Y, Arresis F, Kono Y, Motosaka M (2008) Acquisition and application of observed online earthquake data of the building. Proceedings of AIJ Tohoku chapter architectural research meeting 71:179–182 (in Japanese)
40. Motosaka M et al (2008) Development of an integrated early warning and structural monitoring system to real time earthquake information. AIJ J Tech Design 14:669–674 (in Japanese with English abstracts)
41. Motosaka M et al (2008) Development of an integrated early warning and structural health monitoring system to real time earthquake information—part 1 to part 4. Tohoku J Nat Disast Sci 44:13–34 [in Japanese (Part 1,2) and English (Part 3,4)]
42. Motosaka M et al. (2004) Amplitude dependent dynamic characteristics of an existing building, 13WCEE, CD-ROM No. 1023
43. Motosaka M et al. (2011) Change of dynamic characteristics of a damaged building before, during, and after the 2011 Off Pacific Coast Tohoku Earthquake summaries of technical papers of annual meeting. Architectural Institute of Japan, Structures II, pp 45–46, 2011 (in Japanese)
44. Soft K (1993) Newton Atlas Nihon Rettou (Atlas of Japan). Kyoiku Soft, Tokyo (in Japanese)
45. Koike K, Tamura T, Chinzei K, Miyagi T (2005) Nihon no Chikei 3 Tohoku (Geography of Japan 3 Tohoku region). University of Tokyo Press, Tokyo (in Japanese)
46. Kaizuka S, Koike K, Endo K, Yamazaki H, Suzuki T (2010) Nihon no Chikei 4 Kanto (Geography of Japan 4 Kanto region). University of Tokyo Press, Tokyo (in Japanese)
47. Meteorological Research Institute (2011) Analytical results on 2011 Tohoku-Chiho Taiheiyo-Oki earthquake. http://www.mri-jma.go.jp/Topics/press/20110324/press20110324.html. Accessed 15 Feb 2012 (in Japanese)
48. Japan Meteorological Agency (2011) Report on 2011 Tohoku-Chiho Taiheiyo-Oki earthquake (Issue No. 43). http://www.jma.go.jp/jma/press/1105/13b/kaisetsu201105131700.pdf. Accessed 15 Feb 2012 (in Japanese)
49. Geospatial Information Authority of Japan (2011) Aerial photos of the damage area after 2011 Tohoku-Chiho Taiheiyo-Oki earthquake. http://saigai.gsi.go.jp/h23taiheiyo-ok/photo/photo_dj/index.html. Accessed 15 Feb 2012 (in Japanese)

50. Geospatial Information Authority of Japan (2011) Inundation area map of 2011 Tohoku-Chiho Taiheiyo-Oki earthquake. http://www.gsi.go.jp/kikaku/kikaku60003.html. Accessed 15 Feb 2012 (in Japanese)
51. Tanaka H, Nguyen XT, Roh M, Nguyen XD (2011) Tsunami propagation into rivers in Tohoku district during the 2010 Chile Earthquake. J Hydraul Eng JSCE 55(B2–67):S1627–S1632 (in Japanese)
52. Sugawara D, Imamura F, Matsumoto H, Goto K, Minoura K (2011) Reevaluation of 869 Jogan Earthquake Tsunami utilizing geological data. J Jpn Soc Nat Disast Sci 29(4):501–516 (in Japanese)
53. Tanaka H, Mano A, Udo K (2011) Beach morphology change induced by the 2011 Great East Japan earthquake tsunami. J Jpn Soc Civ Eng Ser B2 (Coast Eng) 67(2):I_571–I_575
54. Geospatial Information Authority of Japan (2011) Estimated inundation area in each local government (tentative summary). 5th report, April 18, 2011. http://www.gsi.go.jp/common/000059939.pdf. Accessed 15 Feb 2012 (in Japanese)
55. The Statistics Bureau and the Director-General for Policy Planning of Japan (2011) Population and households in the inundation area based on the 2010 population census, 2011/4/25. http://www.stat.go.jp/info/shinsai/zuhyou/data0127.xls, 2011. Accessed 15 Feb 2012 (in Japanese)
56. Japan Meteorological Agency (2011) Observed tsunami height data in Hachinohe. http://www.jma.go.jp/jma/press/1105/27b/kaisetsu201105271730.pdf. Accessed 15 Feb 2012 (in Japanese)
57. Port and Airport Research Institute (2011) GPS wave height data of tsunami after the Off the Pacific Coast of Tohoku earthquake. http://www.pari.go.jp/info/tohoku-eq/20110328pari.html. Accessed 15 Feb 2012
58. Port and Airport Research Institute (2011) Executive summary of urgent field survey of earthquake and tsunami disasters. http://www.pari.go.jp/en/files/items/3496/File/20110325.pdf. Accessed 15 Feb 2012
59. Japan Weather Association (2011) Tsunami heights, inundation heights, and inundation areas from Aomori Prefecture to Fukushima Prefecture. The third report on the tsunami by the Off the Pacific Coast of Tohoku earthquake. http://www.jwa.or.jp/static/topics/20110422/tsunamigaiyou3.pdf. Accessed 15 Feb 2012 (in Japanese)
60. Japan Weather Association (2011) Tsunami and inundation heights from Hokkaido to off Boso Peninsula. The second report on the tsunami by the Off the Pacific Coast of Tohoku earthquake. http://www.jwa.or.jp/static/topics/20110407/touhokujishin2_110407.pdf. Accessed 15 Feb 2012 (in Japanese)

Chapter 3
Damage to Timber Buildings

Naohito Kawai, Hiroshi Isoda, and Takahiro Tsuchimoto

Abstract The damage to timber buildings occurred over a wide swath from Tohoku down to northern Kanto. The cause of damage is divided either by ground motion or by tsunami. The damage due to ground motion is categorized by the main cause of damage, either the ground deformation including the liquefaction or the vibration of the superstructure.

In this chapter, the result of the reconnaissance on timber buildings damaged due to the ground motion is shown. Then, the result of the time history analysis is conducted to clarify the relationship between the characteristics of ground motion and the damage of wood houses.

Finally in this chapter, the result of the reconnaissance on timber buildings damaged due to the tsunami is summarized. Many wood houses were swept away by tsunami wave within areas of deep inundation. However, there were a few cases where wood houses with high grade structural specifications and shield against tsunami impact remained in place area.

Keywords Ground deformation • Liquefaction • Time history analysis • Tsunami damage • Tsunami wave force • Vibrational damage

---

N. Kawai (✉)
Kogakuin University, Tokyo, Japan
e-mail: kawai-nk@cc.kogakuin.ac.jp

H. Isoda
Shinshu University, Nagano, Japan
e-mail: hisoda@shinshu-u.ac.jp

T. Tsuchimoto
National Institute for Land and Infrastructure Management,
Ibaraki, Japan
e-mail: tsuchimoto-t92ta@nilim.go.jp

Architectural Institute of Japan (ed.), *Preliminary Reconnaissance Report of the 2011 Tohoku-Chiho Taiheiyo-Oki Earthquake*, Geotechnical, Geological and Earthquake Engineering 23, DOI 10.1007/978-4-431-54097-7_3,

3.1 Overview of Investigation/Introduction

The Managing Committee on Timber Structures of the Architectural Institute of Japan (AIJ), in cooperation with Tohoku and Kanto Chapters of AIJ, conducted earthquake damage reconnaissance on Timber buildings, mainly in the Prefectures of Iwate, Miyagi, Fukushima, Ibaraki and Tochigi.

The damage is divided either by the ground deformation or by the vibration of the superstructure.

The damage of timber buildings including collapse occurred in a wide area along with from Tohoku to northern Kanto, although the severe damage is dotted about the area and the number is limited. The type of damage is extremely diverse due to differences in site amplification of the ground motion characteristics. There also occurred collapse or other severe damage as a result of landslides or failure of retaining wall on slopes, and tilting or sinking of entire structures as a result of the liquefaction of sandy soils.

Many wood houses were swept away by tsunami wave within areas of deep inundation. There were cases, however, of wood houses that, because they were shielded from direct tsunami impacts by relatively large surviving structures, remained in place. There were also a few cases where wood houses presumably having excellent structural specifications remained in place, although they suffered severe damage to their walls or frames.

3.2 Damage due to Ground Motion

The cause of damage is divided by either the damage due to ground deformation or the damage due to the vibration of the superstructure.

In this section, the damage due to the vibration of the superstructure is categorized as minor damage (damage to finishing materials only), moderate damage (structural damage), and collapse. The use of timber buildings are houses, stores, apartments, barns, warehouses, school buildings, public buildings, shrines, gymnasiums, and other large buildings. Their construction method, number of floors, size, and shape are different.

The earthquake caused extensive damage including collapse throughout northeastern Kanto to Tohoku in broad area. In areas such as river basins where the ground is soft, timber buildings are damaged due to the resonance phenomenon. This is consistent with the results of the time history analyses using strong motion records (see Sect. 3.3). In addition to serious structural damage, the minor damage in roof tile and finishing exterior and interior wall occurred in extensively throughout northern Kanto to the Tohoku area.

In addition to vibrational damage, a large amount of damage occurred due to ground deformation. Houses on slopes collapsed or were heavily damaged due to landslides and the destruction of retaining walls. Prominent damage throughout the area was due to slope or land subsidence due to liquefaction of the upper structure

Fig. 3.1 Collapse of wood house in Osaki

of the sandy ground. The whole superstructure was often tilted by the liquefaction but did not collapse. The cause of the damage is not clear due to ground deformation or vibration in the area where damage occurred on soft soil.

3.2.1 Vibrational Damage to Houses

Serious damage to houses has reported in many communities where are Ishinomaki, Osaki, Kurihara, Kesennuma, Sendai and Misato, in Miyagi Prefecture, Sukagawa in Fukushima Prefecture, Nasu Karasuyama, Nasu, Takanezawa and Haga in Tochigi Prefecture and Mito, Ibaraki, Sakuragawa, Naka, and Hitachiota in Ibaraki Prefecture. In addition, the cases of sporadic collapse in dilapidated housing have been reported in other cities such as Katori, Chiba Prefecture and Kounosu Saitama Prefecture. Figures 3.1, 3.2, 3.3, and 3.4 illustrate some of this damage. Most of the extensive damage occurred in relatively old houses, often obsolete buildings, with decay and termite damage. In particular, the building members in Fig. 3.1 showed extensive termite damage.

The houses typically have large openings facing the street, which is inadequate for earthquake resistance. Exterior walls often showed chipped mortar and glass as well as termites in the wood sheathing, which sometimes spread into the structural members. Therefore, the damage of most dwellings was amplified due to the termite damage.

In contrast, vibration damage built in one distinct typical construction is found in Fig. 3.5, which shows the partial collapse of the high stone foundation near the city of Nasu Karasuyama and the town of Ichigai in Tochigi Prefecture. This vibrational failure of the foundation was often found and would lead to the collapse of the wood superstructure.

Fig. 3.2 Partial collapse of wood house in Kurihara

Fig. 3.3 Collapse of wood house with store attached in Misato

3.2.2 Vibrational Damage to Storehouses

A wide array of damage to storehouse buildings was observed from Kanto to Tohoku, many of which have attached dwellings. Examples of storehouse damage are in Figs. 3.6, 3.7, and 3.8.

Fig. 3.4 Severe damage to wood house with store in Osaki

Fig. 3.5 Collapse of wood house with high stone foundation

The primary type of damage was in the form of fallen roof tiles or chipped exterior wall coatings. Major damage, demonstrated in Fig. 3.6, was evident in the form of residual inclination of the building. The group of storehouses in Osaki, as seen in Fig. 3.7, was over 200 years old and had obvious structural deficiencies.

Fig. 3.6 Permanent inclination of a storehouse

Fig. 3.7 Damage to older storehouses in Osaki

Fig. 3.8 Retrofitted storehouse in Osaki

Although no deformation to the columns and other structural members is observed, the termite decay is extensive within the members and would be difficult to repair. Figure 3.8 shows a building on the same property as the warehouses. A retrofit was performed on the interior wooden shear walls and reinforced concrete masonry foundation inside. The structural timbers were still intact but the roof collapsed.

Figures 3.9 and 3.10 show the damage of residential storehouses. Residual deformations were small and the damage was primarily in the form of cracks and spalling of the mud coatings on the wall as well as falling roof tiles.

3.2.3 Vibrational Damage to Barns

In residential areas, many structures such as barns and small storage buildings collapsed. Adequate seismic structural design was not considered in their buildings. Examples of such damage are in Figs. 3.11 and 3.12.

3.2.4 Vibrational Damage to Schools

There is a two construction method of school buildings. One is old timber construction methods and the other is new laminated timber structures. A schoolhouse using

Fig. 3.9 Damage to storehouse with heavy clay walls in Tsukuba

Fig. 3.10 Damage to three storehouses with heavy clay walls

older timber construction was reported to have a partial second floor collapse and large deformations in Osaki city. Traces of sand boiling in the schoolyard were observed because of liquefaction of the sandy ground. Figure 3.13 shows the markings indicating sand boiling.

Fig. 3.11 Collapse of shed in Kurihara

Fig. 3.12 Collapse of shed in Kurihara

School buildings built using laminated wood construction, such as an elementary school gymnasium in Naka-city, only experienced minor damage to the concrete foundation and timber braces.

Fig. 3.13 Damage to school building and markings of sand boiling in Osaki

3.2.5 Vibrational Damage to Temples and Shrines

Damage to shrines is widely reported across the Northeast of Japan with a notable collapse of the main building of a temple in Takanezawa town, seen in Fig. 3.14.

In addition, Fig. 3.15 shows the severe damage of the gate to a shrine in Osaki-city. Marks around the site seem to indicate sand boiling caused by the liquefaction of the sandy ground. These shrines were constructed by traditional methods and were located on soft soils.

3.2.6 Vibrational Damage to Other Buildings

In addition to other building types, a renovated theater in Kurihara-city experienced large residual deformation, evident in Fig. 3.16.

Other examples of damaged timber buildings included houses that used logs assembly. Although this is a rare construction method, there are reports of damage at the intersection of the logs in houses in Nasushiobara-city.

Fig. 3.14 Collapse of the main building of temple in Takanezawa

Fig. 3.15 Severe damage to the gate of a shrine in Osaki

3.2.7 Building Damage due to Landslides

This earthquake caused many landslides and destruction of retaining walls, leading to the destruction of many houses built on the slopes.

Fig. 3.16 Severe damage to old cinema in Kurihara-city

Areas of intense damage were found in Aoba-ku Sendai-city, Izumi-ku Sendai-city, Wakabayashi-ku Sendai-city, Fukushima-city, Kagamiishi town, and Nasu-Karasuyama-city in Tochigi Prefecture. Examples of such damage are found in Figs. 3.17, 3.18, 3.19, and 3.20.

3.2.8 Damage due to Liquefaction

Soil liquefaction occurred rapidly in a wide range of places due to the sandy soil. Ground subsidence leads to serious damage or a tilt of the superstructure.

This damage was observed in Kashima, Kamisu, Itako, Future city, Inashiki, Kawachi-machi, in Ibaraki Prefecture, Asahi city, Urayasu, Funabashi, Narashino, Chiba, Abiko, Katori in Chiba Prefecture and Kuki city, Saitama Prefecture. The structural damage to the superstructure due to liquefaction was rather small if it used a reinforced concrete foundation. Damage was typically in the form of structural inclination and subsidence.

3.2.9 Ground Deformation and Other Damage

In flat ground, even though liquefaction does not occur, the soft ground results in land and soil cracks. Damage to the superstructure from differential settlement is

Fig. 3.17 Collapse of hotel due to landslide in Sendai-city

Fig. 3.18 Severe of a house due to landslide in Sendai-city

reported in Sakura, and Narita in Chiba Prefecture. Similar damage happened in the other regions. This type of the damage occurred to old houses, so the cause of the damage may not be clearly due to ground deformation or vibration.

Fig. 3.19 Severe damage to a house due to landslide in Sendai-city

Fig. 3.20 Severe damage to a house due to landslide in Sendai-city

3.3 Time History Analysis

3.3.1 Analytical Model

A time history analysis was conducted on a typical two-story wood house. The seismic shear coefficient (C_0) used for the first floor was 0.2 while the second floor is a $C_0 = 0.3$ for a standard model. The restoring force characteristics to consider the

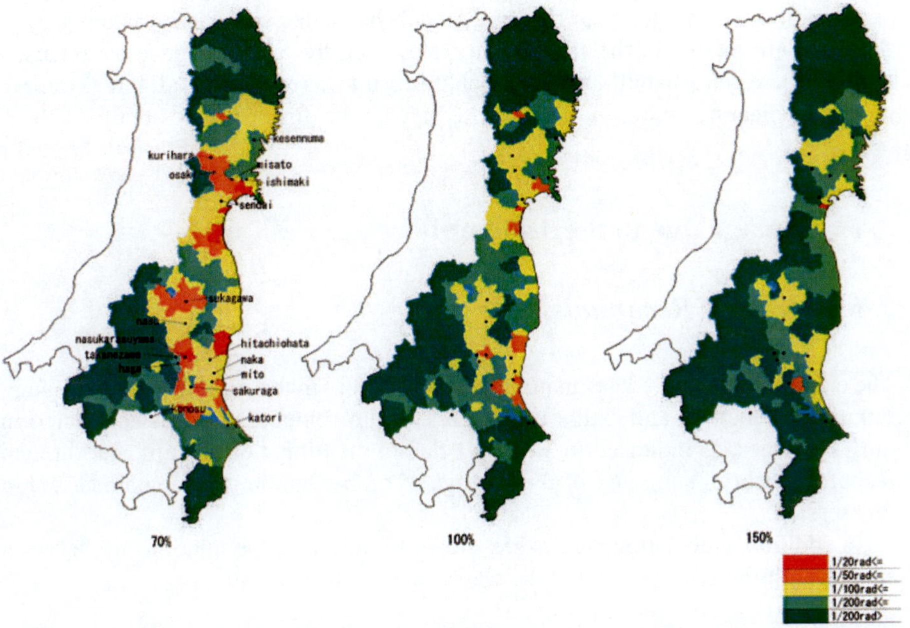

Fig. 3.21 Contour of maximum analytical responses

long duration of ground motion is the modified evolutionary parameter hysteretic model [1] for cyclic behavior. Three models are analyzed to compare the damage depending on the effective strength of the structure. The level of the strength is 70%, 100%, and 150% of the standard seismic resistance model for both the first and second story. The houses are modeled with two lumped masses and two shear springs. 143 strong motions, recorded from Iwate to Chiba by NIED network (K-net), were used for the analysis.

3.3.2 *Analytical Results*

Figure 3.21 shows the results of the analysis. Displacement response of the floors is separated into five categories: greater than 1/20 radians, 1/20–1/50 radians, 1/100–1/200 radians, and less than 1/200 radians. Ishinomaki, Kurihara, and Sendai all experienced large deformation responses in the analyses. However, the models designed for 70% of the adequate seismic resistance lead to large deformations beyond 1/9 radians, which would indicate collapse. Considering the actual damage in this area, the calculated response is considered somewhat excessive in this case. In contrast to that, the models indicate a deformation response of 1/131 radians to buildings in Kesennuma, an area that experienced major damage. The results of these analyses are in correlation with the actual damage. These inconsistent results are possibly due to the effects of the local ground motions.

In addition, the majority of building models had a first story deformation greater than the deformation of the second story. However, five analyses gave the reversed results. There were no buildings in which the result was reversed in all three strength models, indicating higher mode response may not be distinct in the area.

3.4 Damage due to the Tsunami

3.4.1 Sites of Reconnaissance

The earthquake triggered a tsunami that flooded areas in the Aomori, Iwate, Miyagi, Fukushima, Ibaraki, and Chiba Prefectures. The reconnaissance was conducted in only two areas, as indicated in Table 3.1 due to the limitations of time and human resources. A brief summary of the damage of timber buildings is contained within this session.

In addition, calculation results are shown to introduce the relationship between strength of timber buildings and the wave force of the tsunami.

3.4.2 Damage in Plain Areas

The plain areas, often without shield from tsunami force, had extensive damage on reaching locations far inland. The following provides as an overview of this damage.

3.4.2.1 Wakabayashi Ward, Sendai City, Miyagi Prefecture

With a few exceptions, most wood houses were completely washed away by the tsunami. Figure 3.22 demonstrates one of the few wood houses left standing, due to a location downstream of a reinforced concrete building. Inundation depth estimated from remaining reinforced concrete buildings reached 6–8 m.

Table 3.1 Reconnaissance in damage areas due to tsunami

Classification	Municipality	Study dates
Plain area	Wakabayashi-ku, Sendai, Natori, Iwanuma, Watari, and Yamamoto in Miyagi Prefecture	April 6th, 2011–April 8th, 2011
Sloping land	Otsuchi, Kamaishi, Ofunato, and Rikuzentakata, in Iwate Prefecture and Kesennuma Minamisanriku, Onagawa and Higashi-Matsushima in Miyagi Prefecture	May 25th, 2011–May 27th, 2011

Fig. 3.22 Wood house remained downstream of RC building

Fig. 3.23 Wood houses remained in a line

In an area approximately 1.5 km away from the shoreline in Arahama district, many low-rise buildings were washed away. In this area, some houses situated in a line the direction of the water flow withstood damage, as evident in Fig. 3.23. Inundation depth in this area was estimated to be about 4–5 m, from the remaining houses being low-rise and non-wooden. On the other hand, there were several cases identified that no buildings were left standing in the direction of the tsunami waves.

Figure 3.24 demonstrates one of the few timber buildings left standing, with damage limited to the walls in the direction of floodwater. The method of construction was relatively new with adequate structural fasteners.

3.4.2.2 Yuriage, Natori City, Miyagi Prefecture

This area experienced widespread catastrophic damage. Figure 3.25 shows an aerial photo with the positions of investigated buildings.

Initial investigation was conducted on the house in Fig. 3.26 at location (A), which had been carried away with the foundation attached. The junction of the foundation to the steel pile failed due to the force of the tsunami, leaving the steel pipe piles in the sand since in Fig. 3.27 at location (B). This would indicate that the tsunami wave was moving east to west-southwest.

Fig. 3.24 Wood building escaping complete devastation

In the same area, the temple and steel building at locations (C) and (D), respectively, were severely damaged but remained in place. The shop with dwelling at location (E) only received slight damage presumably due to being downstream to the temple and steel building.

Figure 3.28 shows a wood built house in location (J) that withstood the tsunami force without the protection from trees at location (G). This is probably due to the relatively new construction of the house, which was built in the last few years.

A few wooden structures remained around location (K) and (L) without large buildings in the inundation direction, as shown in Figs. 3.29 and 3.30. Inundation depth is estimated at about 3.5 m in this area.

Figure 3.31 in location (N) shows a group of houses that remained downstream of a large RC apartment building. The buildings were older and newly constructed wood houses. In addition, the factory building in location (O) protected wood houses. This is expected to be due to the relatively new construction of these houses that had superior structural specifications. These houses are seen in Fig. 3.32 taken from location (P). Another wood house, at location (Q), was spared from catastrophic damage presumably due to superior construction.

3.4.2.3 Arahama, Watari Town, Miyagi Prefecture

The Arahama, Watari Town area is surrounded by the Pacific Ocean on the east and a harbor to the south, as seen in Fig. 3.33. In the area between the Pacific Ocean

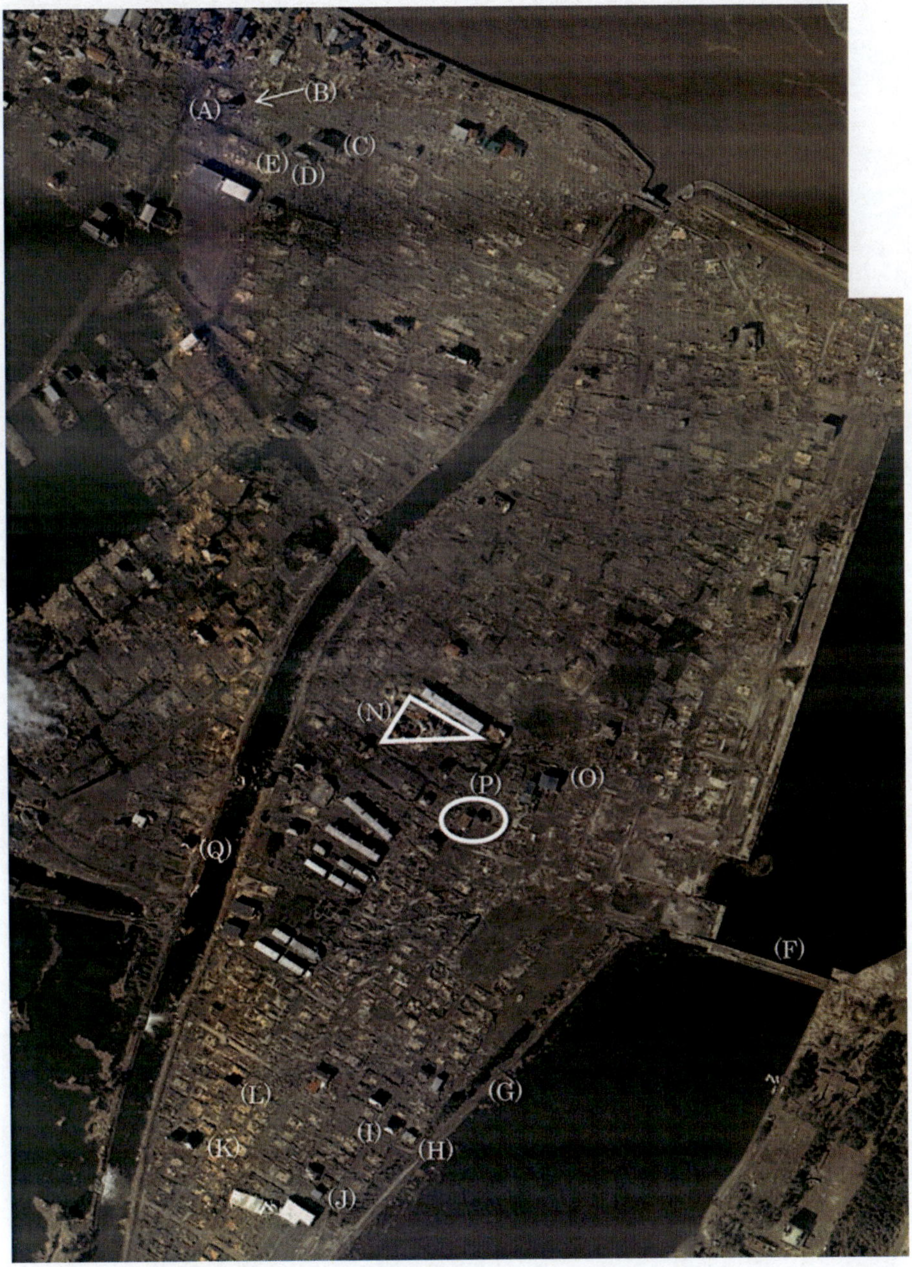

Fig. 3.25 Aerial photograph and the position of investigated buildings in Yuriage, Natori City

Fig. 3.26 Wood house moved away with foundation

Fig. 3.27 Steel pile left of the original position in Fig. 3.26

Fig. 3.28 Wood house remaining without the protection of the forest (location (J))

Fig. 3.29 Japanese conventional post and beam house remaining (location (K))

and the harbor, the tsunami washed away many wood houses, leaving only the foundation.

The extent of the severe damage was mostly on the eastern peninsular region. One timber building, as pictured in Fig. 3.34, was previously "L" shaped in plan. The part of the house that had the largest amount of area exposed to the tsunami

Fig. 3.30 One of few light frame construction houses remaining (location (L))

Fig. 3.31 Group of wood houses escaped from damage downstream of the RC structure (location (N)), Factory building (location (O))

wave washed away while the shorter side remained. Judging from the joint fastener in Fig. 3.35, the completion of construction is estimated after 2000.

Figure 3.36 shows a building that withstood the tsunami force. This building consists of reinforced concrete structure on the first floor and wood construction on the second floor. Figure 3.37 shows a one-story glulam framed structure in which the structural members remained intact, but the building was filled with wreckage.

Fig. 3.32 Wood houses moved away from downstream of the factory building (location (P))

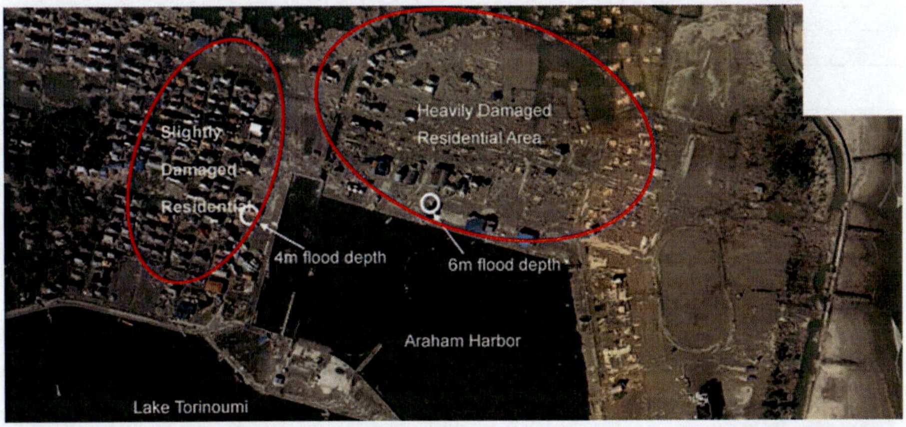

Fig. 3.33 Aerial photograph in Arahama, Watari-machi

Buildings in north of the harbor, estimated to have been in 6 m of floodwater, mostly consisted of wooden residential houses. Figure 3.33 illustrates the impacted area, with very few houses remaining on the north side of the harbor. One of the few remained houses, pictured in Fig. 3.38, was a three-story wood house with superior seismic protection. The lateral strength of three-story house was larger than that of two-story house. West of the harbor, estimated to have been in 4 m of floodwater, had a number of buildings withstand the tsunami force. One of the houses, pictured in Fig. 3.39, also withstood contact from a ship. This is probably due to newly construction of the house, which presumably has superior structural specification.

Fig. 3.34 A portion of a wood house washed away

Fig. 3.35 Fastener of joint in the house of Fig. 3.34

Fig. 3.36 Composite structure consisting of RC on the first floor and wood construction on the second floor

Fig. 3.37 Glulam frame structure remaining

3.4.2.4 Summary of Damage in Plane Areas

The following conclusions were drawn about wood building in the plane areas:

- Damage was quite severe in plane area.
- If there are effective shields such as reinforced concrete building, the wood buildings survived from tsunami wave.

Fig. 3.38 Three-story wood house remaining

Fig. 3.39 Wood house remaining even after collision with ship

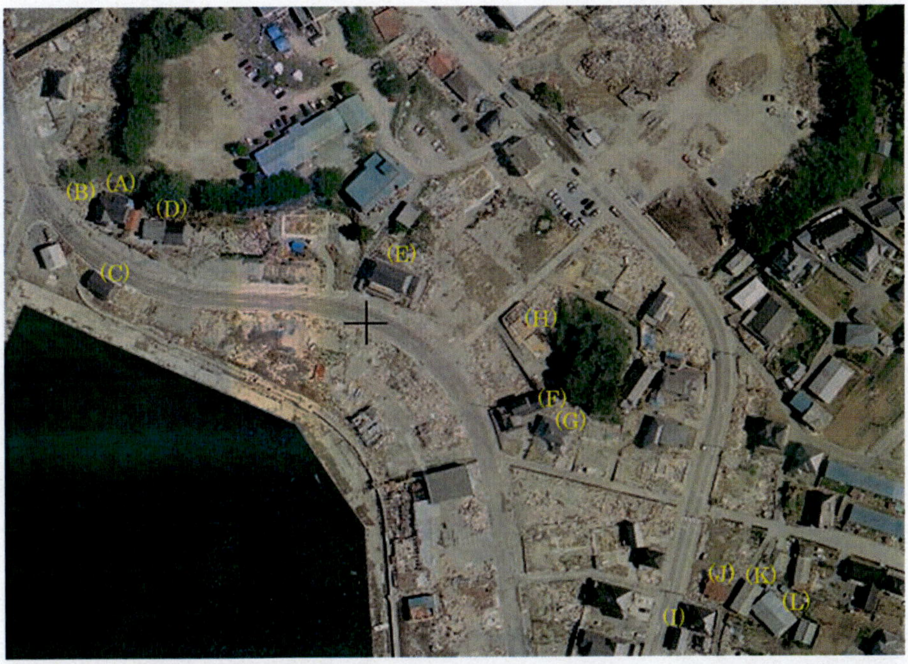

Fig. 3.40 Aerial photograph of Akasaki-cho, Ofunato City

- Even if there was no building blocking the force of the tsunami, there are a few examples of low-rise timber buildings that remained intact due to newer construction specifications. However, the buildings that remained had severe damage to the lower level walls perpendicular to the tsunami direction.

3.4.3 Damage in Sloping Lands

In this section, the damage reconnaissance is only given for Akasaki-machi, Ofunato City, a village on the east side of the bay of Ofunato situated on a gradual slope. Figure 3.40 shows the aerial photograph with the position of the investigated buildings.

A house in location (A) (Fig. 3.41), according to residents, was almost completely submerged with inundation depth over 7 m. The house suffered some damage but was largely spared overall. An adjacent work shed in location (B) was disconnected from the concrete foundation and moved. The two-story wood house across the street in location (C) on the ocean side received very little damage (Fig. 3.42). The hill to the north of these buildings probably reduced the magnitude of the tsunami force.

Figure 3.43 shows a traditional Japanese wood house in location (E) that survived the inundation depth of about 5–6 m and it was located 1.5 m above the road level. Although the walls were destroyed the structure was intact, and there

Fig. 3.41 Wood house intact with 7 m wave depth

Fig. 3.42 Wood house without damage with over 7 m wave depth

Fig. 3.43 Survived traditional Japanese wood house under 5–6 m wave depth

Fig. 3.44 Survived wood house with flood damage up to the first floor roof

seemed to be little residual deformation in the house. Figure 3.44 in location (F) was about 2 m above the road level. The house was spared despite floodwater levels reaching 6 m, damaging parts of the roof as well as the house. A relatively new house to the northeast was washed away. This house was equipped with significant tie-down fastener and large partial corrosive pores were observed. This indicates that the tie-down fastener was not sufficient to fix the house to the foundation under the tsunami force with about 5–6 m high.

Along the prefecture road up the hill from these buildings, a lightweight steel frame house in location (I) withstood floodwaters up to 5 m. Adjacent to this building in location (J), a small one-story building that had minor structural specifications survived, pictured in Fig. 3.45. The house was confirmed to have one brace with

Fig. 3.45 Survived house in spite of minor structural specifications

Fig. 3.46 Survived warehouse in spite of minor structural specifications

nailed joints and anchor bolts into the foundation. In addition, despite a 3.8-m inundation depth, a dilapidated warehouse in location (K) remained (Fig. 3.46). Just to the East in location (L), old houses were rotated about 30° clockwise (Fig. 3.47). The fact that some of these older buildings withstood these floodwaters suggests that the sloping land reduced the force of the tsunami in contrast to the plain areas.

Fig. 3.47 Survived and rotated old house

3.4.4 Relationship Between the Horizontal Tsunami Force and the Lateral Strength of Timber Building

After the damage due to the Indian Ocean tsunami in 2004, the Cabinet office published the *Guidelines Pertaining to Tsunami Evacuation of Buildings* in June 2005 [2]. This document is meant to promote the spread of tsunami evacuation buildings in areas that are difficult to escape from tsunami waters. Specifications and other precautions were prepared for the structural requirements of the tsunami evacuation buildings.

3.4.4.1 Calculation of the Tsunami Wave Pressure

Tsunami wave pressure distribution used for design was three times the normal design water height in the hydrostatic pressure distribution.

$$q_z = \rho g\,(3h - z) \tag{3.1}$$

where q_z: wave pressure in the direction of travel (kN/m^2); ρ: density of water (kg/m^3); g: acceleration due to gravity (m/s^2); h: Flood water depth for design (m); z: Height above ground level of the portion ($0 \leqq z \leqq 3$ h) (m).

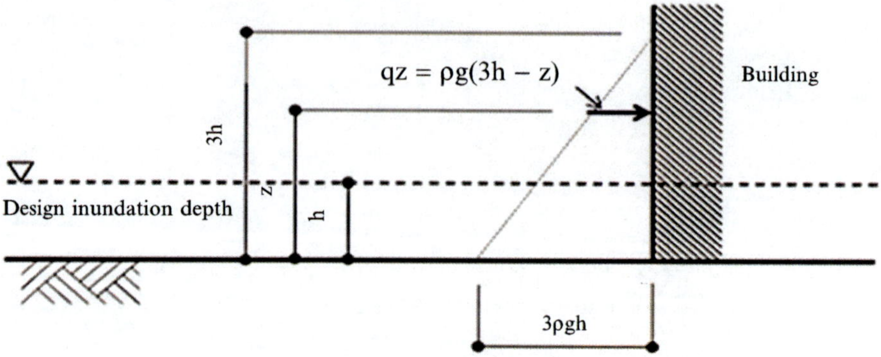

Fig. 3.48 Calculation of wave pressure acting to building

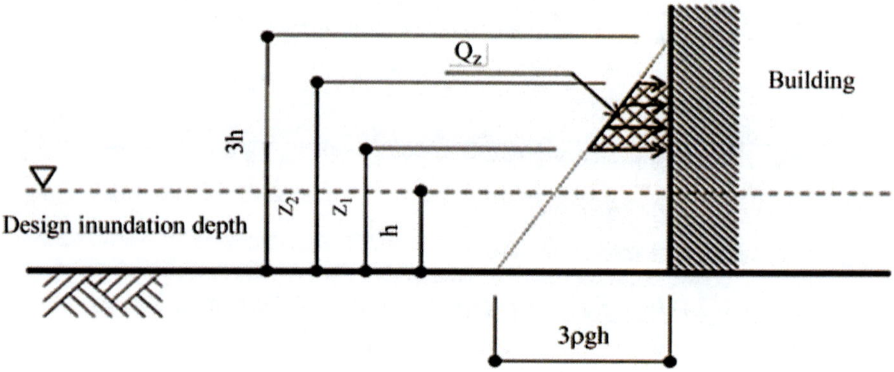

Fig. 3.49 Calculation of wave force acting to building

Values obtained from the hydraulic model experiment assumed a structure close to the seawall. In this case, the wave pressure in the direction of tsunami travel is calculated by the following formula.

3.4.4.2 Tsunami Wave Force Calculation

The tsunami wave pressure (3.1) in the direction of travel is assumed to be acting with the triangular wave distribution shown in Fig. 3.48. The following equation is obtained by integrating the value of the pressure over the area receiving pressure from the tsunami as shown in Fig. 3.49.

$$Q_z = B\int_{z_1}^{z_2} q_z dz = \text{ñ}gB\int_{z_1}^{z_2} (3h - z)dz \tag{3.2}$$

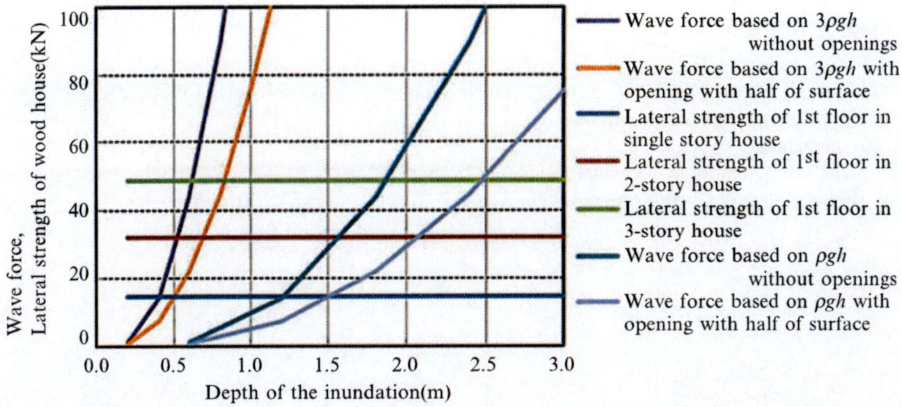

Fig. 3.50 Relationship between lateral strength of wood house and tsunami wave force

where Q_z: tsunami wave force for structural design (kN); B: width of wall portion (m); z_1: minimum height of pressure receiving surface ($0 \leqq z_1 \leqq z_2$) (m); z_2: maximum height of pressure receiving surface ($z_1 \leqq z_2 \leqq 3$ h) (m).

The relationship between tsunami wave force and strength for single, two-story, and three-story wood detached houses was simulated, with the same pressure-receiving surface of 4.55 × 7.28 m. The lateral strength was determined as the allowable strength (3.3) based on the assumption that the existing shear wall is equal to the required shear wall amount as indicated in *Article 46 Enforcement Ordinance of Building Standard Law*. The relationships between the wave force of the tsunami and the determined lateral strength of the wood houses are shown in Fig. 3.50. In addition, strength was assumed up to 1.5 times the allowable tolerance.

$$\text{Holding Strength} = \text{Wall Volume} / 100 \times 1.96\ \text{kN}/\text{m} \times 1.5 \qquad (3.3)$$

If we assumed the wave pressure distribution to be 3 times the depth of the inundation, the wave force of the tsunami at 1 m is higher than the lateral strength of wood houses, regardless of the number of stories.

If the wave velocity was zero, the wave force based on the wave pressure distribution is equal to the static water pressure. Assuming that one half of the pressure receiving area (wall) was an opening, wooden detached houses can withstand tsunami with the wave height of 1.5 m for single story, 2 m for two storied and 2.5 m for three storied houses. If the tsunami wave was higher, wood houses may be swept away.

Furthermore, this results of simulation don't correspond to the damage survey proved that wood houses could withstand tsunami waves from 4 to 6 m.

Acknowledgments This chapter is based on the reconnaissance report by the Managing Committee on Timber Structures of the Architectural Institute of Japan (AIJ), in cooperation with Tohoku and Kanto Chapters of AIJ. Contribution of the reconnaissance team members, listed below, is greatly appreciated: Naoyuki Itagaki (Akita Prefectural University); Masahiro Inayama (University of Tokyo); and Takafumi Nakagawa, Yasuhiro Araki (Building Research Institute).

References

1. Isoda H, Kawai N et al. (2009) Hysteresis model of wood shear wall considering of large displacement and deterioration of cyclic load. J Struct Construct Eng 74(646):2239–2306, AIJ (in Japanese)
2. JOC/Task Committee under the Japanese Cabinet Office (2005) Design guidelines for tsunami shelters, http://www.bousai.go.jp/oshirase/h17/tsunami_siryo2.pdf (in Japanese)

Chapter 4
Damages to Reinforced Concrete Buildings

Toshimi Kabeyasawa, Yoshiaki Nakano, Masaki Maeda, Kazuhiro Kitayama, Daisuke Kato, Toshikatsu Ichinose, Akira Tasai, Koichi Kusunoki, Hitoshi Shiohara, Nobuyuki Izumi, Toshikazu Kabeyasawa, Masanori Tani, and Kazuhiro Watanabe

Abstract Typical damages to reinforced concrete buildings, both by the ground motions and the tsunami waves, are outlined in this section. The report herein is based primarily on the field survey of the AIJ members and collected through the

T. Kabeyasawa (✉) • Y. Nakano • H. Shiohara
University of Tokyo, Tokyo, Japan
e-mail: kabe@eri.u-tokyo.ac.jp; iisnak@iis.u-tokyo.ac.jp; shiohara@arch.t.u-tokyo.ac.jp

M. Maeda
Tohoku University, Sendai, Japan
e-mail: maeda@archi.tohoku.ac.jp

K. Kitayama
Tokyo Metropolitan University, Tokyo, Japan
e-mail: kitak@tmu.ac.jp

D. Kato
Niigata University, Niigata, Japan
e-mail: dkato@eng.niigata-u.ac.jp

T. Ichinose
Nagoya Institute of Technology, Nagoya, Japan
e-mail: ich@nitech.ac.jp

A. Tasai • K. Kusunoki
Yokohama National University, Yokohama, Japan
e-mail: tasai@ynu.ac.jp; kusunoki@ynu.ac.jp

N. Izumi
Chiba University, Chiba, Japan
e-mail: nobuyuki.izumi@faculty.chiba-u.jp

T. Kabeyasawa • M. Tani
Building research Institute, Tsukuba, Japan
e-mail: tosikazu@kenken.go.jp; mtani@kenken.go.jp

K. Watanabe
Yokohama, UR, Japan
e-mail: k-watanabe07@ur-net.go.jp

Architectural Institute of Japan (ed.), *Preliminary Reconnaissance Report of the 2011 Tohoku-Chiho Taiheiyo-Oki Earthquake*, Geotechnical, Geological and Earthquake Engineering 23, DOI 10.1007/978-4-431-54097-7_4, © Springer Japan 2012

members of the Reinforced Concrete Steering Committee of AIJ. The typical damages to reinforced concrete buildings are described below through the reports on the buildings selected and available for detailed survey. The damage reports below are translated selectively from the Chap. 6, Sect. 6.2 of AIJ report in Japanese version, which contains more photos and figures, especially on the structural plans as well as the damage rate evaluation of members. The reports on the buildings below are arranged as from the Japanese version: (1) school buildings located from north to south (Iwate, Miyagi, Fukushima and Kanto areas), (2) government offices, (3) residential buildings.

Keywords Post-earthquake damage inspection • Residual seismic capacity ratio • School buildings • Seismic capacity index • Seismic strengthening/retrofit

4.1 Outline of Damage Survey

4.1.1 Introduction

The content of this chapter is based on the field survey and other sources provided by the members of the Reinforced Concrete Steering Committee of AIJ.

The typical reinforced concrete buildings in Japan are categorized by the use of the buildings into four groups as follows:

1. School and public hall
2. Municipal (mainly for offices)
3. Residential
4. Commercial

As so many damaged buildings were widely distributed in a broad area from Tohoku region to northern Kanto region, it was virtually impossible to coordinate reconnaissance teams which covers all the important severely damaged buildings. So it is challenging to draw a comprehensive conclusions at this moment. Thus, a number of reinforced concrete buildings subjected to detailed survey are selected in this chapter and typical damages are outlined as examples observed after 2011 Tohoku-chiho Taiheiyo-oki Earthquake in the subsequent sections. The type of damage here are due to the ground motions or tsunami inundation. The buildings are listed in the order: (1) school buildings located from north to south (Iwate, Miyagi, Fukushima prefectures and Kanto areas), (2) municipal offices, and (3) residential buildings.

The description of damaged buildings in this chapter are excerpts from the Chap. 6, Sect. 6.2 of AIJ reconnaissance report in Japanese. More photos and figures, in particular on the structural plans as well as the results of damage rating of structural members are shown in the report.

4.1.2 *Target of Field Survey*

As to buildings for school and public collective houses, it is not so difficult to obtain the list of damaged buildings. Fortunately, detailed information on the design, construction, age, and structural drawings were found to be available for the most of the public buildings. Particularly for school buildings, results of seismic vulnerability assessment are also available. Professional engineers carries out the assessments upon the requests of municipal administrators. The most popular procedure for the seismic assessments in Japan is described in the Standard for Seismic Evaluation of Existing Reinforced Concrete Buildings (2001 Japanese version and 2004 English version) [1]. The result of seismic assessment is valuable information to investigate the correlation between the observed damage level and the results of the assessment to verify the reliability of the procedure adopted in the standards.

On the other hand, access to the detailed information for the buildings for commercial use, offices of private company or private collective houses were limited. It is also basically difficult to get acceptance for survey of such buildings by the owner. So survey of such buildings were carried out based on the availability.

4.1.3 *Seismic Intensity*

JMA scale for seismic intensity is defined by Japan Meteorological Agency. The JMA intensity scales at the site of the damaged buildings are referred in this chapter such as 5-lower 5-upper, 6-lower, 6-upper or 7, which approximately correspond to intensities, 8-lower, 8-upper, 9, 10 and 11 defined by the Modified Mercalli intensity (MMI) scale.

4.1.4 *Post-earthquake Damage Evaluation Standards*

Recently the importance of post-earthquake damage evaluation of buildings has been recognized for smooth recovery after earthquake disasters. One of such standards in Japan is "The Japanese Standard of Damage Evaluation in Post-earthquake Inspection" [2]. The standard is made for damage grading after earthquake to assist owner's decision of demolishing or repairing of a building suffered various damages. Results of the evaluation are shown for the most buildings in this chapter. So the method of the damage evaluation standard is herein briefly explained. The objective of the standard is to grade the structural damage of a building into six grades: (Grade 0) no damage, (Grade 1) slight, (Grade 2) minor, (Grade 3) moderate, (Grade 4) severe or heavy, or (Grade 5) collapse (or near collapse). In the standard, the grade is defined as tabulated values according to residual capacity ratio R. Once the value of R is estimated, the damage grade is given as (0) $R = 100\%$—Grade 0, (1) $95\% < R < 100\%$—Grade 1, (2) $80\% < R < 95\%$—Grade 2, (3) $60\% < R < 80\%$—Grade 3,

(4) $R<60\%$—Grade 4, or (5) R is nearly zero—Grade 5, respectively. The residual capacity ratio R is usually estimated as the minimum value among the calculation for each story and direction of a building.

The residual capacity ratio R with respect to a particular story and a direction is estimated using the following assumptions. *Assumption 1*: At the field survey, all the vertical structural elements in a story are classified into one from five categories: (1) shear column, (2) flexure column, (3) wall without boundary columns, (4) wall with a boundary column and (5) wall with boundary columns at both ends. Then, ratios of lateral capacity of each type from (1) through (5) can be simply assumed to be 1:1:1:2:6, respectively, despite of the difference in the dimension of the member and the amount of reinforcement. *Assumption 2*: The damage of each vertical member is rated into none [0] or one of five levels from [I] to [V] by the observation of crack width, with or without crushing of concrete, with or without buckling of steel reinforcement etc., based on the tabulated criteria given in the standard. The ratios of the residual capacity of each member can be assumed to be 1.0:0.95:0.6:0:0 for shear columns and the other wall members, while the ratios of the residual capacity can be assumed to be 1.0:0.95:0.75:0.5:0.1:0 for the flexural column, in accordance with the damage rates of [0] to [V], respectively. *Assumption 3*: The residual capacity ratio R can be calculated as a sum of the residual capacity of all the vertical structural elements divided by the sum of the total original capacity of the members. By these assumptions, the residual capacity ratio R can be simply and quickly estimated for any buildings without detailed information such as structural drawing or material properties.

4.1.5 Summary

Result of the survey on typical examples of damage to reinforced concrete buildings from the 2011 Earthquake may be summarized as:

1. A number of buildings suffered very serious damages, such as collapse or severe, which requires demolishing and reconstruction for restoration, although the rate was very low.
2. Some buildings suffered moderate or minor damages, by which they were not functional in continuous use and requires repair or strengthening.
3. Most of the damaged buildings were constructed before 1981 when the current code was put into practice.
4. Shear failure of columns, especially of short columns, was observed. It is probably due to insufficient rate of hoops, concentration of earthquake load to captive columns, or high varying axial load.
5. Relatively heavy damages are observed at the joint concrete with steel and timber members in concrete structures combined with steel or timber members.
6. Most of the new buildings designed by the current code or older buildings retrofitted to the current required level suffered relatively slight or minor damages.

7. Non-structural walls and ceilings were severely damaged in many cases, not only with old, but also with new buildings.
8. Damages estimated to be increased by the amplification of ground motions due to soils and hills.
9. Settlement of base foundations and inclination of buildings were observed due to damages to piles or soil liquefaction.
10. Not only non-structural but also structural failures due to the tsunami waves were observed in RC buildings, from which the rehabilitation seemed to be difficult in most cases.

4.2 K Middle School in Kamaishi City

This school is located in the flatland of Unosumai riverside, 1 km distant from the river mouth, and consists of a three-story classroom building, a four-story special classroom building, a single-story administration building, a single-story technical building, and a gymnasium as shown in Figs. 4.1 and 4.2. Although the seismic capacity indexes of some buildings were identified lower than required, they were not yet seismically retrofitted at the time of the event.

The gymnasium suffered most serious damage by tsunami that flowed over the riverbank in the northeast. In the building, steel frames supporting a roof completely collapsed southwestward by tsunami as shown in Fig. 4.3. Although the buildings had no structural damage by tsunami and/or ground shaking except the gymnasium, the interior was remarkably damaged due to sand deposit and debris flow. The damage in the third floor of classroom building is shown in Fig. 4.4. Tsunami trace was observed at the height of 10 cm from the third floor level in the stair hall of the classroom building (7.9 m from the ground level).

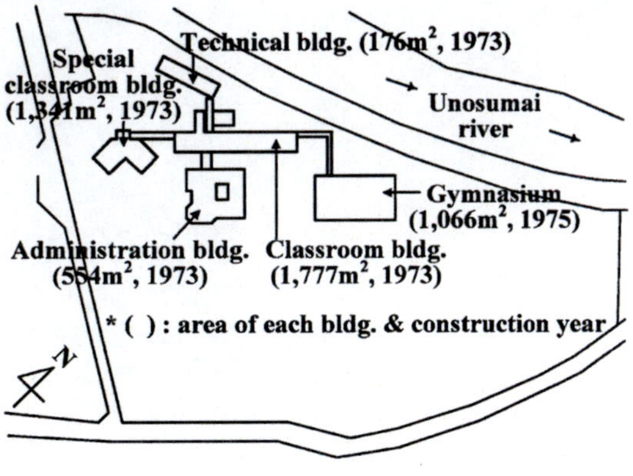

Fig. 4.1 Location of buildings

Fig. 4.2 General view of buildings

Fig. 4.3 Damage to gymnasium

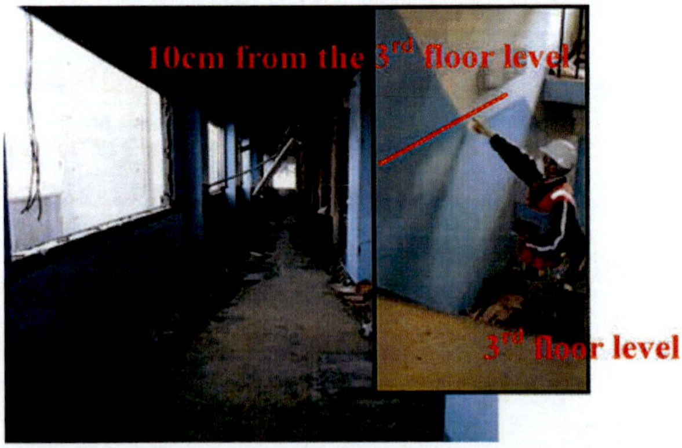

Fig. 4.4 Damage to classroom building

4.3 Y Elementary School in Ichinoseki City

This school is located in the flatland near Iwai River, which is a branch of Kitakami River. It consists of two classroom (west wing and east wing) buildings, an administration building, a special classroom building, and a gymnasium as shown in Fig. 4.5. All the buildings except the gymnasium are three-story buildings. Figure 4.6 shows the general view of the buildings.

The west part of the west wing building and the administration building suffered the most serious damage due to ground shaking, while the other buildings had no remarkable damage. Some columns of the west wing building were rated as damage rate of IV or V. The first story had the most serious damage in the west wing building, and the damage grade was identified as moderate. Shear cracks were observed also in the non-structural walls as shown in Fig. 4.7. In the administration building, a column of the west part in the first story was rated damage class V due to shear failure as shown in Fig. 4.8, and the damage level of the building was identified as moderate. Shear cracks were found in a large number of RC walls in the transverse direction of the building.

4.4 S Junior High School in Shichigahama Town

S junior high school is located near Shichigahama Town Hall where seismic intensity of 5-upper was recorded. Two buildings (east and west building) were connected in L-shape plan as shown in Fig. 4.9. The buildings were three-story reinforced concrete structure constructed in 1966 and designed according to the old seismic code. Structural system in longitudinal direction is a moment frame, whereas RC shear walls are installed between classrooms in transverse direction. Seismic

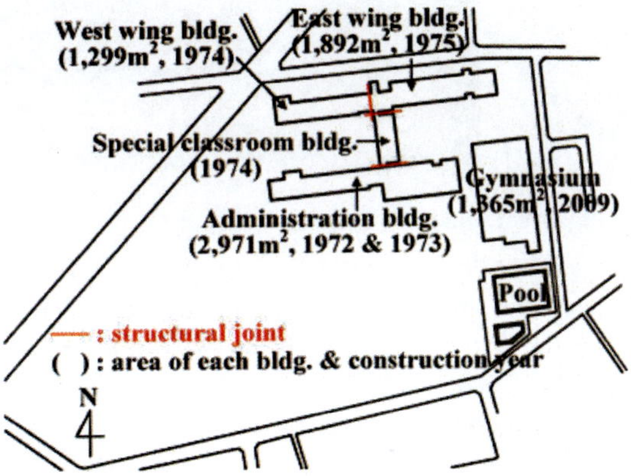

Fig. 4.5 Location of buildings

Fig. 4.6 General view of buildings

Fig. 4.7 Damage to nonstructural wall

evaluation had been carried out and the seismic capacity indices I_s were insufficient for the criteria, seismic retrofit had not yet been applied to the buildings at the time of the quake.

Structural members in the first story suffered most severe damage in NS direction. Shear failure of columns was observed accompanied with spall-off and crushing

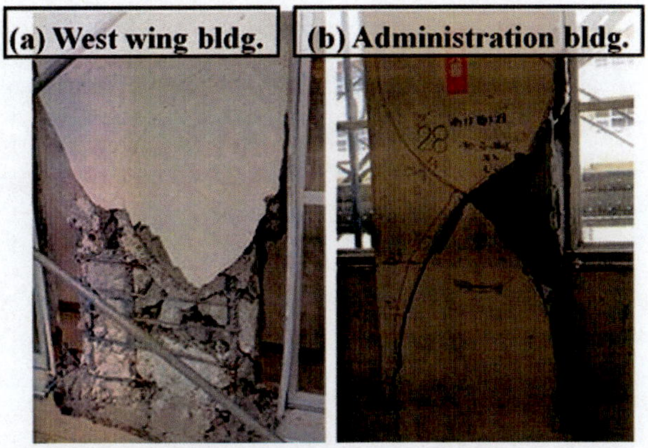

Fig. 4.8 Damage to columns

Fig. 4.9 Overall view of S junior high school building

of concrete as shown in Fig. 4.10. Most of the RC walls in the first story also failed in shear and concrete in the wall panel crushed as shown in Fig. 4.11.

The residual seismic capacity ratios R of the east and west buildings were 25% and 16%, respectively, and the damaged level is graded as "severe."

Fig. 4.10 Shear failure of a RC shear wall

Fig. 4.11 Shear failure of a column at the first story

4.5 TH Elementary School in Sendai City

TH elementary school is located in Sendai City where the JMA seismic intensity of 6-upper was recorded. The buildings are divided into the east building and the west building as shown in Fig. 4.12. Both of the east and the west buildings were three-story reinforced concrete structures constructed in 1973 and 1974, respectively.

Fig. 4.12 Overall view of TH elementary school building

Fig. 4.13 Seismic retrofit with framed steel brace in east building

The first and the second stories of the east building had been seismically retrofitted by the addition of framed steel braces (see Fig. 4.13), and RC shear walls in the longitudinal direction. On the other hand, the seismic retrofit had not been applied to the west building, as the seismic capacity index I_s of the building was larger than 0.7, which is the criterion set by the municipality of Sendai City.

Fig. 4.14 Shear failure of short columns in the west building

The damage to the first and the second story of the east building was slight as it was retrofitted. Flexural and shear cracks were observed in the columns in the third story. But the damage rating was relatively minor to the west building.

As for the west building, severe shear failures were observed in the short columns and non-structural walls in the north frame as shown in Figs. 4.14 and 4.15, respectively. However, the other columns demonstrated ductile behavior which prevented from a collapse of the structure. The residual seismic capacity ratio *R* of the west building was calculated 67%, corresponding to damage grade of moderate.

4.6 N Junior High School in Sendai City

N junior high school is a rare example of damage grade of moderate in new R/C buildings built after the seismic code revision in 1981. The damaged building shown in Fig. 4.16 was four story reinforced and prestressed concrete (PC) structure built in 1984. The school is located at a flat site in the northeast of Sendai City where the JMA intensity of the main shock was 6—and is supported by a spread foundation mounted on unreinforced concrete block with a height of 4 m. The building consists of RC 12 bay frames for the longitudinal direction and 1 bay of cast-in-place PC frames for the transverse direction to achieve a long span of 18.4 m. The building has an open ceiling space over the full height in the middle. RC shear walls in the transverse direction are placed only at the ends of the building while light steel partitions are placed between the classrooms.

Fig. 4.15 Shear failure of non-structural wall with opening in the west building

Fig. 4.16 Overall view of school building

A shear wall with an opening at the second floor in the longitudinal direction failed in shear as shown in Fig. 4.17, the thickness of the wall was 150 mm. Many flexural cracks of damage rate II were observed for RC beams of all floor levels. Cover concrete spalled off at the end of the beams. Shear cracks of damage rate II

Fig. 4.17 Shear failure of R/C wall with opening at second floor

occurred at a plastic hinge region of the beams. RC walls located at an out-of-frame failed remarkably in shear, accompanied with spall-off of concrete and buckling of steel bars. The residual seismic capacity rate *R* was estimated to be 69% for the second story in the longitudinal direction and the building damage was graded as "moderate."

For the transverse direction, no significant damages were observed for beams, columns and shear walls, except for slight flexural or shear cracks of damage rate of I.

4.7 F College in Fukushima City

Collapse of a building, so-called pancake crush, was observed at a college building located north of Fukushima city, which was a three-story reinforced concrete building for administration constructed in 1965. The building is located in the campus of the north area of Fukushima city, which is on a shallow hill with an adjacent drain on north end of the site. Relatively heavy damages were observed in the neighborhood area, which are now crowded with wooden houses, but were marshes in the old days.

The building had a Y-shaped plan with a stair hall in the center and three longitudinal blocks spread to north, east and west directions as equivalent triangular as shown in Fig. 4.18. The foundation bases are PC piles with the depth of 6 m. A continuous shear wall is located only at each end of three blocks, while the walls along the stair in the center are not continuous and the other walls are non-structural partitions. The columns are located with 11 m in the span and 5 m in the longitudinal direction, respectively.

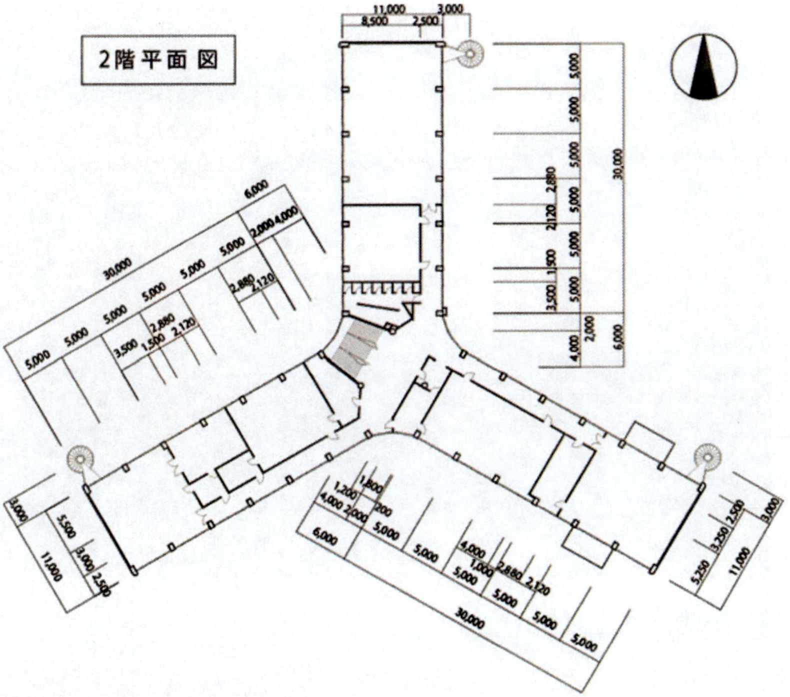

Fig. 4.18 First floor plan with columns and walls in the second story

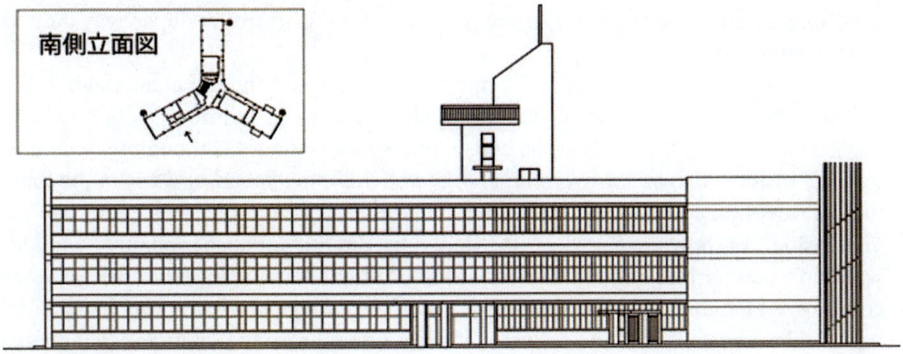

Fig. 4.19 South elevation

The typical columns in the second story had the sectional size of 500 × 750 mm, 12-D25 + 2-D16 main bars, the 9 mm hoop bars set with the spacing of 250 mm. The story height was 3,800 mm, though the inner height of the columns in the longitudinal direction was 2,000 mm due to the spandrel walls, as shown in an elevation in Fig. 4.19, while the column inner height in the span direction was longer, up to 3,000 mm. Therefore, it is estimated that the concentration or unequal distribution of the lateral earthquake loads might have occurred among the three blocks, due to

Fig. 4.20 South overview of west wing

the difference of the inner column heights, because when one of the blocks is in the longitudinal, the others are mostly in span directions, both skewed in 30°, where the shear distribution ratios might have been relatively smaller. The unequal shear distributions might have been one of the reasons why the severe collapse was induced on this building.

As shown in Figs. 4.20 and 4.21, most of the slabs around the center stair hall in the third floor had fallen down due to the shear and axial failure of columns in the second story, while the walls at the ends could support the slabs though they inclined towards inside. Also as shown in Figs. 4.22 and 4.23, the spandrel walls in the longitudinal direction could support the floor slabs so that there are spaces enough to survive inside. As shown in Fig. 4.24, the first-story did not collapse probably owing to a wall in the center hall and the amount of longitudinal bars increased in the first-story columns. At the time of the earthquake, most of the college staffs working in the building could evacuate before the collapse. It is estimated from the interview that the collapse might have occurred at the end of the main shock, probably around 2 minutes after they first felt the shaking. Three of them were confined but rescued several hours later so that no casualties were reported. A two-story reinforced concrete building, attached nursery school, in the adjacent area suffered relatively severe damage with axial compression at the column base. The other buildings in the college have been constructed with the current code or retrofitted, which had no damages, slight or minor. The collapsed building was the only one to be retrofitted in the near future.

Fig. 4.21 South overview of east wing

Fig. 4.22 Axial collapse of a column in the second story

Fig. 4.23 Collapse of the second story

Fig. 4.24 First-story column

4.8 M-Junior High School in Motomiya City

Motomiya city is in the middle area of Fukushima prefecture where the JMA seismic intensity recorded as 5-upper. M junior high school in the city is located in a wide field area with a shallow hill north of the school. Figure 4.25 shows a plan view of the site. The school consists of two reinforced concrete buildings, one steel gymnasium and one steel works. Figure 4.26 shows two reinforced concrete buildings,

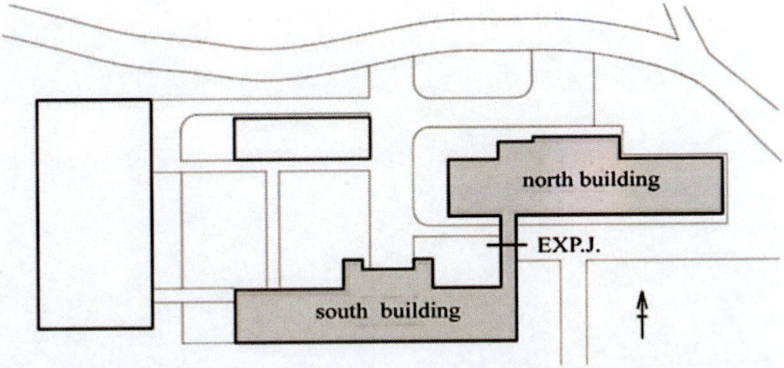

Fig. 4.25 Plan view

Fig. 4.26 South-east panoramic view of south building and north building

both are three-story buildings built in 1966. One building located in the south of the site, called south building, nearly collapsed and the other building located in the north, called north building, suffered heavy damage due to the earthquake.

The south building has typical floor plan as school buildings in the age with five classrooms in a floor. The foundation bases are piles. The structure in the longitudinal direction is moment resisting frame without shear walls. On the other hand, continuous shear walls from base to top are located as partitions between classrooms in the span direction except for two frames of the first floor. In other words, the shear walls of these two frames are not continuous and isolated columns of the

Fig. 4.27 Southern view of south building

first floor support all axial load generated by walls of upper floors, which could become very high during earthquakes.

The columns are located only at the corners of each classroom with 9 m in the longitudinal and 7 m in the span direction, respectively. The typical columns in the first floor have the sectional size of 750 × 500 mm and 14–22φ (longitudinal south frame) or 20–22φ (longitudinal middle frame) main bars (SR235). Nine millimeter round hoop bars (SR235) with 90° hooks are arranged for shear reinforcement with the spacing of 250 mm (column center) or 150 mm (column end).

Figure 4.27 shows southern view of south building. As shown in the figure, the slabs around the center of the building in the second floor had fallen due to the shear and axial failure of columns of the first floor. These columns were isolated columns in the first floor supporting upper shear walls. Therefore, the failure was supposed to be caused by very high axial load generated by walls of upper floors. Figure 4.28 shows the damage of the isolated column located in the corridor side of the classroom. Although the inner height of the column was long, the column suffered severe damage showing crush of concrete and broken hoop reinforcement around the center of the column with hoop spacing of 250 mm. On the other hand, Fig. 4.29 shows the damage of a side column of a barbell type shear wall in the first floor. As shown in the figure, the column failed in shear severely in the longitudinal direction but the damage was slight compared to the isolated columns due to the presence of the shear wall in the span direction. Figure 4.30 shows the damage of a column with a wing wall of the northern frame in the longitudinal direction. The inner height of the column was short due to the spandrel walls and the column failed in shear severely. The damage grade of the south building may be regarded as near collapse, while the grade of north building was rated as severe from the residual seismic capacity ratio.

Fig. 4.28 Shear failure of isolated column of first floor supporting walls of upper floors

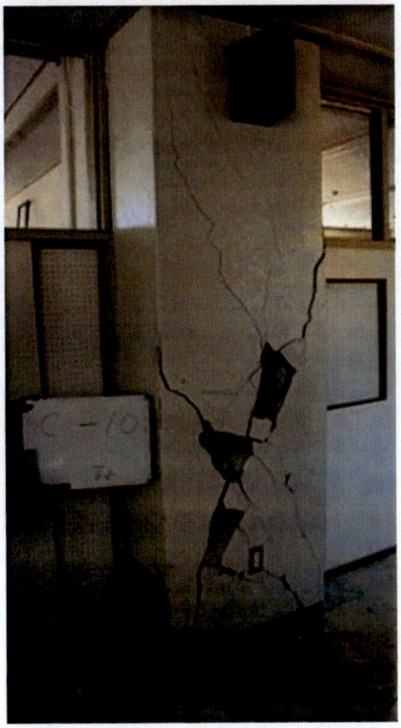

Fig. 4.29 Shear failure of side column of barbell type shear wall of first floor

Fig. 4.30 Shear failure of column with side walls of first floor

4.9 S Primary School in Sukagawa City

The appearance and the site plan of S primary school are shown in Figs. 4.31 and 4.32. The school consists of four RC buildings with three stories, all of which were built in 1965, which were connected or isolated structurally using expansion joints. Among them, A- and B-buildings were severely damaged. The JMA seismic intensity was 6-upper at the station approximately 1 km east from the school. The damage to A-building is described here as an example.

The damage of the building was mainly observed in the longitudinal direction. The I_s-values of the first and second stories in the longitudinal direction had been evaluated as 0.46 and 0.53, respectively. The ultimate base shear coefficient ($C_T S_D$) was 0.47.

In the first story of the building, the damages of the columns with wing-walls were serious as shown in Fig. 4.33. These members failed in shear and vertically deformed, the damage rate of the members was V. The space of hoop reinforcements was approximately 300 mm. In addition, about 70% of columns without wing-wall also failed in shear as shown in Fig. 4.34. Damage rate of these columns were between II and IV. The residual seismic capacity ratio was estimated to be 39% of the original capacity and the damage was graded as severe.

Fig. 4.31 An overview

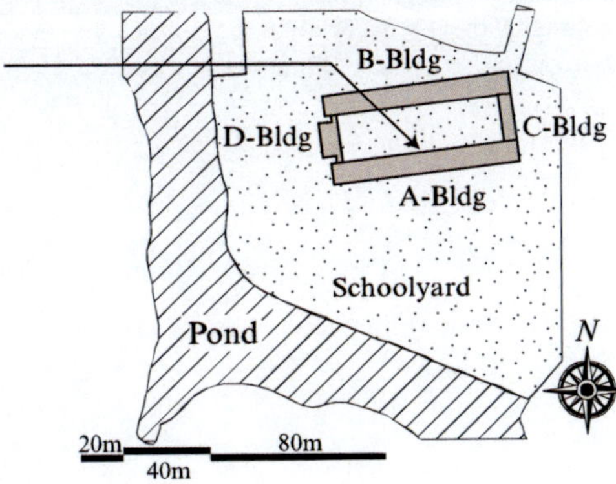

Fig. 4.32 Site plan

4.10 S Elementary School in Iwaki City

S elementary school is located in the southern coast of Iwaki City, and the site is on a plateau locally. The investigated classroom building is a reinforced concrete building of the west side part of the buildings shown in Fig. 4.35. The building is connected to the east side building through an expansion joint. The investigated building with three stories and (2,502 m^2 total floor area) was constructed in 1974. The JMA seismic intensity of the site was 6-lower.

Fig. 4.33 Damage to the column with wing-wall (Grade V)

Fig. 4.34 Damage to the column without wing-wall (Grade III)

The longitudinal direction of the building is composed of moment resisting frames including non-structural walls. In this direction, minor cracks of damage rate of I or II were observed in many columns and their wing walls. The transversal

Fig. 4.35 South side view of the building

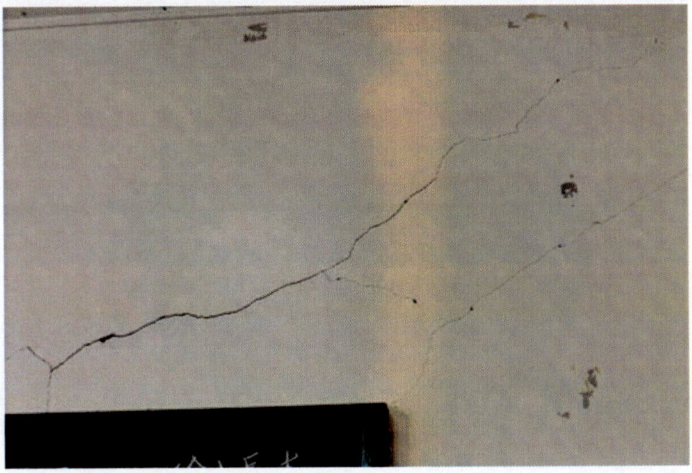

Fig. 4.36 Shear cracks of a wall in span dir

direction of the building is rigid-frame structure with seismic walls of 100–120 mm in thickness. Diagonal cracks of damage rate III were observed in the shear walls at the first floor as shown in Fig. 4.36.

4.11 I High School in Iwaki City

Layout plan of the buildings in I high school is shown in Fig. 4.37. Buildings for classroom (Buildings A and B) are located in the north and south. Two connecting corridor buildings (Buildings C and D) are connected to Buildings A and B with expansion joint. Building A is four-story reinforced concrete (RC) building, while Building B is three-story, and the Buildings C and D both two-story for connecting corridor and classrooms. These were constructed from 1973 to 1975, portion to portion over the years, as was typical process in the age.

The seismic intensity around the school was reported as 5-lower in JMA scale. However, the school is on a small hill with height of 30–40 m from the plain land, and all four buildings suffered moderate or severe damages. The damage grade seems to be much higher than the other school buildings located just at the foot of the hill. The amplification of the ground motions due to the hill zone effect is estimated from the damages though the quantity should be investigated further.

An overview of the Building A is shown in Fig. 4.38. The north frame of the building was severely damaged, and the columns of the special classroom at the west end of the building in the first floor failed in shear up to the damage index of V. The vertical shortening of the columns was estimated obviously from the deformation of the window frames between them, as shown in Fig. 4.39. On the other hand, the shear wall at the west end of the building suffered no serious damage, as shown in Fig. 4.40. Some girders with standing wall in the north frame failed in shear. The floor level of Building A became 15 cm lower than that of Building C.

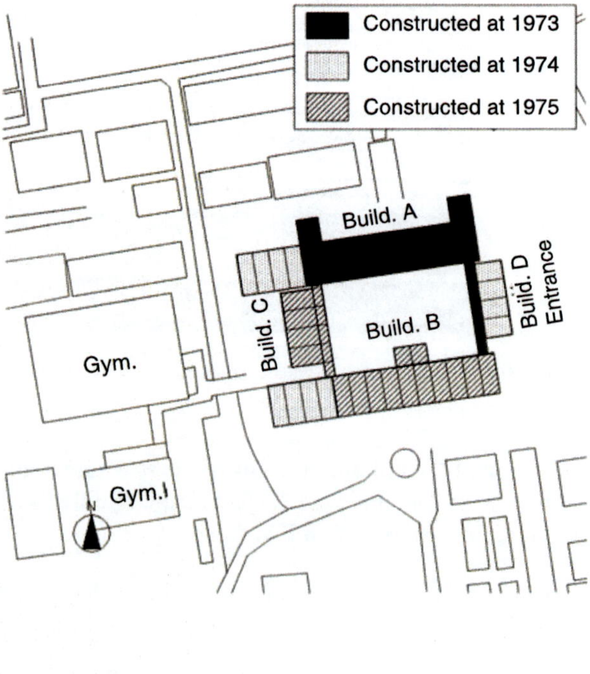

Fig. 4.37 Location of buildings

Fig. 4.38 Overview of Building "A"

Fig. 4.39 Shear failure of the first story column at north frame of Building "A"

The residual seismic capacity index was calculated as R = 33.3% for the longitudinal direction in the first story, by which the damage was classified into the severe damage.

Some of the columns in the Building B failed in shear, which were classified as the damage rate of IV as shown in Fig. 4.41. The joint between Building B and D was damaged due to pounding. The floor level of Building B was lower than that

Fig. 4.40 West side view of Building "A"

Fig. 4.41 Shear failure in the first story column at south frame of Building "B"

of Building D by 12 cm. The residual seismic capacity ratio was calculated as R = 47% for the longitudinal direction of the first story. The damage was classified also into the severe damage.

Columns of the east frame in the first story of the Building C failed in shear evaluated as V with vertical shortening. The residual seismic capacity ratio R=72% for the longitudinal direction of the first story, by which the damage was classified into the moderate damage. In Building D, shear cracks classified as III were observed in some columns, from which the residual seismic capacity ratio was 73% also classified into the moderate damage. The ground in front of the entrance subsided much.

4.12 M Senior High School in Mito City

The buildings of M senior high school located in a site near the center of Mito city, Ibaraki prefecture, on a hill between Senba-ko Lake and Nakagawa River flowing from west to east. The site is flat and surrounded by streets on three sides as shown in Fig. 4.42. The buildings include, a gymnasium and three reinforced concrete buildings for administration, classroom and rooms for particular subjects. The buildings No. 1 with five stories and No. 3 with three stories are relatively new and completed in 1980s and had little damage, while the building No. 2 is the oldest and had extensive damage as reported below.

The building No. 2 is four-story reinforced concrete framed structure with shear walls in the transverse direction, an overview is shown in Fig. 4.43. The building consists of two parts separately completed in 1969 and 1970 while they are structurally jointed. It longitudinally spans 22 m and typical bay length is 4.5 m. The distance of the longitudinal frames is 9.7 m and there are some shear walls in the transverse direction. The building is supported by concrete piles. A set of four piles with diameter of 350 mm and length of 5 m supports each footing beneath each column. The story height is 4.2 m for the first and 3.6 m for the upper stories.

The first bay through 14th bay of frame A and the 15th bay through 23rd bay of the frame B is facing to south and have larger windows in height. Thus the clear height of these columns in the frames are large and approximately 2.0 m, while the clear height of the most columns in the frames on the northern side are 1.35 m short

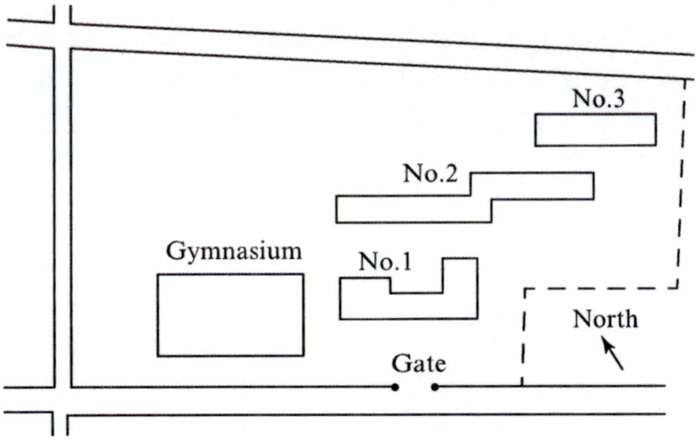

Fig. 4.42 Site plan of M senior high school

Fig. 4.43 Northern side view of the building No. 3

columns with reinforced concrete spandrel walls. The typical dimensions of the columns are 800 mm in depth and 500 mm.

The lowest values of Seismic Indices (I_s) for longitudinal and transverse direction are evaluated as 0.26 at the second story and 0.38 at the second story, respectively. The second and the third story did not satisfy the standard target performance of I_s>0.6, while the first and fourth story satisfied the target.

There are two factors of the high seismic vulnerability particular to this building. One is the building age of 1969 and 1970. The other is deficient concrete with the lowest strength of 11 MPa tested in 2009 with sampled concrete cylinders taken from the part completed in 1969, while concrete cylinders taken from the part completed in 1970 exceeded the design strength of 17.6 MPa.

The photos of the northern view of the building and the most extensively damaged columns are shown in Fig. 4.44. It is observed that the columns of second and third story in the northern frames had extensive shear cracks. The number of columns with damage index of IV, III and II were 1, 7 and 8, respectively, for the part constructed in 1969 in the northern frame. The damage in the northern frame is heavier than that of the southern frame due to the smaller clear height. It is noted that the damage of the columns constructed in 1970 is relatively minor, probably owing to the higher concrete strength, because the reinforcement design of the columns in both parts are to be similar.

The damage level of this building is calculated as "moderate damage" based on the evaluation standard. The relatively old building with high seismic vulnerability suffered more extensive damage and new building suffered slight or no damage. In the same building with similar structural design, the damage levels of the columns with insufficient concrete strength are obviously severer than those with sufficient concrete strength in a building sharing a common diaphragm.

Fig. 4.44 Shear cracks observed on the columns on second and third story (from the sixth bay through the eighth bay of the B frame)

4.13 S Junior High School in Chiba

S junior high school is located in Katori city in the area of the Tone River basin where the JMA seismic intensity scale was 5-upper. Damage to foundation was observed in two three-story reinforced concrete school buildings as shown in Fig. 4.45, which were built in 1970–1971and supported by PC piles with the depth of 16 m. The soil liquefaction was observed around the buildings and the peripheral soils settled down by 30–50 cm as shown in Fig. 4.46. The inclination of the building due to the settlement of the base foundations was observed as shown in Fig. 4.47. The maximum inclination of the building was around 1/70 rad, by which the damage was classified into the severe damage to the foundation. Compression failure was identified at the top of the PC piles reportedly from the excavation survey conducted later on.

4.14 U High School in Chiba

U high school is located in Urayasu city on the Tokyo Bay area where the JMA seismic intensity scale of 5-upper was recorded. No significant structural damage was observed in the four-story reinforced concrete buildings supported by piles, though the damage to the plumbing systems due to the liquefaction was observed as shown in Figs. 4.48 and 4.49.

Fig. 4.45 Overall view of school buildings

Fig. 4.46 Liquefaction of ground around the building

4.15 T Municipal Office Building in Iwate

T municipal office building is a three-story reinforced concrete city hall completed in 1963 and located in T city in the middle south of Iwate prefecture. An overview of the building is shown in Fig. 4.50. The floor plan is 3×6 bay, and some of the inner columns shifts from regular allocation. In the region, the seismic intensity was 5-upper in JMA scale, and the maximum acceleration of 469 cm/s^2 was recorded at a K-NET site nearby.

Fig. 4.47 Inclination of buildings

Fig. 4.48 Settlement of ground by liquefaction

The observed damages were shear failure of columns and walls in the first story and in longitudinal direction, including two short columns in the south section (Fig. 4.51), four columns in the north section, and a structural wall. Shear cracks were also observed in other columns with one-side spandrel wall or without spandrel wall.

Fig. 4.49 Settlement of ground by liquefaction

Fig. 4.50 Overall view of the building

The short columns failed in shear had been temporarily strengthened by steel columns and repaired with concrete after Sanriku-Minami Earthquake 2003. It was not clear whether the strengthening could have been effective or not: the steel columns located in the side surface as well as the front of the concrete column seemed to have reduced the damage compared with the damage to un-strengthened short columns, though the base concrete for anchorage of steel columns showed compression failure at the bottom of the column. Cracks were not observed on the columns

Fig. 4.51 Shear failure of short columns in south section

in the middle frames and continuous structural walls in longitudinal direction in the first story. Shear cracks were also observed in a few columns in the second story.

The post-earthquake damage evaluation was conducted for the longitudinal direction of the first story. The numbers and damage rates of the vertical members were ten shear columns (V:6, IV:1, III:2, 0:1), eight flexural columns (IV:1, II:2, 0:5), and three structural walls (V:1, 0:2), by which the damage was graded as "severe" with the residual seismic performance ratio *R* of 58%.

4.16 East-Building of F Municipal Office in Fukushima

F municipal office is a reinforced concrete (RC) and steel-reinforced concrete (SRC) building with six stories above ground level and one basement floor. An overall view of the building is shown in Fig. 4.52. The building was constructed in 1967 and had been used as a university building until 1990. After that it has been used as a branch of the municipal office. The structural type is SRC from the basement level to the middle of the third story and RC in the upper stories. The change of the column from SRC to RC had been preferred in practice and seen in old medium-rise buildings in Japan. The structural system is moment frames with 13 spans in the longitudinal direction with span length of 4.5 or 9 m and 3 spans of frames and bearing walls of 7.5 m in the transverse direction. Each story height is 4 m, though the measured clear height of the exterior columns was 1.57 m due to spandrel walls in the longitudinal direction. The dimensions of the exterior columns section on the second and the third story are 700 × 800 and 700 × 700 mm, respectively.

Fig. 4.52 Overview of the building

Fig. 4.53 Column in the third story

The third story suffered more damage than the other stories, where minor flexural cracks were observed in the columns and two RC short columns suffered wide shear cracks in the third story as shown in Fig. 4.53. It is estimated that the strength of RC columns in the third story was much smaller than that of SRC columns in the second

Fig. 4.54 Non-structural wall in the third story

story, which might cause heavy damage to the third story. Many non-structural walls located between the exterior columns suffered severe damage in shear failure as shown in Fig. 4.54. The residual seismic capacity ratio R was estimated to 81% in the longitudinal direction of the third story and the damage was graded as "minor."

4.17 S Municipal Office Building in Fukushima

S municipal office is a four-story reinforced concrete city hall building built in 1970 and located in S city in the middle corridor of Fukushima prefecture. The building is connected to annex buildings with a passage on the second floor. The floor plan and the elevation is regular and has two center cores of continuous walls. The plan has 3 by 12 bays of moment resisting frames in the peripheral and two core walls of square shape, by which the most of the earthquake loads are to be carried in both directions.

Axial shear failure of short columns due to spandrel walls was observed in the first story (damage rate of V, Fig. 4.55). The neighboring column in the next span is missing irregularly, which might have caused relatively higher axial load on the damaged column compared with the others. Twenty-five millimeter round bars were used as longitudinal bars in the column, some of which were interrupted and anchored by 180° hook in the middle of the column. The shear crack spreads from the middle level, and the shear failure might have affected by the reinforcing detail and also by insufficient amount of hoops. Several other short columns in the peripheral frames showed shear cracks especially in the east and south corners. Buckling of the longitudinal bars was observed at the bottom of the boundary corner columns

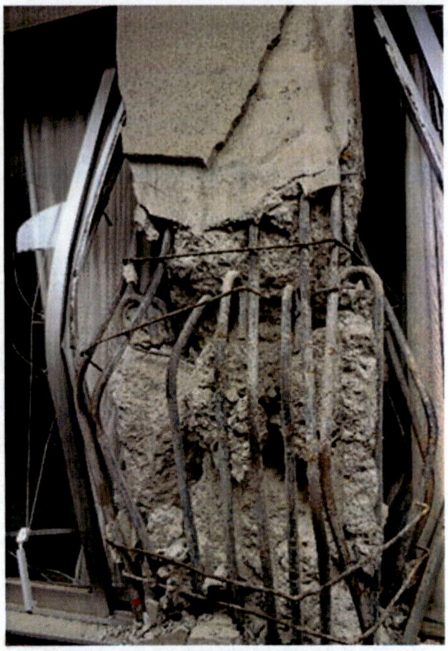

Fig. 4.55 Shear failure of short column

Fig. 4.56 Shear failure of wing walls

in the core walls, and also in the top corner of the walls without boundary columns in the transverse direction. Numbers of shear cracks exceeding 0.2 mm were observed in the wall panels in both directions.

The shear failure and cracks were observed in the columns with wing walls in the north and east section on the second story (damage level III, Fig. 4.56).

Fig. 4.57 Shear failure of structural walls

The structural walls including center core walls on the second floor showed shear failure in both directions (damage level V, Fig. 4.57). Non-structural walls were also heavily damaged in a passage and east section.

The building suffered heavy damages in the first and second stories where the residual seismic performance ratios *R* were evaluated as 50% and 38% for the first story and the second story, respectively, thus the damage level is graded as "severe" for the both of them.

4.18 Damage to K Municipal Office in Ibaraki

K municipal office is a three-story reinforced concrete building with one basement floor constructed in 1964 and located on a hill. An overview of the building is shown in Fig. 4.58. This building had an irregular plan mixed with parts of different number of stories from one to three.

The typical dimensions of columns in the first story are 550×600, 550×550 and 500×500 in mm for three-, two- and one-story frame, respectively. The longitudinal reinforcing steel in the columns were round bars of 10–22φ or 14–22φ, for example. The hoops are 9φ@250 mm in common. Maximum accelerations of 968 gal in NS and 596 gal in EW direction were recorded at K-NET station 200 m away from the building.

Damages were observed in the RC members of each story, especially severe damage on the first floor. Typical damage was shear failure of short columns with

Fig. 4.58 An overview

Fig. 4.59 Shear failure of short column in first floor

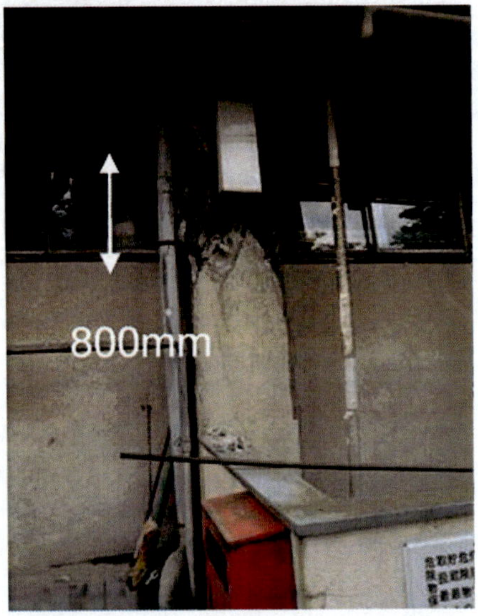

spandrel walls as shown in Figs. 4.59 and 4.60, a column with wing wall in Fig. 4.61 and a wall with opening in Fig. 4.62. Horizontal dislocation at the top of the walls on the third story was observed as shown in Fig. 4.63.

Fig. 4.60 Shear failure of short column in first floor

Fig. 4.61 Column with wing wall

Although the recorded acceleration in NS direction was larger, the structural damage was much heavier in the longitudinal (EW) direction, probably due to the short columns in the longitudinal (EW) direction and also due to the effects of some walls in the transverse (NS) direction. The residual seismic capacity ratio R was estimated to be 38% for the longitudinal direction of the first story, which was graded as "severe."

Fig. 4.62 Wall in the first story

Fig. 4.63 Wall in the third story

4.19 N Housing Complex in Sendai City

N housing complex is a nine-story RC building located in Taihaku ward of Sendai City built in 1969. The first and second floors are for the use of commercial facilities partially with the underground floor level. The third through ninth floors are for

Fig. 4.64 Overview

Fig. 4.65 Bottom of column

dwelling houses of an inside passage type with 84 units. An overview is shown in Fig. 4.64. Observed damages are flexural failure at the bottoms of four columns in the first story as shown in Fig. 4.65, and shear failures of structural shear walls in the first and second story as shown in Fig. 4.66, non-structural walls around the doors from the third through ninth story as shown in Fig. 4.67, and also external walls between the

Fig. 4.66 Shear wall

Fig. 4.67 Damage to non-structural walls

Fig. 4.68 Damage to external walls. (example of damaged reinforced concrete housing complex)

windows as shown in Fig. 4.68. The similar types of relatively moderate damage or shear failure of non-structural walls were quite frequently observed in many other medium-rise buildings in Sendai City, even if they were built after 1981 in accordance with the current code. The damage caused serious troubles in continuous use for dwelling, although the buildings were safe enough against collapse.

4.20 K Municipal Housing Complex

K housing complex is ten-story reinforced concrete frame building located in Aoba ward of Sendai City. The floor plan is shown in Fig. 4.69, while an overview is shown in Fig. 4.70. The building had been strengthened and retrofitted with viscous dampers attached to external frames. Toggle amplification mechanism, a proprietary damping system was used as shown in Fig. 4.71. However, shear failure and shear cracking were observed in non-structural walls adjacent to opening such as doors and window on the corridor side throughout most of the stories including the strengthened stories with dampers, as shown in Figs. 4.72 and 4.73.

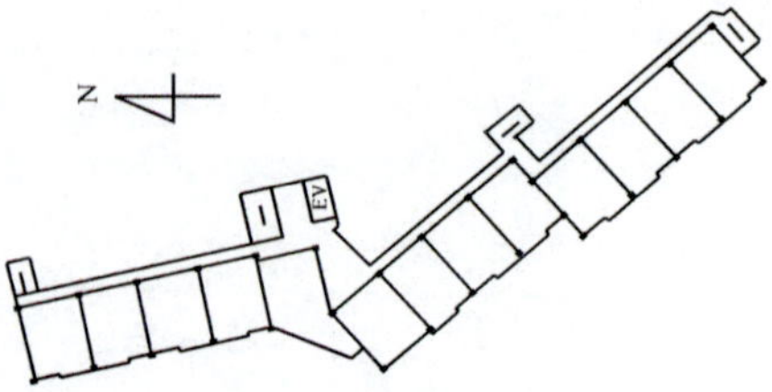

Fig. 4.69 Typical floor plan

Fig. 4.70 Overview of the building

Fig. 4.71 Oil damper brace with toggle amplification mechanism

Fig. 4.72 Damages to non-structural walls between openings

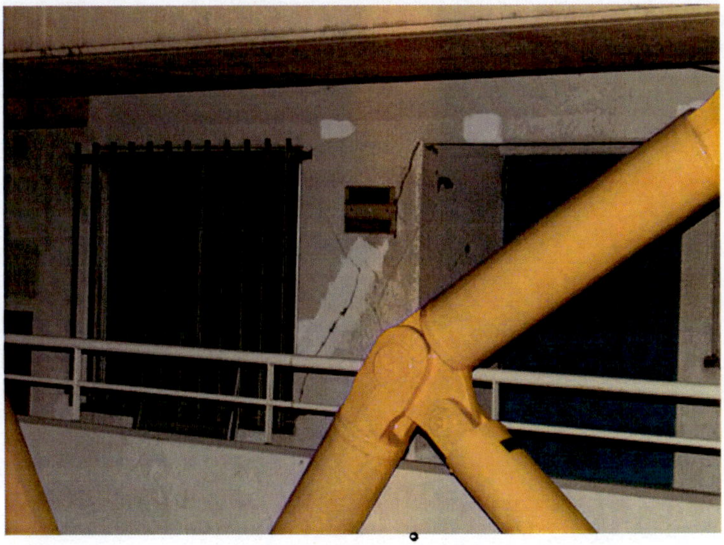

Fig. 4.73 Damages to non-structural walls between openings near the damping system

Acknowledgements The field survey was conducted as the reconnaissance activity of Architectural Institute of Japan, from April to June 2011, part of which was requested from the local governments through MEXT, Ministry of Education as combined with the rehabilitation procedure on the damaged school buildings. The voluntary activities of the committee members of AIJ, the clerical supports of the staffs in the local governments and MEXT and the technical support of practical engineers on the field survey are gratefully acknowledged.

References

1. JBDPA (2001) Standard for Seismic Evaluation of Existing Reinforced Concrete Buildings (2001 Japanese version and 2004 English version). Japan Building Disaster Prevention Association
2. JBDPA (2001) Standard for Damage Level Evaluation and Guidelines for Recovery Technologies of Buildings Damaged by Earthquake (in Japanese). Japan Building Disaster Prevention Association, 360 pp

Chapter 5
Damage to Steel Reinforced Concrete Buildings

Akihiko Kawano, Junichi Sakai, Mareyasu Doi, Hiroshi Kuramoto, Takashi Fujinaga, and Teruhisa Tanaka

Abstract Steering Committee on Steel-Concrete Composite Structure (SCCS), Architectural Institute of Japan, conducted a survey on the damage of Steel Reinforced Concrete (SRC) buildings in Sendai City. The SRC buildings suffered damage to nonstructural members in principle and little damage was observed in structural members. However, some buildings designed by old seismic design codes suffered severe damage to structural members. But they are not such serious damage as story collapse at the middle story of SRC buildings, which were found after the 1995 Great Hanshin earthquake. This chapter briefly describes damage to four SRC buildings which showed distinct types of damage.

Keywords Earthquake damage • Nonstructural walls • Pile foundation • Sendai City • Shear failure • Steel reinforced concrete buildings • Structural members

A. Kawano (✉)
Kyushu University, Fukuoka, Japan
e-mail: kawano@arch.kyushu-u.ac.jp

J. Sakai • T. Tanaka
Fukuoka University, Fukuoka, Japan
e-mail: sakaij@fukuoka-u.ac.jp; sttanaka@fukuoka-u.ac.jp

M. Doi
Niigata University, Niigata, Japan
e-mail: mare@cc.niigata-u.ac.jp

H. Kuramoto
Osaka University, Suita, Japan
e-mail: kuramoto@arch.eng.osaka-u.ac.jp

T. Fujinaga
Kobe University, Kobe, Japan
e-mail: ftaka@kobe-u.ac.jp

Architectural Institute of Japan (ed.), *Preliminary Reconnaissance Report of the 2011 Tohoku-Chiho Taiheiyo-Oki Earthquake*, Geotechnical, Geological and Earthquake Engineering 23, DOI 10.1007/978-4-431-54097-7_5, © Springer Japan 2012

5.1 Introduction

Steel Reinforced Concrete (SRC) composite structure had evolved independently in Japan as a means to enhance the seismic resistance of ordinary reinforced concrete structural member. The 1978 Miyagi-ken oki earthquake shook Sendai City, Miyagi Prefecture, with the JMA seismic intensity of 5. This was the first major earthquake to test the seismic performance of SRC buildings in Sendai City.

It was reported in the reconnaissance report of AIJ [1] that the rate of the structures with some damage in Sendai City, Miyagi Prefecture was 13% of the SRC buildings. Most frequently observed damage were shear cracks or shear failures of non-structural walls and damage to the parts adjacent to expansion joints. Many apartment buildings also experienced the problem of inoperable doors at the entrance of each household. Importance of the design of the non-structural walls to decrease inoperable doors was recognized. Damage reported to structural elements was minimal.

Typical detailing of steel encased in SRC beams and columns in 1970s was open-web type, which are classified into (a) batten-plate type or (b) lattice type, both of which are assembled with L-shapes and plates (see Fig. 5.1a). The performance of SRC structural members of open-web type was regarded to be sufficient at that time.

After the 1978 Miyagi-ken oki earthquake, a new type of SRC member; full-web type was developed, which uses H-shape steel for encased steel (see Fig. 5.1b). Then the open-web type was replaced with the full-web type, because it was revealed by tests that full-web type is structurally superior to the open-web type in seismic performance.

The 1995 Great Hanshin earthquake extensively damaged the SRC structures, bringing into question on the expected superiority in earthquake-resisting performance. It was reported in the reconnaissance report of AIJ [2] that the collapse of the intermediate story were observed in 32 SRC buildings with SRC members of open-web type while no collapse of SRC buildings of full-web type was observed. Anyway, typical damages to SRC buildings were (1) shear failure of column, (2) fracture of the bolts connecting the column to the base plate, both of which make columns unable to support vertical load.

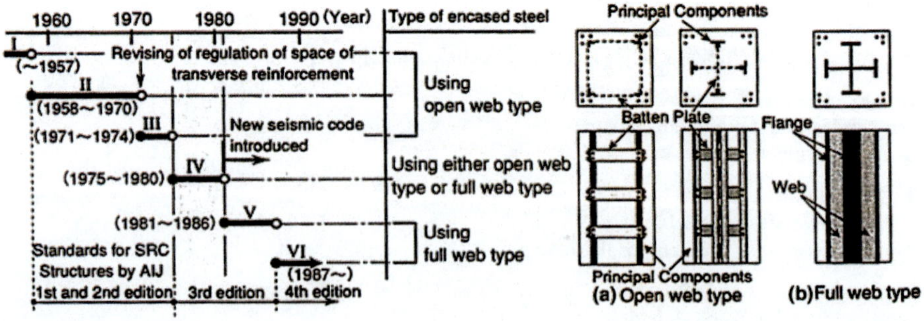

Fig. 5.1 Type of encased steel

Table 5.1 Number of SRC buildings classified by ward

	Sendai-city					
Area	Aoba	Izumi	Miyagino	Wakabayashi	Taihaku	Total
Number of buildings	2	9	16	13	3	43

In March 2005, an moderate earthquake hit the west of the Fukuoka Prefecture on northern part of Kyusyu Island. Shear cracks on shear walls, damage adjacent to expansion joints, and shear failure of non-structural walls were observed to SRC members with full-web type [3]. It was revealed that the lessons learned were not utilized in the design, although the 1995 Great Hanshin earthquake caused significant damage to such structural members.

Sendai City, Miyagi Prefecture have experienced three major earthquakes, the 1978 Miyagi-ken oki earthquake (JMA intensity of 5-upper), 2005 Miyagi earthquake (JMA intensity of 5-upper), and 2008 Iwate-Miyagi Inland earthquake (JMA intensity of 5-upper) before the 2011 Tohoku-chiho Taiheiyo oki earthquake. In order to understand the extent and feature of the damage to SRC building, Steering Committee on Steel-Concrete Composite Structure, Architectural Institute of Japan (AIJ SCCS Steering Committee), coordinated and sent reconnaissance teams consisting of the member of the committee. This chapter reports the findings of the reconnaissance by the team.

5.2 Method of Survey

AIJ SCCS Steering Committee have sent three teams to reconnaissance damages of SRC buildings from April to June in 2011. Both low-rise and high-rise SRC buildings were included in the survey. Table 5.1 lists the number of SRC buildings subjected to the survey.

The visited areas are Aoba Ward, Izumi Ward (JMA intensity of 6), Miyagino Ward (JMA intensity in 6-upper), Wakabayashi Ward (JMA intensity of 6), and the Taihaku Ward (JMA intensity in upper 5-upper) in Sendai City, Miyagi Prefecture, where JMA intensity larger than 5-upper is recorded. Table 5.1 shows the number of buildings in each area while Fig. 5.2 shows each location of the SRC building investigated.

5.3 Typical Damage Pattern

Non-structural damage was extensive for the SRC buildings while the damage to structural members were very little. However, some damage to the structural members were found in the buildings built before 1981. No collapse of the intermediate story was found.

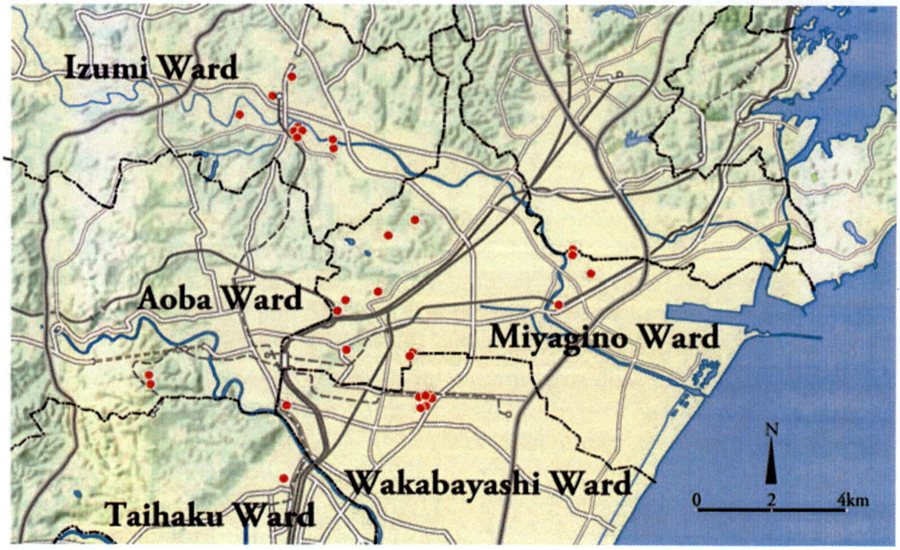

Fig. 5.2 Distribution of investigated SRC buildings

Notable damages to structural and non-structural members found in SRC buildings in Sendai City, Miyagi Prefecture are:

1. Structural Members
   (a) Flexural cracking and shear cracking of columns
   (b) Shear failure of column bases
   (c) Flexural cracking and shear cracking of beams
   (d) Flexural yielding of columns
   (e) Shear failure and bond splitting failure of coupling beams
   (f) Cracks on concrete brace
   (g) Shear cracks and crushing of concrete of shear walls and boundary elements
   (h) Flexural failure of multistory shear wall
   (i) Damage of elevator shaft
   (j) Damage to pile foundation
   (k) Penthouse damage
2. Non-structural Members
   (a) Shear failure and shear cracks of non-structural wall
   (b) Cracks of horizontal construction joint
   (c) Damage to door frame
   (d) Chipped tiles
   (e) Expansion joint damage
   (f) Ceiling panel damage

Fig. 5.3 Southside view of the apartment building S1

(g) Autoclaved lightweight concrete panels (ALCs) damage
(h) Windowpanes damage
(i) Corridor railing damage
(j) Ground subsidence

The field survey on the SRC buildings was completed on June 14, 2011. The total number of buildings studied was 43, 6 of which received no damage. Four buildings are selected from the inventory of surveyed buildings to show typical damage of SRC buildings in the following sections.

5.4 Apartment Building S1

An apartment building S1 is a 14 story residential building of a condominium in Miyagino Ward. Construction of the building completed in 1976 and is a kind of typical 1970s architecture. It experienced the 1978 Miyagi-ken oki earthquake after 2 years of completion and survived the earthquake.

Figure 5.3 shows the photo of overview of the building. The floor plan of the building is of "L" shape while two parts (north and south tower) are structurally separated by expansion joints. The corridors of the building meets at the expansion joints. The south tower is oriented to the east–west direction while the north tower

Fig. 5.4 Shear failure of non-structural reinforced concrete wall

is oriented to the north–south direction. The dimensions and type of the encased steel or the arrangements of reinforcing bars are not known. But it is an SRC structure with open-web type presumably according to the construction age. The 1978 earthquake had caused no damage to structural elements such as columns, beams, or shear walls. Ground subsidence and inclination of the building was also not evident after the event. Non-structural walls and foundation piles beneath the footing were found to be damaged. So they were repaired in 1979 [4].

After the earthquake in 2011, the south tower was identified to be inclined to the south two hundredth radians approximately. The type of damage and extent are similar to those in 1978. The inclination of the first floor slab is so large, so it is presumed that the pile might have some damage again. In addition, shear cracks were observed on the first story column. Non-structural walls experienced shear failures throughout all floors in both parts (see Figs. 5.4, 5.5, 5.6, 5.7, 5.8, 5.9, 5.10, 5.11, and 5.12). The cover concrete spalled off severely and the vertical and the horizontal reinforcing bars were visible. Damage to non-structural elements also occurred adjacent to the expansion joints, so the two parts of the building might have collided. The most extensive damage is seen at the top floor. Pipes were destroyed and it caused ceiling panels to fall off.

Fig. 5.5 Shear failure of non-structural reinforced concrete wall

Fig. 5.6 Shear failure of non-structural reinforced concrete wall

Fig. 5.7 Shear failure of non-structural reinforced concrete wall

Fig. 5.8 Shear failure of non-structural reinforced concrete wall

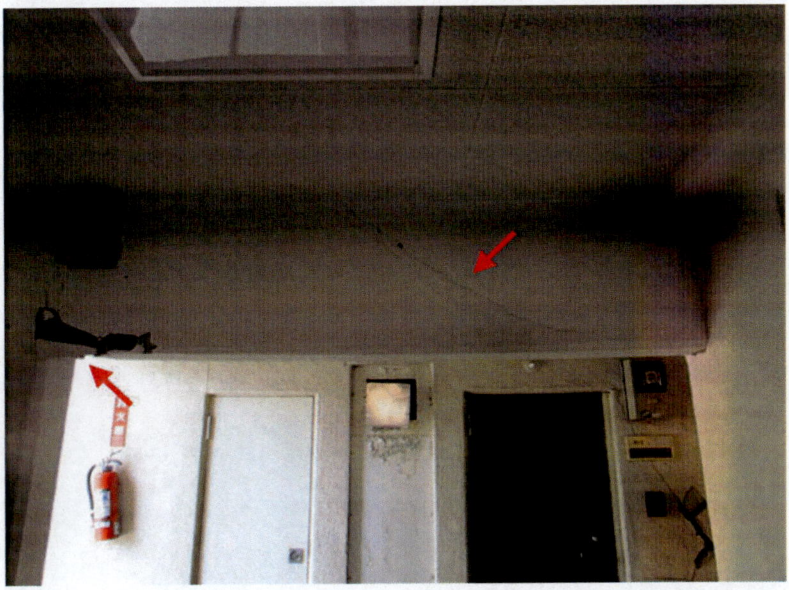

Fig. 5.9 Shear crack of short span beam

Fig. 5.10 First floor shear crack (shear crack of column on first story)

Fig. 5.11 Expansion joint damage

Fig. 5.12 Ground subsidence around building

5.5 Apartment Building K1

An apartment building K1 is a eight story residential building of a condominium in Izumi Ward. It was built in 1996. The floor plan is of "L" shape comprised of two parts, north building and south building, which are structurally separated by expansion joints. Figure 5.13 shows the photo of the overview of the building. The corridors in the two parts meet at the expansion joints. The north building orients to east–west direction. The columns and beams from the first through the six stories are of SRC while those of the seventh and eighth story are of RC. The south building orients to the north–south direction and all the columns and beams are of SRC.

Figures 5.13, 5.14, 5.15, 5.16, 5.17, 5.18, 5.19, 5.20, 5.21, and 5.22 show the close up of the various damage of the building. In the south building, the beams adjacent to an elevator shaft, walls, penthouse, and stairs leading to the roof floor were severely damaged. In addition to the elevator, evident shear cracks were seen on three of five RC shear walls with 150 mm thickness below the sixth story. Concrete crushed and reinforcing bars were exposed at the bottom of the walls.

In the north building, the shear wall at the second floor was damaged at the bottom edge of the RC walls and reinforcing bars were exposed. The staircase between the ground floor and the second floor was also damaged. Overall damage extent of the north building is much slighter than that of the south building.

Fig. 5.13 Overall view of the apartment building K1

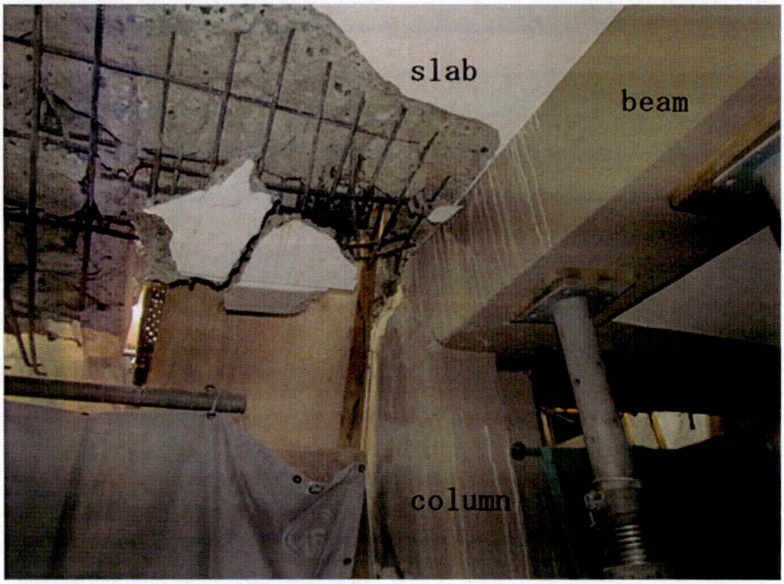

Fig. 5.14 Serious damage of beam with slab

Fig. 5.15 Shear crack of RC wall

Fig. 5.16 Crushed RC wall of a *lower* end part

Fig. 5.17 Damage to connecting corridor

Fig. 5.18 Damage to RC non-structural wall to handrail connection

Fig. 5.19 Shear failure of RC non-structural wall

Fig. 5.20 Crushed RC non-structural walls

Fig. 5.21 Damage to RC non-structural wall at the stairs

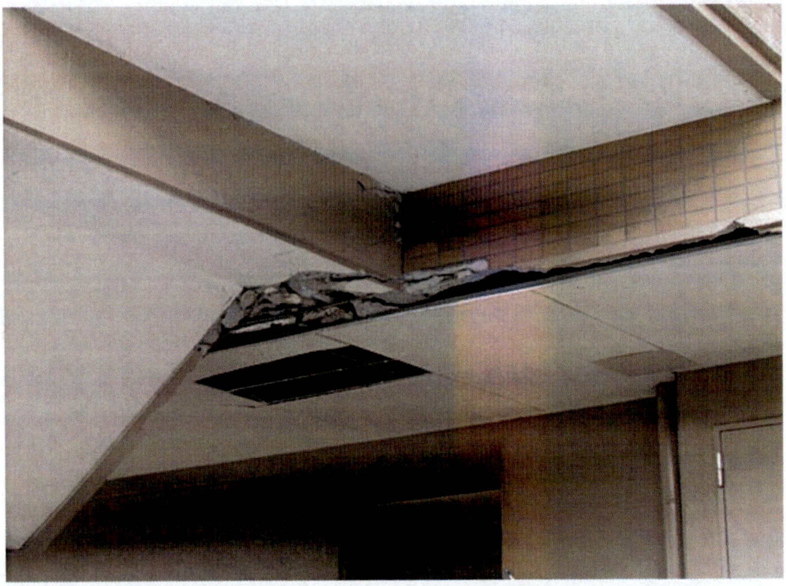

Fig. 5.22 Damage to ceiling panel

Shear cracks and shear failures were observed in non-structural walls in the both buildings. In particular, the severest damage were seen from the third story through five story of the south building.

5.6 University T-1

The University T-1 is a eight story building for higher education and research in a university campus in Aoba Ward. It was built in 1966. The floor plan is of seven spans by three spans in longitudinal and transverse direction, respectively. Figure 5.23 shows the photo of the overview of the building. This building was retrofitted for seismic resistance in 1996. Dimensions of the encased steel and reinforcement detailing are not known but steel type is estimated as open-web type by the age of construction. Multistory shear walls with coupling beams experienced a shear failure as well as splitting bond cracks (see the close up shown in Figs. 5.24 and 5.25). In addition, it was reported that a cab of the elevator fell during the shaking [4].

5.7 University T-2

The University T-2 is a nine story building in the same campus of the university described above. It was built in 1969. The floor plan is ten spans and two spans in the longitudinal direction and the transverse direction, respectively. Two independent

Fig. 5.23 Overall view of University T-1 building

Fig. 5.24 Shear failure of coupling beams

Fig. 5.25 Splitting bond failure of coupling beams

two story buildings are attached to the two sides of the nine story building. Figure 5.26 shows the photo of the overview of the building.

Similar to the other buildings, dimensions of the encased steel, and reinforcement detailing is not known but estimated by the age of construction. This building was retrofitted for seismic resistance in 2000, where steel braces with frame were inserted and shear walls were added in the longitudinal and the transverse direction, respectively.

The earthquake caused extensive damage to the column base at the floor level of the third story. Shear failures, peeling of concrete, buckling of rebar, and rupture of shear reinforcement were observed. It is believed that the extensive damage of the columns is due to a large difference in the fundamental modes of vibration between the two buildings. Figures 5.27, 5.28, 5.29, and 5.30 show the close up of the damages.

5.8 Future Investigation

The plan for future research is listed below:

1. Further analysis of the SRC buildings repaired due to the 1978 Miyagi-ken oki earthquake
2. Damage survey of the SRC buildings in the other main cities besides Sendai (Fukushima, Koriyama, and Iwaki)
3. Statistical analysis of SRC damage in buildings in Tohoku area (including buildings that were undamaged)

Fig. 5.26 Overall view of University T-2 building

Fig. 5.27 Shear failure of perimeter column on the third story

Fig. 5.28 Shear failure of perimeter column on the third story

Fig. 5.29 Shear failure of perimeter column on the third story

Fig. 5.30 Shear failure of perimeter column on the third story

4. Structural investigation of other steel-concrete composite structures (other than SRC)
5. Survey of the steel-concrete composite structures affected by the tsunami
6. Develop future measures to protect against damage of typical SRC buildings

Acknowledgements This chapter is based on the damage reconnaissance reports prepared by the Steering Committee on Steel-Concrete Composite Structures, Architectural Institute of Japan. Contributions of the reconnaissance team members, listed below, are greatly appreciated: Juan Jose Castro, Hisatoshi Kashiwa, Masato Sakurai, Suguru Suzuki (Osaka Univ.); Shintaro Matsuo (Kyushu Univ.); Masayuki Handou, Tomomi Fujita (Sendai National College of Tech.); Yo Kuratomi (Fukuoka Univ.); Koichi Minami (Fukuyama Univ.). Part of this report was written in reference to the content of public housing damage survey conducted by the Steering Committee on Wall Construction, Architectural Institute of Japan.

References

1. Architectural Institute of Japan, Report on the Damage Investigation of the 1978 Miyagiken-oki Earthquake, pp 595–629, February 1980 (in Japanese)
2. Architectural Institute of Japan, Report on the Hanshin-Awaji Earthquake Disaster, Building Series vol 2, pp 100–502, August 1998 (in Japanese)
3. Architectural Institute of Japan, Report on the 2005 Fukuoka-ken Seiho-oki Earthquake, September 2005 (in Japanese)
4. The Nikkei BP, Nikkei Architecture, No.951, pp 12–23, May 2011 (in Japanese)

Chapter 6
Damage to Reinforced Concrete Box-Shaped Wall Buildings

Eiichi Inai, Shinji Tokita, Yoshio Inoue, Masayoshi Iizuka, and Takahiro Sasaki

Abstract Reinforced concrete box-shaped wall structure is one of the typical structural systems for public housing in Japan. The earthquake and tsunami damage to reinforced concrete box-shaped wall buildings with various types is described in this chapter. Damage is mainly based on surveys of public housing in Sendai City and the coastal areas of Miyagi Prefecture.

Keywords Mid-sized ribbed thin concrete panel buildings • Precast pre-stressed reinforced concrete shear wall buildings • Public housing • Reinforced concrete box-shaped wall buildings • Sendai City • The post-earthquake damage evaluation • Tsunami

E. Inai (✉)
Graduate School of Science and Engineering, Yamaguchi University, Yamaguchi, Japan
e-mail: inai@yamaguchi-u.ac.jp

S. Tokita
Urban Renaissance Agency, Tokyo, Japan
e-mail: s-tokita@ur-net.go.jp

Y. Inoue
UR Linkage Co., Ltd, Tokyo, Japan
e-mail: inoue-y@urlk.co.jp

M. Iizuka
The Japan Prefabricated Construction Suppliers & Manufacturers Association, Tokyo, Japan
e-mail: carlos_fudo@yahoo.co.jp

T. Sasaki
Rescohouse Co., Ltd, Tokyo, Japan
e-mail: takahiro.sasaki@rescohouse.co.jp

Architectural Institute of Japan (ed.), *Preliminary Reconnaissance Report of the 2011 Tohoku-Chiho Taiheiyo-Oki Earthquake*, Geotechnical, Geological and Earthquake Engineering 23, DOI 10.1007/978-4-431-54097-7_6,

6.1 Introduction

Reinforced concrete box-shaped wall buildings are a category of engineered buildings being constructed in line with "the Standards for Structural Design and Structural Calculation of Reinforced Concrete Box-Shaped Wall Structures (Architectural Institute of Japan 2003)" developed in 1960s and used for more than 5 decades. This chapter describes the earthquake and tsunami damage to reinforced concrete box-shaped wall buildings, precast pre-stressed reinforced concrete shear wall buildings, and mid-sized ribbed thin concrete panel structures popularized in the 1960s due to the demand for mass production of public housing. The summary includes earthquake damage of public housing in Sendai and tsunami damage in the coastal areas of the Miyagi Prefecture.

6.2 Damage Caused by Ground Motion

6.2.1 Outline of Damage Survey

6.2.1.1 Building Inspection

There are total 64 housing estates in the city of Sendai which include the following: (a) Miyagi Prefectural Municipal Housing, (b) Housing Owned by Miyagi Prefecture Housing Supply Corporation, (c) Sendai Municipal Housing, and (d) Housing of Urban Renaissance Agency. Total number of residential buildings in these estates is 638. Table 6.1 shows the number of investigated structures classified by the type of structure and district. Out of total 638 buildings, 500 buildings were 2–5 story reinforced concrete box-shaped wall buildings (WRC buildings) built from cast-in-site concrete, 47 buildings were 4–5 story precast pre-stressed reinforced concrete shear wall buildings (WPCa PS buildings) and 91 buildings were 2 story public houses made of ribbed concrete panels (ribbed panel buildings). The investigated residential buildings do not include any of the types classified as box-shaped reinforced concrete block wall construction and reinforced masonry constructions with concrete block or ceramic shell. The investigation of 634 buildings in 63 housing estates was conducted on April 21 through April 24, April 29 through May 2, and on June 26, 2011. The remaining four buildings of WRC construction type haven't been investigated yet.

6.2.1.2 Methodology

The post-earthquake damage evaluation on the residential buildings was carried out by the method described in "Guidelines for inspection of post-earthquake damage evaluation for reinforced concrete buildings," issued by The Japan Building Disaster Prevention Association. In the evaluation of damage of WPCa PS buildings, and

Table 6.1 Number of box-shaped wall type public apartments at each area in Sendai City

Type of structure	Aoba Ward	Miyagino Ward	Wakabayashi Ward	Taihaku Ward	Izumi Ward	Total
WRC buildings 2F	2	0	0	0	0	2
WRC buildings 3F	35	9	13	7	30	94
WRC buildings 4F	39	43	4	25	16	127
WRC buildings 5F	24	101	1	55	96	277
WPCa PS buildings 4F	6	8	3	3	6	26
WPCa PS buildings 5F	0	12	2	1	6	21
Ribbed panel buildings 2F	0	70	0	0	21	91
Total	106	243	23	91	175	638

ribbed panel buildings special attention was made to the damage to vertical connections and horizontal connections between precast concrete elements. Only visual inspection has been conducted and the level of damage was noted for each building.

The seismic intensity of each housing estate is estimated from the nearest observation station operated by Japan Meteorological Agency, Building Research Institute [1], Tohoku University Disaster Control Research Center [2], National Research Institute for Earth Science and Disaster Prevention [3] and Strong Motion Array of Local Lots by the Tohoku Institute of Technology Area Network [4]. Figure 6.1 shows the location of apartment complexes investigated in Sendai as well as the measured seismic intensities.

6.2.1.3 Summary of Findings

Summary of damage investigation conducted on 634 building structures is presented in Table 6.2 in which buildings are classified by the type of structure. Investigation of WRC structures showed that 97.4% of investigated buildings had only slight or no damage, while some structural damage is observed on 2.6% of investigated structures. Moderate damage is observed only on one building of this structural type. Severe damage or collapse is not observed.

All of 47 investigated WPCa PS structures in Sendai were only slightly damaged or had no damage at all. However, as shown in Sect. 6.2.3, one building of WPCa PS structural type located in Natori was severely damaged at the foundation structure.

Out of 91 ribbed panel-building structures, 72.5% experienced only slight, or no damage. Among the remaining 27.5%, 11 buildings were moderately damaged, while only 1 building was severely damaged.

The relationship between distribution of earthquake intensity and the level of damage on three considered structural types are shown in Figs. 6.2, 6.3, and 6.4. It can be observed that seismic intensity of 7 did not cause structural damage on WRC structures. Seismic intensity of more than 6 caused only slight damage on most of WRC and WPCa PS structures, while 27.5% of ribbed panel structures had minor, moderate or severe damage. The reason ribbed panel structures had more damage is because the majority of them were designed in line with the old seismic design code prior to 1981.

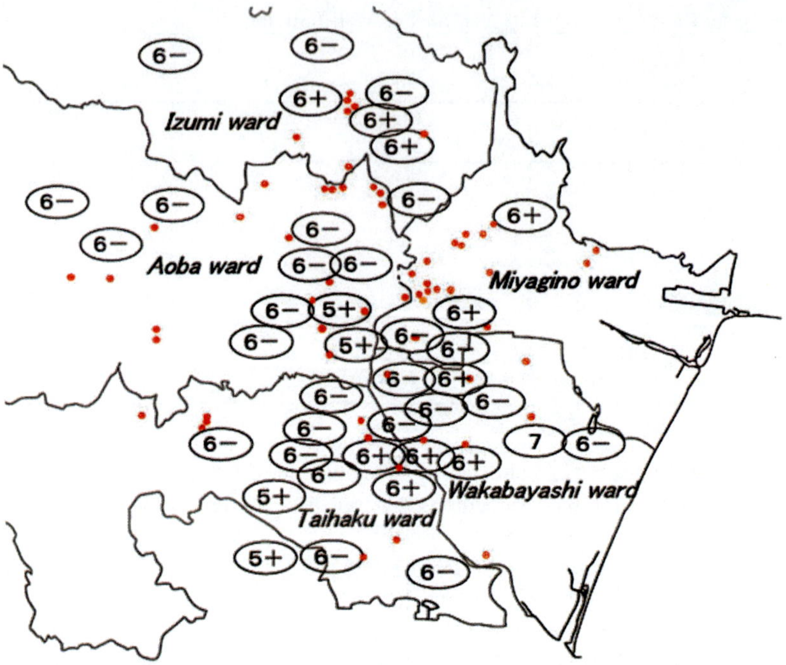

Fig. 6.1 Locations of investigated apartment complexes and seismic intensities in the City Sendai

Table 6.2 Level of earthquake damage on investigated residential buildings in Sendai

	Damage level of super-structure					
Type of structure	No damage	Slight damage	Minor damage	Moderate damage	Severe damage	Collapse
WRC buildings	369	114	12	1	0	0
WPCa PS buildings	31	16	0	0	0	0
Ribbed panel buildings	43	23	13	11	1	0
Total	443	153	25	12	1	0

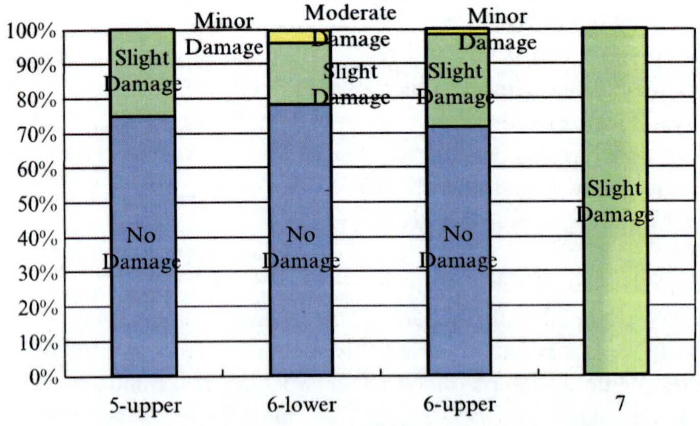

Fig. 6.2 Distribution of earthquake damage level of WRC apartment buildings and seismic intensity

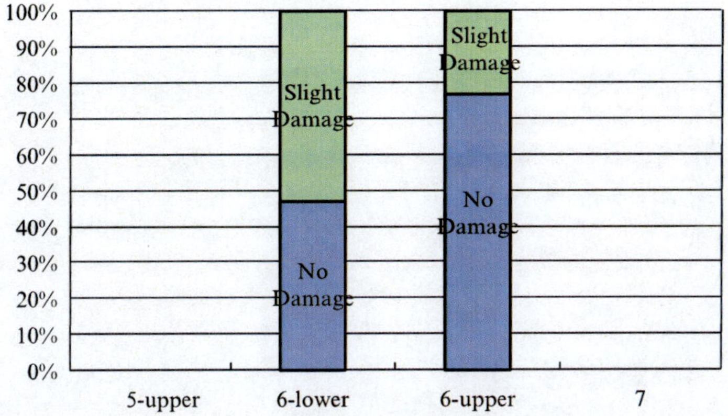

Fig. 6.3 Distribution of earthquake damage level of WPCa PS apartment buildings and seismic intensity

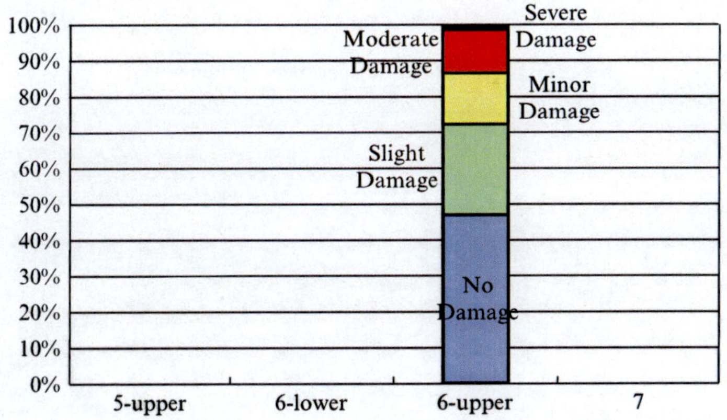

Fig. 6.4 Distribution of earthquake damage level of ribbed panel structures and seismic intensity

6.2.2 Characteristics of Damage Observed on WRC Building Structures

6.2.2.1 Apartment Building Complex A Located in Aoba Ward in Sendai

Apartment Building Complex A is located on sloped terrain and consisted of six Y shaped buildings (Building Nos. 1–6 called the Star-House) and one rectangular shape building (Building No. 7). The observed seismic intensity at the location of Complex A was 6-lower. While Building No. 7 was not damaged, three Star-House buildings had only slight damage, two Star-House buildings had minor damage, and one is moderately damaged. Figure 6.5 shows the locations of these residential buildings. Figure 6.6 gives an overview of the Star-House apartment building. Figure 6.7 shows a plan view of the Star-House apartment buildings.

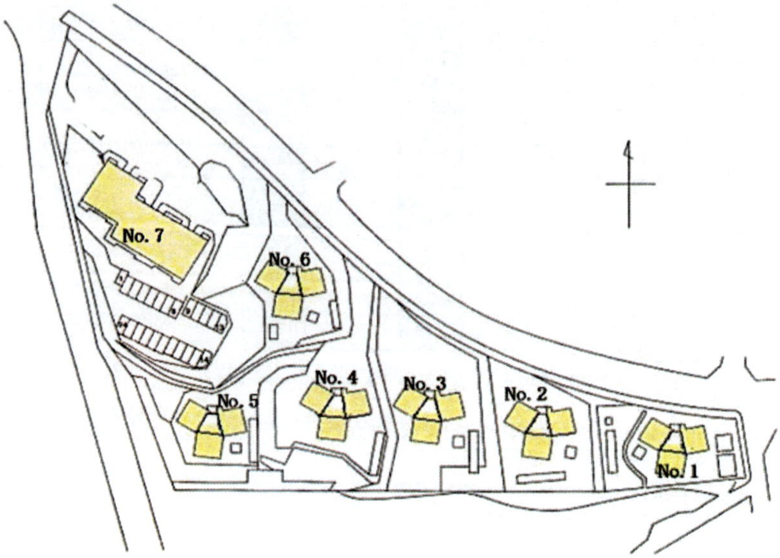

Fig. 6.5 Site plan with locations of Star-House apartment buildings

Fig. 6.6 The Star-House apartment building

Shear failure in structural wall was observed on the east side of the first floor. The location of the wall with shear failure is shown with dotted lines on the plan view of the building (Fig. 6.7), while cracks in the structural wall are shown on Figs. 6.8 and 6.9. Also, it should be mentioned that ground subsidence has been observed along the slope south of the building complex.

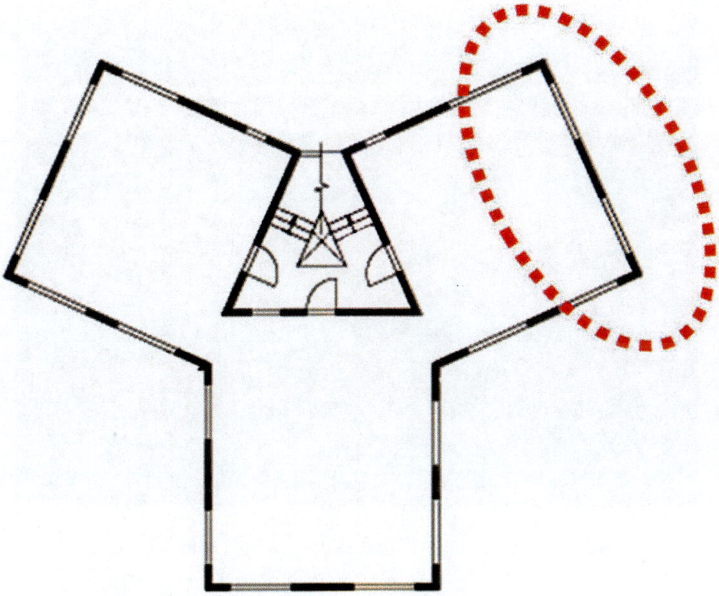

Fig. 6.7 Plan view of typical Star-House apartment building

Fig. 6.8 Cracks on structural walls at east side

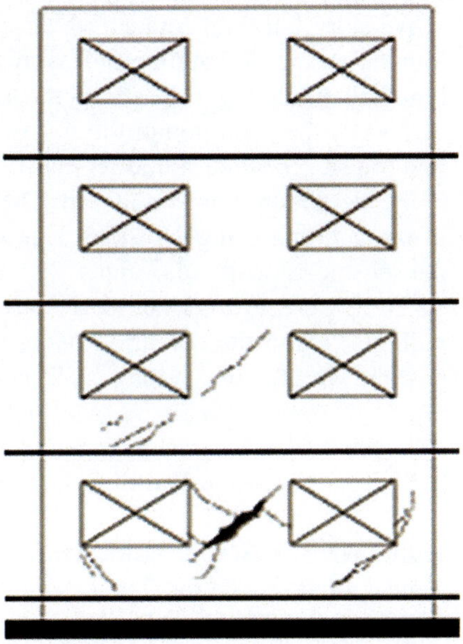

Fig. 6.9 Shear failure of structural wall of the first floor at east side

6.2.2.2 Apartment Building Complex B Located in Miyagino Ward in Sendai

Because of the ground subsidence in the vicinity of Apartment Building Complex B a five-story building rotated which caused serious damage to the retaining wall, foundations, and building infrastructure. Damaged retaining wall is shown in Fig. 6.10, while Fig. 6.11 shows the building in a rotated state. Also, Figs. 6.12 and 6.13 show the condition of the soil surrounding the building. The angle of building rotation in the short side direction is 22/1000 rad, while the rotation angle in the long side direction is 9/1000 rad. The inclination was measured at all four corners of the building using 1-m-long inclinometer. This buildings complex experienced the seismic intensity of 6-upper.

There were no cracks observed on external wall on the balcony side of the rotated building, while the condition of other exterior walls is unclear because they are covered with the insulation panels as shown in Fig. 6.14.

6.2.2.3 Apartment Building Complexes Located in Izumi Ward in Sendai

Izumi Ward in Sendai is located in a valley surrounded by hills north from the Sendai JR train station. The following nine building complexes were considered: Complex C1–C5, Complex D1–D2 and Complex E1–E2. There are 76 five-story reinforced concrete box-shaped wall buildings and 8 five-story precast pre-stressed reinforced concrete shear wall buildings in these nine building complexes.

Fig. 6.10 View of rotated building and damaged retaining wall

Fig. 6.11 View of rotated building

These nine building complexes are located 200–800 m from the location where the earthquake intensity of 6-upper was observed. The level of earthquake intensity at the actual location of the buildings is unknown. Table 6.3 gives the overview of the level of damage on the WRC building structures. Overall, the level of damage in WRC-type apartment building in north area of Sendai is very small.

Fig. 6.12 Ground deformation around the building

Fig. 6.13 Ground deformation around the building

As it can be found from the table minor damage is reported on three buildings in Complexes E1–E2, Buildings A–C. More detailed overview of site conditions and damage on these three buildings is discussed below

Fig. 6.14 Outside panel for insulation

Table 6.3 Earthquake damage level of reinforced concrete box-shaped wall buildings at Izumi-Ward Shogen area in City Sendai

Complex name	No damage	Slight damage	Minor damage	Moderate damage	Severe damage	Collapse
Complex C1–C5	41	1	0	0	0	0
Complex D1–D2	12	0	0	0	0	0
Complex E1–E2	9	10	3	0	0	0
Total	62	11	3	0	0	0
Ratio (%)	81.6	14.5	3.9	0	0	0

Description of Terrain

Building Complexes E1–E2 are located about 600 m south from the location where seismic intensity of 6-upper was observed. There is a retaining wall with 5 m high at the south end of the property of Complex E2. The building site is on gentle slope toward the east, near Shogen-numa lake, which is located in north-east of the property.

Overview of Damage to Building A

The site of Building A is 5 m higher than the level of the street, and it is laterally supported by retaining wall running in east–west direction. The distance of the buildings to the retaining wall is about 8 m.

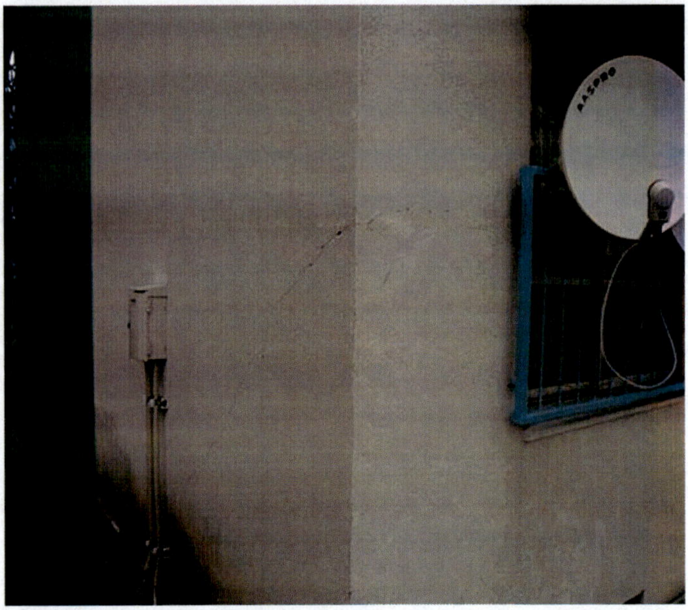

Fig. 6.15 Shear crack in structural wall at west side

The following damage is observed:

- Diagonal cracks of 0.25 mm wide, can be observed on a structural wall running in the longitudinal direction
- A diagonal crack of 1 mm width on the west structural wall of the building (Fig. 6.15)
- Spalling of finishing mortar on the connection of structural wall and footing beam (Fig. 6.16)
- A 15 mm wide gap between building wall and washing space at the footing of stairway entrance on the north side of the building
- A 3 mm gap between foundation beam and ground on the north side of the building

Overview of Damage to Building B

Observed damage to residential Building B located western from the Building A is:

- Cracks of second damage degree are observed at the east structural wall in the short side direction (Figs. 6.17 and 6.18)
- Diagonal cracks of first damage degree are observed at the corner of the spandrel wall above an entrance door
- A gap between foundation beam of the west gable wall and ground
- A gap between foundation beam of the north side and ground

Fig. 6.16 Spalling of finishing mortal of foundation beam at west side

Fig. 6.17 Flexural crack in structural wall at east side

Overview of Damage to Building C

Residential Building C is located approximately 15 m from the north wing of Building B on the flat ground. The following damage is observed:

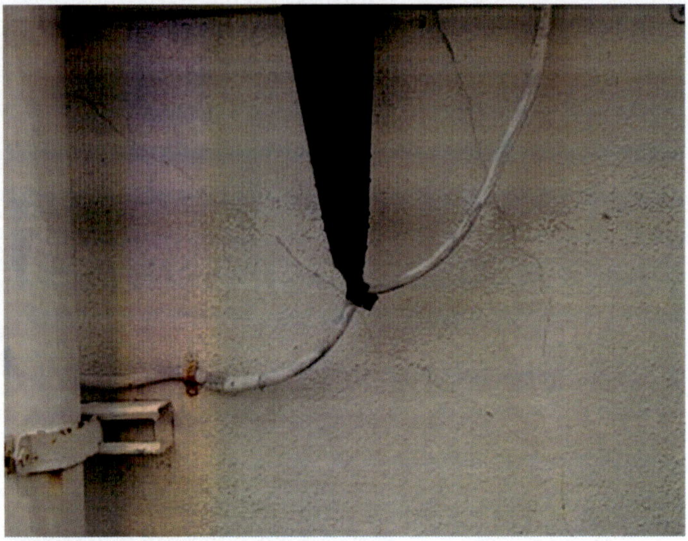

Fig. 6.18 Shear cracks on structural wall at east side

Fig. 6.19 Shear cracks on structural wall at south side

- Diagonal cracks of first to third damage degree in structural walls at southern balcony side in the longitudinal direction (Fig. 6.19, 6.20, 6.21, and 6.22)
- Diagonal cracks of first damage degree above the top of the entrance doorway
- Spalling of the finishing layer of mortar on foundation beam of the west gable wall
- Diagonal cracks on the corners of RC balconies at the locations of handrails

Fig. 6.20 Shear cracks in structural wall at south side (magnification of Fig. 6.19)

Fig. 6.21 Shear cracks in structural wall at south side

Fig. 6.22 Shear cracks in structural wall at south side

6.2.3 *Damage to Precast Pre-stressed RC Shear Wall Buildings*

6.2.3.1 Precast Pre-stressed RC Shear Wall Buildings Near Sendai

Precast pre-stressed RC shear wall buildings (WPCa PS buildings) in Natori city were inspected, in addition to 47 WPCa PS buildings in the city of Sendai. They include 1 building of Urban Renaissance Agency, 24 buildings of Miyagi Prefectural Municipal Housing, and 31 buildings of Sendai Municipal Housing.

While WPCa structures are precast reinforced concrete box-shaped wall structures in which precast concrete plain panels are connected to construct, in WPCa PS structures use precast concrete units with special configuration. Typical units are three dimensionally panels with T, L, or I-shaped cross section integrated in two directions. These 2D panels and precast beams are stacked together and post-tensioned by vertical PC bars to complete the connections. WPCa PS building complex and typical WPCa PS building for this region are shown in Figs. 6.23 and 6.24, respectively.

6.2.3.2 Building D in Building Complex F in Natori

Examination of residential buildings of WPCa PS construction type typically shows that superstructure was affected by the ground deformation.

Fig. 6.23 WPCa PS building complex

Fig. 6.24 Typical WPCa PS building

Overview of the Building

Building Complex F consists of two five-story buildings of WPCa PS structural type, Building D on the north with 30 units, and Building E on the south side of the complex with 20 units. The construction of these buildings started in October 1981 and finished in March 1982. Figure 6.25 shows Building D.

Fig. 6.25 Building D in Natori city

Summary of Damage

Relative settlement of 43 mm between axis X8 and X9 (distanced 5,100 mm) was observed on the south side of the building (balcony side), which caused rotation of the structure for 1/118 rad. Figure 6.26 shows this situation from the east side.

Settlement of ground under X9 axis caused cracks on the foundation beams as well as beams on floors 2 through 5 in between axis X8 and X9, as schematically shown in Fig. 6.27. In addition, Fig. 6.28 shows a photo of cracks in foundation beam at axis X9, and Fig. 6.29 shows cracks in beam on the second floor at axis X9.

Classification of Damage

The level of damage in the superstructure was determined to be minor damage as a result of examination of the degree of damage to the first-floor structural walls on the balcony side and second-floor floor beams where damage was visually confirmed. Subsequent detailed surveys found that there was a high possibility of piles failing and that settlement of more than 100 mm occurred. Then, the level of damage to the foundation was determined to be severe damage.

Fig. 6.26 Ground subsidence around apartment Building D

Damage to Precast Concrete Units

Cracks and chippings on brackets supporting precast concrete plank were observed. However, it is unclear if the earthquake caused this damage.

6.2.4 Damage to Mass-Produced Precast Thin Ribbed Concrete Panel Building Structures

6.2.4.1 Overview of Damage

Public housing buildings made of precast thin ribbed concrete panels (ribbed panel structure) bolted to each other showed to be relatively simple to build with good fireproof characteristics. This construction type was developed in 1962, and it has been used for almost 50 years.

Ninety-one public, residential, two-story, ribbed panel building structures were investigated in the city of Sendai. The level of seismic intensity in locations of investigated buildings was 6-upper.

Inspection of these buildings was performed in line with modified inspection procedure for box-shaped wall structures modified by AIJ Committee. These provisions were based on the method of post-earthquake damage evaluation of reinforced concrete and SRC composite buildings, issued by The Japan Building Disaster Prevention Association.

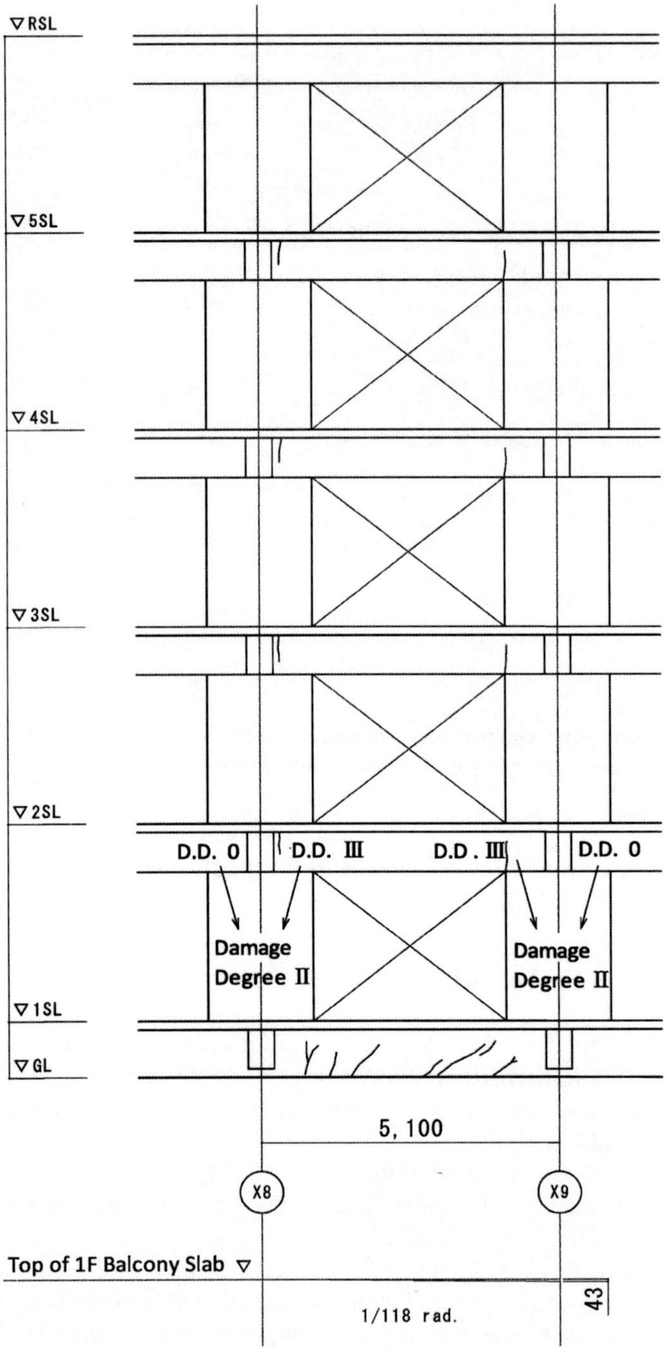

Fig. 6.27 Schematic view of cracks on Building D

Fig. 6.28 Cracks on foundation beam of Building D

Fig. 6.29 Crack in second floor beam of Building D

The levels of the damage of the 91 investigated buildings are classified as no damage for 43 buildings (47%), slight for 23 buildings (25%), minor for 13 buildings (14%), and moderate or severe for 12 buildings (13%) (see Table 6.4).

Most of the investigated ribbed panel buildings were designed and constructed in line old seismic design standards prior to 1981. Only 20 out of 91 investigated

Table 6.4 Earthquake damage level of mass production type public housing investigated

Complex name	Year of build	No. of buildings	No damage	Slight damage	Minor damage	Moderate damage	Severe damage	Collapse
Complex G (Miyagino)	1969–1970	36	12	10	9	5 (5)	0	0
Complex H (Miyagino)	1970–1971	14	5	1	2	5 (5)	1 (1)	0
Complex K (Izumi)	1975	21	9	11	0	1	0	0
Complex L (Miyagino)	1985	20	17	1	2 (1)	0	0	0
Total		91	43	23	13	11	1	0
Ratio (%)			47.3	25.3	14.3	12.1	1.1	0.0

( ) Cause of damage is relating to embankment and retaining walls

Fig. 6.30 Damage to mass production type public apartment buildings and masonry stone retaining wall

buildings were designed using new seismic buildings codes. Among these 20 buildings, 17 of them (85%) had no damage.

Figure 6.30 shows that a building sitting on the ground supported by masonry stone retaining wall damaged as well as that the superstructure itself was severely damaged.

It can be observed from the figure that the deformation of the building changed the shape of its elevation view.

In addition, ten more buildings built on embankment were moderately damaged due to deformation of the embankment.

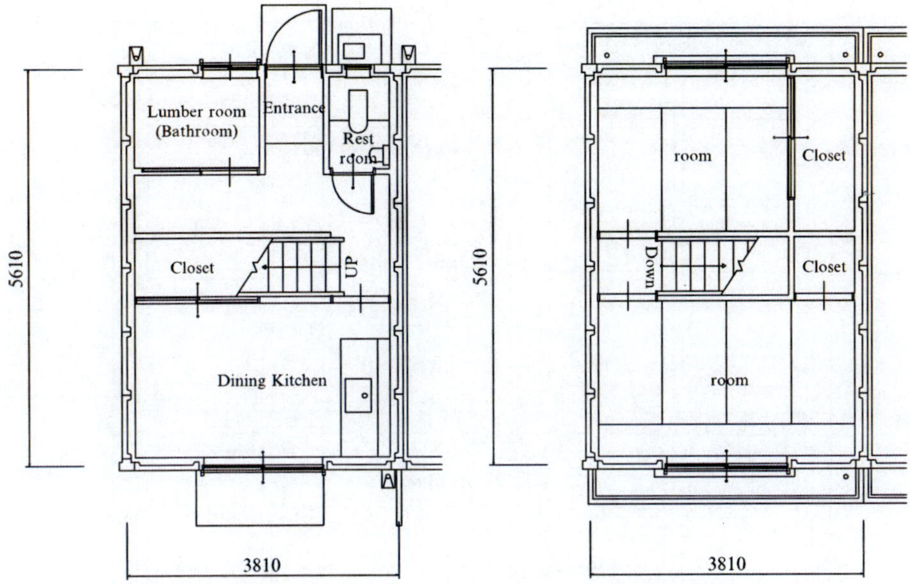

Fig. 6.31 Plan view of mass production type public apartment buildings

The floor plan of the buildings of Sendai Municipal Housing resembles to the standard floor plan (Fig. 6.31) proposed by the Japan Prefabricated Construction Suppliers and Manufacturers Association [5]. A floor consists of four to eight housing units. It is observed that damage is limited to the structural elements in longitudinal direction.

6.2.4.2 Types of Structural Damage

Figure 6.32 shows an elevation view of the structure made of precast ribbed concrete panels. Observed structural damage on ribbed panel structures is:

- Vertical crack in joints between structural wall and hanging wall (Fig. 6.33)
- Vertical cracks on the joint surface of structural wall and spandrel wall
- Horizontal cracks near the base of a structural wall at the boundary where the concrete thickness abruptly changes from 40 to 120 mm
- Shear crack in the structural wall (Fig. 6.34)
- Horizontal cracks of structural wall at the level of the top of window
- Cracks along vertical connections between structural walls and column form (Fig. 6.35)
- Vertical cracks on the sides of footing and foundation beams accompanied by settlement and fissure on the ground (Fig. 6.36)

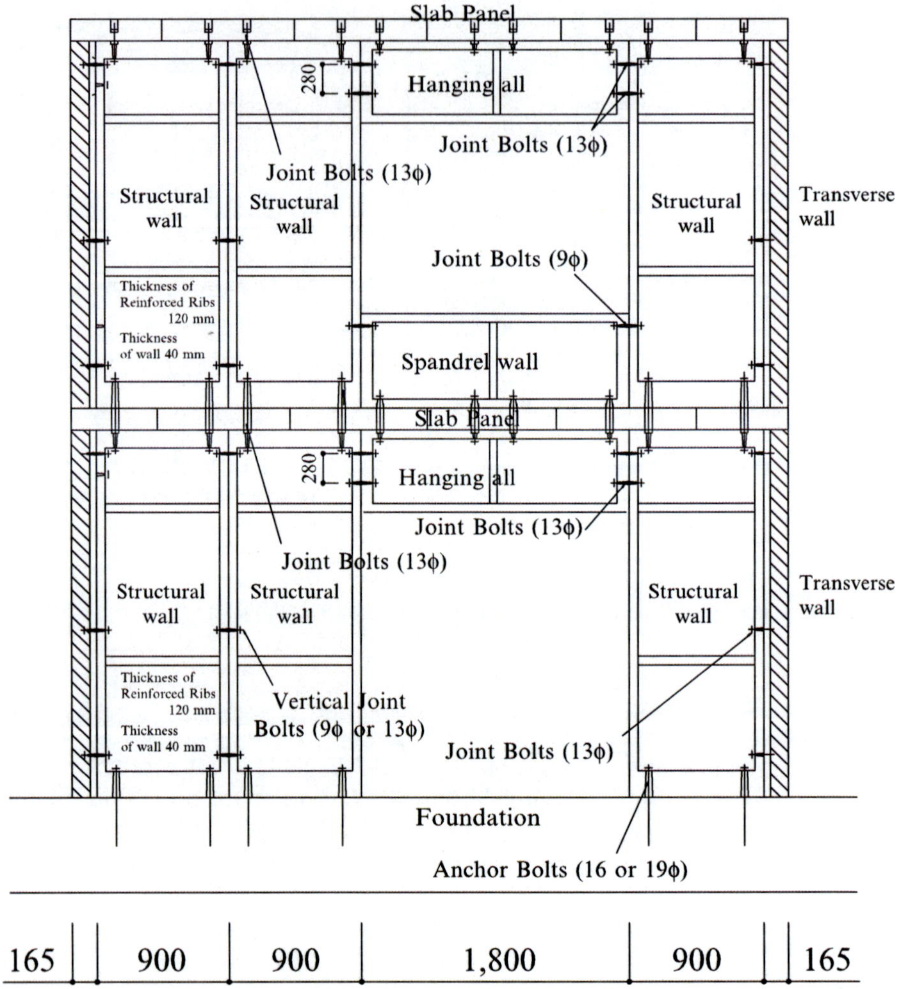

Fig. 6.32 Ribbed panel structural system

In the buildings constructed after the enforcement of the current code 1981, improvements have been made in detail and damage is incurred less frequently.

6.3 Damage Caused by Tsunami

Damage caused by tsunami on exterior walls of apartment buildings in the coastal area of Miyagi Prefecture was investigated on June 5 and June 6, 2011. Figure 6.37 shows damage of mass-produced two-story apartment building in Natori where severe

Fig. 6.33 Crack in vertical joint between structural wall and hanging wall

Fig. 6.34 Shear cracks on structural wall

damage can be observed on the first floor. Tsunami damage in the case of three-story residential building of WPCa PS structural type is shown in Fig. 6.38. It can be observed that there is no major damage to the exterior walls. Figure 6.39 shows a four-story WRC residential building in Onagawa affected by tsunami. Temperature isolation panels on gable wall and balconies were damaged. Figures 6.40 and 6.41

Fig. 6.35 Vertical crack between structural wall and column form

Fig. 6.36 Crack in foundation by ground fissures

show damage on three-story residential WRC buildings in Minamisanriku. In most buildings in this area tsunami reached the top floor of the buildings, and in some cases leaving floating objects on the roofs. Damage can be observed on the balconies, while structural walls and beams did not have major damage.

Fig. 6.37 Mass production type public apartment building in Natori city damaged by tsunami

Fig. 6.38 WPCa PS apartment building in Natori city damaged by tsunami

Fig. 6.39 WRC apartment building in Onagawa damaged by tsunami

Fig. 6.40 WRC apartment building in Minamisanriku damaged by tsunami

Fig. 6.41 WRC apartment building in Minamisanriku damaged by tsunami

Acknowledgments The damage surveys on reinforced concrete box-shaped wall buildings described in this chapter were conducted under the control of Box-shaped Wall Structure Committee of Architectural Institute of Japan. The authors acknowledge with sincere thanks the cooperation of the Housing Division, Civil Engineering Department, Miyagi Prefectural Government; Miyagi Prefecture Housing Supply Corporation; Municipal Housing Division, Sendai City Government and Urban Renaissance Agency in the surveys of public apartment buildings in Sendai City. Contributions of members of survey teams, listed below, are greatly appreciated: Akita Prefectural University team (Tetsuya Nishida), Oita University team (Kenji Kikuchi, Masayuki Kuroki and Koji Nishino), Osaka University team (Hiroshi Kuramoto, Juan Jose Castro, Hisatoshi Kashiwa, Masato Sakurai and Taku Suzuki), Tokyo Institute of Technology team (Yo Hibino), Toyohashi University of Technology team (Yasushi Sanada, Yuta Sato and Sohei Matsubara), Nagoya University team (Masaomi Teshigawara, Ippei Maruyama and Motoyuki Nakamura), Japan Prefabricated Construction Suppliers & Manufacturers Association team (Masayoshi Iizuka, Takahiro Sasaki, Toshiyuki Masubuchi and Hiroki Yasuda), Mie University team (Shigemitsu Hatanaka and Naoki Mishima),Yamaguchi University team (Eiichi Inai, Nanako Marubashi, Junji Ozaki and Ken Harayama), Urban Renaissance Agency team (Shinji Tokita and Shiro Suzuki), UR Linkage team (Yoshio Inoue, Motoyuki Tanaka, Masumi Hishikura, Takashi Kitahori, Yasuhiko Kitano, Yoichi Wakasugi and Naoto Shiraishi), Yokohama National University team (Akira Tasai, Koichi Kusunoki, Yuichi Hatanaka and Hidekazu Watanabe) and Green Design Office team (Masatsugu Iwata, Kimiko Tamura and Rie Kawahara).

References

1. Building Research Institute, http://smo.kenken.go.jp/smreport/201103111446
2. Tohoku University Disaster Control Research Center, https://sites.google.com/site/tohokuuni-vdcrc/home/jishin
3. National Research Institute for Earth Science and Disaster Prevention, http://www.k-net.bosai.go.jp/k-net/

4. Strong Motion Array of Local Lots by the Tohoku Institute of Technology Area Network, http://smweb.tohtech.ac.jp/smalltitan/japanese/smalltitanhome.html
5. The Japan Prefabricated Construction Suppliers & Manufacturers Association, Two decade history of the Japan Prefabricated Construction Suppliers & Manufacturers Association

Chapter 7
Damage to Masonry Buildings

Kenji Kikuchi, Masayuki Kuroki, Toshikazu Hanazato, Kazuya Koga, Katsuya Kawakami, and Noriyuki Mita

Abstract This chapter describes the earthquake and tsunami damage to partially grouted concrete masonry building, fully grouted concrete masonry buildings, and concrete masonry nonbearing walls provided in reinforced concrete moment resisting framed buildings. These are all used in small housing, commercial and office buildings. Damage caused by the earthquake and tsunami on concrete masonry and stone masonry garden walls are also reported in this chapter.

Keywords Masonry garden wall • Masonry nonbearing wall • Reinforced masonry

K. Kikuchi (✉) • M. Kuroki
Department of Architectural Engineering, Oita University, 700 Dannoharu, Oita 870-1192, Japan
e-mail: kikuchi@oita-u.ac.jp; mkuroki@oita-u.ac.jp

T. Hanazato
Division of Architecture, Graduate School of Engineering, Mie University, 1577 Kurimamachiya-cho, Tsu 514-8507, Japan
e-mail: hanazato@arch.mie-u.ac.jp

K. Koga
Department of Architecture, Faculty of Science and Technology, Tokyo University of Science, 2641 Yamazaki, Noda 278-8510, Japan
e-mail: koga@rs.noda.tus.ac.jp

K. Kawakami
Department of Architecture, Oyama National College of Technology, 771 Nakakuki, Oyama 323-0806, Japan
e-mail: kawakatu@oyama-ct.ac.jp

N. Mita
Department of Architectural System Engineering, Polytechnic University, 4-1-1 Hashimotodai, Midori-ku, Sagamihara 252-5196, Japan
e-mail: mitanori@uitec.ac.jp

Architectural Institute of Japan (ed.), *Preliminary Reconnaissance Report of the 2011 Tohoku-Chiho Taiheiyo-Oki Earthquake*, Geotechnical, Geological and Earthquake Engineering 23, DOI 10.1007/978-4-431-54097-7_7, © Springer Japan 2012

7.1 Introduction

In 1952, since some types of the reinforced masonry structures had been developing, AIJ established structural design standards for respective masonry structures as well as unreinforced masonry structures. Current AIJ Standards for Structural Design of Masonry Structures [1] contains a total of six standards. The first one is for the reinforced hollow concrete masonry building with three or less stories whose walls are partially grouted. The second and third ones are for the reinforced fully grouted concrete masonry buildings with three or less stories and four or five stories, respectively. The forth one is for the concrete masonry nonbearing walls provided in the moment resisting framed buildings. The fifth one is for the concrete masonry garden walls with 2.2 m or less height. The last one is for the unreinforced masonry buildings.

Also, AIJ established specification for the masonry work called as JASS 7 in 1956, and the latest revision was done in 2009 [2].

Japan Association for Building Research Promotion established a guideline for structural design of reinforced masonry (RM) structures [3] based on the results of a US-Japan cooperative research project during 1984–1988, which is one of the fully grouted concrete masonry structures.

7.2 Damage to Concrete Masonry Buildings and Walls

7.2.1 Summary of Investigation

This section gives an overview of the damage on reinforced masonry buildings made of concrete masonry units and nonbearing masonry walls for partition in reinforced concrete framed buildings.

The types of reinforced concrete masonry construction are classified as (a) reinforced hollow concrete masonry, which is usually partially grouted, (b) reinforced fully grouted concrete masonry and RM, which should be fully grouted.

The investigations of buildings as well as walls of these types were carried out Sendai City, Minami-Sanriku Town, and Kesennuma City, on April 23, 24 and 28–30.

7.2.2 Reinforced Hollow Concrete Masonry Buildings

A one-story building of this type with dimensions of 14.2 × 7.4 m in a plan view in Wakabayashi ward in Sendai is shown in Fig. 7.1. It is a building for boiler room in a junior high school. The thickness of the hollow concrete block is 150 mm and no obvious deterioration to the masonry structure was observed. This building also survived the 1978 Miyagiken-oki earthquake without damage [4].

Fig. 7.1 A reinforced hollow concrete masonry building without any damage

Fig. 7.2 Damaged two-story reinforced hollow concrete masonry houses

Two buildings of reinforced hollow concrete masonry stand side by side in Arahama district in the city of Sendai located 100 m from a coastal seawall (Fig. 7.2). Plan view dimensions of the building in the front on the figure are 6.0 × 5.6 m, while the masonry units are decorative. Dimensions of the masonry units are 450 mm in length, 150 mm in height and 150 mm in thickness. Although the building was rotated because the foundation was washed away, there was no crack observed on the masonry wall.

Fig. 7.3 A reinforced hollow concrete masonry building damaged by tsunami

The building that can be seen behind the mentioned one on Fig. 7.2 is a reinforced masonry with hollow concrete blocks with finishing mortar. Dimensions in the plan view are 14.8 × 9.4 m. This building suffered inundation and the soil beneath it was partially washed away, but no rotation or structural damage was observed in this case. Most probably, the building in front of this one took almost all the impact from the tsunami waves.

Figure 7.3 shows a one-story reinforced hollow concrete masonry building located in proximity of the ocean. Visual investigation did not reveal any structural damage. It can be seen on the right side of Fig. 7.3 that the concrete masonry garden wall stands near this building toward the ocean. This wall has buttress walls on its west side and therefore did not collapse. However, it did rotate due to tsunami impact.

A reinforced masonry building made of hollow concrete blocks (150 mm thick) in northern Minami-Sanriku Town, shown in Fig. 7.4, was also affected by the tsunami. The building is located 400 m from the inclined embankment. Base dimensions of the building are 7.5 × 3.9 m. Tiled roof is formed from timber truss elements. There is an interior wall on the first floor in the span direction, while there is no wall on the second floor. Although its foundation is 6.0 m above the sea level, the tsunami affected the building up to the height of the second floor ceiling. Cracks occurred in the wall on the first floor. Collapsed timber buildings in the vicinity of the present structure suggest better performance of concrete block buildings against tsunami attacks.

Figure 7.5 shows a reinforced masonry structure with hollow concrete blocks located in Motoyoshi district in Kesennuma, one of the areas heavily affected by the devastating tsunami. This overturned structure existed 200 m from the Tsuya River. The powerful tsunami completely overturned the structure, as shown in Fig. 7.5, having caused separation of the roof structure from the rest of the building. Most of the timber houses were washed away by the tsunami, leaving large amount of rubble in the surrounding area.

Fig. 7.4 A reinforced hollow concrete masonry building slightly damaged by tsunami

Fig. 7.5 A reinforced hollow concrete masonry building overturned due to tsunami wave

7.2.3 *Fully Grouted Concrete Masonry Buildings*

Figure 7.6 shows a two-story reinforced fully grouted masonry residential building built in 1980 in Nagamachi district in Sendai. Plan view dimensions of the building are 15.0×12.4 m. Thickness of concrete blocks is 190 mm. Although this building is located in the area that experienced seismic intensity of 6-upper on JMA (Japan Meteorological Agency) scale, the building remains undamaged.

Fig. 7.6 A reinforced fully grouted concrete masonry house without any damage

Figure 7.7 shows a three-story RM building, which is located in Aoba ward in Sendai. Plan dimensions of the building are 11.5×6.1 m, while the thickness of concrete blocks is 190 mm. Earthquake ground motion with seismic intensity of 5-upper did not cause structural damage to this building.

7.2.4 Concrete Masonry Nonbearing Walls

Figure 7.8 shows a two-story building in Izumi ward in the city of Sendai. The building has infilled concrete hollow masonry in the structural frames of reinforced concrete. Although this area experienced seismic intensity of 6-upper, damage to the exterior as well as interior structure was not observed.

Structures with concrete masonry nonbearing walls affected by tsunami are considered in the following. Figure 7.9 shows a bank building located on the intersection of National Road Route 398 and National Road Route 45 in the Minami-Sanriku Town. The whole residential area in the seaside plain was heavily impacted by tsunami and almost all buildings located around this building had been destroyed. The structural system of this two-story building is reinforced concrete frame with four columns and with infill walls of nonbearing concrete masonry. The building had a reinforced masonry extension at the first floor made of hollow concrete blocks without RC columns. The concrete masonry units are concrete hollow blocks with dimensions of 400×200×190 mm. This extension had collapsed and been washed away. It was observed on the reinforcing round bars (8 mm) that they were significantly deteriorated by corrosion.

Fig. 7.7 Undamaged reinforced masonry (RM) building

Fig. 7.8 Undamaged RC building with concrete masonry nonbearing walls

7.3 Damage to Masonry Garden Walls

7.3.1 Summary of Investigation

After the Miyagiken-oki earthquake (June 12, 1978) [4], a variety of measures have been taken in Miyagi Prefecture, ahead of other prefectures, regarding prevention of damage to concrete masonry garden walls. An investigation has been conducted in order to investigate effectiveness of these measures.

Fig. 7.9 Damaged RC building with nonbearing concrete masonry walls

The inspection of the masonry walls was conducted on March 25–28 in Natori City and Shiroishi City, and from April 2–4 in Natori City, Watari Town, Sendai City, Higashi-Matsushima City and Ishinomaki City.

7.3.2 Concrete Masonry Garden Walls

Figure 7.10 shows a concrete masonry garden wall in Aoba ward in Sendai. The wall partially collapsed due to significant deformation of the ground, while the part of the wall next to the entrance gate has relatively minor damage. It is made of six layers of decorative concrete masonry with unit dimensions of 190 (height) × 490 (length) × 100 (thickness) mm. The wall was connected to an intersecting wall with horizontal reinforcing bars spaced at 600 mm.

Figure 7.11 shows another concrete masonry garden wall located in Aoba ward in Sendai. The wall is facing the street which is inclined about 30° in east–west direction. Due to the inclination of the ground, six layers of concrete blocks (190 × 390 × 100 mm) are stacked in a staircase formation to fit the slope. Vertical reinforcement is placed with spacing of 800 mm. However, the vertical reinforcement did not extend into the upper two blocks. Furthermore, there were no horizontal reinforcing bars at the top of the wall. As a result, these blocks fell down.

Figure 7.12 shows a concrete masonry wall overturned and collapsed completely. This wall was located in Natori, where the recorded seismic intensity was 6-upper. The reasons of the collapse are (1) no anchorage into foundation for vertical

Fig. 7.10 Collapse of concrete masonry garden wall due to bad soil conditions

Fig. 7.11 Falling of concrete blocks

reinforcing bars, and (2) insufficient length of the buttress wall as well as insufficient depth of the embedment of the foundation. This wall did not meet the criteria of the current provisions of the Building Standard Law Enforcement Orders.

Figure 7.13 shows a concrete masonry wall located at Shinden district of Miyagino ward in Sendai. The collapse of the wall was caused by soft soil conditions and insufficient anchorage length of vertical reinforcements. Decorative concrete blocks (with dimensions of $140 \times 490 \times 100$ mm) were placed on the continuous

Fig. 7.12 Collapse of concrete masonry garden wall due to bad construction

Fig. 7.13 Insufficient anchorage of vertical reinforcing bars

footing of 300 mm height in six layers. There was no horizontal reinforcement in the wall. While the wall's vertical reinforcement consisted of 9 mm round bars spaced at 1,000 mm.

Figure 7.14 shows a long concrete masonry wall at Uematsu district in Natori. The wall consists of concrete blocks (190 × 390 × 100 mm) stacked in five layers. Since height of the wall does not exceed a limit of 1.2 m, buttress walls were not provided. Observed diagonal cracks are the result of insufficient length of lap splice

Fig. 7.14 Cracks on concrete masonry garden wall

Fig. 7.15 Collapse of old concrete masonry wall

of top reinforcing bars under seismic action. Thickness of the concrete cover around top horizontal bars was about 10 mm. These observations suggest that it is necessary to improve details of top of the walls including coping unit.

Figure 7.15 shows masonry wall built from hollow concrete blocks ($190\times 390\times 100$ mm) stacked in five layers. Vertical round 9 mm reinforcing bars were spaced at 800 mm, while horizontal bars were provided at the top of wall and third layer of concrete blocks. Vertical bars were heavily corroded between the second and third layers of concrete blocks, which caused fracture of bars and wall collapse.

Fig. 7.16 Concrete masonry wall supporting collapsed building

Figure 7.16 shows concrete masonry wall in Wakabayashi ward in Sendai damaged by collapsed building. The wall of about 36 m long was built from hollow concrete blocks (190 × 390 × 100 mm) stacked in five layers with reinforced concrete columns spaced at 4.4 m. The wall was reinforced with round 9 mm steel bars spaced at 800 mm (vertical bars) and 600 mm (horizontal bars). There were no hooks at the end of vertical bars. Corrosion on the reinforcing bars was observed at joint of concrete blocks, as well there were not sufficient filling in concrete blocks. However, the wall was still able to support the collapsed building due to strength of RC columns.

Figure 7.17 shows the concrete masonry wall located in Aoba ward in Sendai. The wall consisted of hollow concrete blocks (190 × 390 × 100 mm) is standing on retaining wall with a maximum height of about 1.80 m. In addition there is a metal fence attached to the top of concrete masonry wall with five layers. As it can be observed on the picture, large cracks occurred along the joints of concrete blocks.

In the following text the summary of damage to concrete masonry walls caused by tsunami is presented. Figure 7.18 shows masonry wall composed of hollow concrete blocks (190 × 390 × 100 mm) with incorporated metal fence located in Naruse district in Higashi-Matsushima. Total height of the wall is about 1.0 m. Although wall remained in one piece, it was overturned by tsunami wave. There are many similar walls in the same area that remained undamaged. Cause for the collapse of this wall is a short anchorage into the ground.

Figure 7.19 shows the remaining of the concrete masonry wall in Watari seriously damaged by tsunami. The wall was made of hollow concrete blocks (190 × 390 × 100 mm) placed in five layers sitting on reinforced concrete foundation. Vertical reinforcement of the wall consisted of 9 mm round reinforcing bars placed at 800 mm intervals, however, they are short as a reaching the second row of

Fig. 7.17 Cracks on concrete masonry garden wall

Fig. 7.18 Concrete masonry wall overturned by tsunami

concrete blocks. The wall collapsed under the impact of the tsunami waves. A number of walls in the same area oriented in the same direction as the discussed wall have not been damaged.

Figure 7.20 shows concrete masonry garden wall located in Yuriage district in Natori, where is seriously damaged by tsunami. Part of the wall made of concrete masonry blocks remained undamaged, but the metal fence was missing. This shows that intersecting walls perform better in the tsunami attack.

Fig. 7.19 Remaining of concrete masonry garden wall damaged by tsunami

Fig. 7.20 Undamaged concrete masonry garden wall by tsunami wave

This paragraph presents the summary of damage on concrete masonry garden walls in Miyagi Prefecture. The level of damage was difficult to estimate because it also included the damage caused by tsunami. Miyagi Prefecture conducted investigation on safety of 8,193 masonry garden walls in 2002. As a result of the investigation, 79% of concrete masonry walls were classified as safe. The rate of destroyed walls in Natori is 16% (4/25). In the 1995 Hyogoken-Nanbu earthquake, 25%

Fig. 7.21 Damage to stone masonry garden wall

of concrete masonry walls collapsed in Kobe [5]. While during the 2007 Niigataken Chuetsu-oki earthquake 17.6% of concrete masonry walls collapsed in Kashiwazaki [6]. This situation shows that measures based on lessons learned from the 1978 Miyagiken-oki earthquake reduced damage of concrete masonry walls.

7.3.3 Stone Masonry Garden Walls

The Tohoku is a famous production district of tuff stone. There are a number of garden walls built by high quality tuff stone.

Figure 7.21 shows collapsed wall made of this kind of stone built with only unreinforced mortar located in Masuda district in Natori. Strength of this wall comes from adhesive strength of mortar and the resistance of dowels used in the wall construction. The earthquake significantly damaged many walls of this type.

On the other hand, Fig. 7.22 shows stone masonry wall retrofitted with a steel structure. It can be observed that the wall remained undamaged owing to the seismic retrofit.

7.4 Summary

In the coastal area of Miyagi Prefecture several hollow reinforced concrete masonry buildings had fallen over or collapsed as a result of tsunami, or tilted as a result of wash out of soil under the foundation. Also, partial collapse of concrete masonry nonbearing walls was observed.

Fig. 7.22 Stone masonry garden wall retrofitted with steel structure

As a result of the investigation in the city of Sendai, in the areas distanced from the coastline, notable damage was not observed on reinforced hollow concrete masonry buildings, reinforced fully grouted concrete masonry buildings, and concrete masonry nonbearing walls.

The following conclusions were made after the investigation of concrete masonry garden walls in Miyagi Prefecture. The observed types of damage were as follows: collapse due to ground deformation, collapse of unreinforced portions of walls due to poor construction, tip-over and collapse due to insufficient anchorage of vertical bars into foundations or due to insufficient embedding of the foundation into ground, collapse due to rebar corrosion, major cracking at lap splice joints along the horizontal top rebar, and tsunami-induced overturn and collapse. The investigation of a certain district in Natori showed that 16% of concrete masonry garden walls and many non-reinforced stone masonry garden walls collapsed.

Acknowledgments The damage investigation of masonry buildings in Sendai was made possible by supports of S-BIC Company, Taiyo Cement Industrial, and Kubota Cement Kogyo. Japan Architectural Concrete Block Industry Association and JABEC are gratefully acknowledged for their cooperation during some of the damage investigation of concrete masonry garden walls.

References

1. Architectural Institute of Japan, AIJ Standards for Structural Design of Masonry Structures, AIJ, 437 pp, 2006.3 (in Japanese)
2. Architectural Institute of Japan, Japanese Architectural Standard Specification, JASS 7 Masonry Work, AIJ, 395 pp, 2009.6 (in Japanese)
3. Japan Association for Building Research Promotion, Guidelines for Structural Design of Reinforced Masonry (RM) Structures, JABRP, 358 pp, 2004.12 (in Japanese)
4. Architectural Institute of Japan, Report on the Damage Investigation of the 1978 Miyagiken-oki Earthquake, AIJ, pp 687–688, 1980.2 (in Japanese)
5. Architectural Institute of Japan, Report on the Hanshin-Awaji Earthquake Disaster, Building Series vol 2, AIJ, pp 613–621, 1998.8 (in Japanese)
6. Architectural Institute of Japan, Report on the Damage Investigation of the 2007 Noto-Hanto Earthquake and the 2007 Niigataken Chuetu-oki Earthquake, AIJ, 356 pp, 2010.3 (in Japanese)

Chapter 8
Damage to Steel Buildings

Mitsumasa Midorikawa, Isao Nishiyama, Motohide Tada, and Takehiko Terada

Abstract The Committee of Steel Structures of AIJ conducted earthquake damage reconnaissance over cities where severe ground shaking was recorded among Miyagi Prefecture and Fukushima Prefecture. Tsunami damage reconnaissance was also performed along the coastlines of Iwate, Miyagi, and Fukushima. The observed damage to steel buildings is categorized into that caused by ground motion and that caused by tsunami. Steel buildings exhibited excellent seismic performance. Buildings that used older cladding construction methods sustained damage to their claddings even though their structural performance was excellent. In some older buildings, severe ground motion caused damage to structural members and connections, which are the same modes of damage observed from past earthquakes. A unique feature is the damage caused by tsunami. In the most extreme cases, buildings were displaced from their original location and completely destroyed. In other cases, fracture of connections and members caused the building to tilt or collapse. In buildings whose claddings were completely washed away by tsunami, structural damage was minor. The extent of tsunami damage varied substantially depending on the locality of tsunami attack.

M. Midorikawa (✉)
Hokkaido University, Sapporo, Japan
e-mail: midorim@eng.hokudai.ac.jp

I. Nishiyama
Building Research Institute,
Tsukuba, Japan
e-mail: nishiyam@kenken.go.jp

M. Tada
Osaka University, Suita, Japan
e-mail: tada@arch.eng.osaka-u.ac.jp

T. Terada
Shimizu Corporation, Tokyo, Japan
e-mail: t.terada@shimz.co.jp

Architectural Institute of Japan (ed.), *Preliminary Reconnaissance Report of the 2011 Tohoku-Chiho Taiheiyo-Oki Earthquake*, Geotechnical, Geological and Earthquake Engineering 23, DOI 10.1007/978-4-431-54097-7_8,

Keywords Seismic damage • Steel building • Tsunami damage

8.1 Overview of Investigation

The Steel Structures Committee of the Architectural Institute of Japan conducted earthquake damage reconnaissance over Miyagi Prefecture and Fukushima Prefecture. Tsunami damage reconnaissance was conducted along the coastlines of Miyagi, Fukushima and Iwate Prefectures.

The observed damage to steel buildings is categorized into that caused by ground motion and that caused by tsunami. Severe ground motion caused damage to beam-to-column connections, buckling of diagonal braces, cracking of concrete overlaying the column base, yielding and fracture of anchor bolts, which are the same modes of damage observed from past earthquakes.

Severe nonstructural damage occurred to ceilings and claddings of structures with large open areas, such as gymnasiums and factories. Widespread damage was observed in external finishes composed of mortar over light-gauge metal lath.

A unique feature of the Tohoku event is the damage caused by tsunami. In the most extreme cases, buildings were displaced from their original location and completely destroyed. In other cases, fracture of connections and members caused the building to tilt or collapse. In buildings whose claddings were completely washed away by tsunami, structural damage was minor. The extent of tsunami damage varied substantially depending on the locality of tsunami attack.

8.2 Damage Caused by Earthquake Ground Motion

The ground motion caused damage to many low- to mid-rise buildings. Judging from the types of members and framing system, the majority of damaged buildings were constructed in an older era preceding the major change in the Building Standard Law in 1981.

8.2.1 Damage to Structural Members

8.2.1.1 Beam-to-Column Connections

Brittle fracture of beam-to-column connections, which was observed extensively after the 1995 Hyogo-ken Nanbu (Kobe) earthquake, has not been reported from the Tohoku event. Figure 8.1 shows damage found in an older structure that employed built-up HSS (hollow structural steel) columns composed of a light W-shape and a pair of cover plates. Such sections were commonly used before cold-formed HSS sections became available.

Fig. 8.1 Yielding of an older built-up column (Koriyama)

Fig. 8.2 Buckling of double-angle brace (Miyagino, Sendai)

8.2.1.2 Braces and Bracing Connections

Brace buckling, net-section fracture, and distortion and fracture of the gusset plates were observed in many brace framed structures (Figs. 8.2, 8.3, 8.4, 8.5, 8.6, 8.7, 8.8, 8.9, 8.10, and 8.11).

Failure of angle-section braces, which were commonly used in older construction, was typically governed by buckling and fracture initiating at the bolt

Fig. 8.3 Local buckling in square-HSS brace (Aoba, Sendai)

Fig. 8.4 Net section fracture of single-angle brace (Miyagino, Sendai)

holes. In contrast, the predominant damage to HSS braces was out-of-plane bending of the gusset plates. Some gusset plates fractured as a result of a large number of repeated bending.

Fig. 8.5 Bending of middle gusset plate in an X-brace (Miyagino, Sendai)

Fig. 8.6 Fracture of gusset plate-to-column weld and spalling of concrete covering an exposed base plate (Ishinomaki)

8.2.1.3 Column Bases

The majority of investigated buildings were low- to mid-rise, where exposed base plate connections are more commonly used than embedded or encased column base connections. The damage to exposed base plates indicates that, unless the anchor bolts fractured, residual drift and structural damage to the building was

Fig. 8.7 Out-of-plane deformation of gusset plate (Miyagino, Sendai)

Fig. 8.8 Yielding of column web near bracing connection (Koriyama)

Fig. 8.9 Out-of-plane deformation and fracture of gusset plates (Miyagino, Sendai)

Fig. 8.10 Out-of-plane deformation of gusset plates (Miyagino, Sendai)

Fig. 8.11 Out-of-plane deformation of gusset plate caused by compression (Miyagino, Sendai)

Fig. 8.12 Spalling of reinforced concrete encasing a steel column base (Wakabayashi, Sendai)

minimal. On the other hand, evidence suggests that fracture of anchor bolts led to dislocation of the column and severe residual drift (Figs. 8.12, 8.13, 8.14, 8.15, 8.16, 8.17, 8.18, 8.19, and 8.20).

Fig. 8.13 Elongation of anchor bolts in an exposed base plate (Miyagino, Sendai)

Fig. 8.14 Elongation of anchor bolts in an exposed base plate (Miyagino, Sendai)

Fig. 8.15 Cracking of asphalt covering a column base (Miyagino, Sendai)

Fig. 8.16 Spalling of concrete covering a column base (Koriyama)

8.2.1.4 Other Damage (Figs. 8.21, 8.22, 8.23, and 8.24)

8.2.2 Damage to Non-structural Elements

Extensive damage was observed in dry-construction elements such as ceilings composed of mortar over metal lath and ALC-panel cladding. Non-structural

Fig. 8.17 Spalling of reinforced concrete foundation supporting a column base (Koriyama)

Fig. 8.18 Fracture of anchor bolts (Miyagino, Sendai)

damage was observed in buildings of all construction ages (Figs. 8.25, 8.26, 8.27, 8.28, 8.29, 8.30, 8.31, 8.32, 8.33, 8.34, 8.35, 8.36, 8.37, 8.38, 8.39, 8.40, 8.41, 8.42, 8.43, and 8.44). Extensive damage was observed in older-type external finishes that place mortar.

Fig. 8.19 Fracture of anchor bolts, spalling of concrete covering a column base (Miyagino, Sendai)

Fig. 8.20 Fracture of anchor bolts, out-of-plane deformation of base plate (Miyagino, Sendai)

Fig. 8.21 Lateral-torsional buckling of crane girder (Miyagino, Sendai)

Fig. 8.22 Bending of stair stringer (Miyagino, Sendai)

Fig. 8.23 Collapse of sky bridges (Koriyama)

Fig. 8.24 Collapse of shear tab in a parking ramp (Rifu)

Fig. 8.25 Fallen ceiling grid and boards (Fukushima)

Fig. 8.26 Fallen claddings; damaged ceiling boards (Miyagino, Sendai)

Fig. 8.27 Fallen ceiling, partitions damaged by impact (Aoba, Sendai)

Fig. 8.28 Fallen ceiling grid and boards (Aoba, Sendai)

Fig. 8.29 Partly fallen ceiling (Koriyama)

Fig. 8.30 Fallen ceiling on the outside of a building (Wakabayashi, Sendai)

Fig. 8.31 Cracking of external finish (Aoba, Sendai)

Fig. 8.32 Fallen cladding in a penthouse (Aoba, Sendai)

Fig. 8.33 One side of a building lost cladding entirely (Wakabayashi, Sendai)

Fig. 8.34 Failure of timber lath-and-mortar cladding (Miyagino, Sendai)

Fig. 8.35 Cracking of cladding panels (Miyagino, Sendai)

Fig. 8.36 Failure of metal lath-and-mortar cladding (Miyagino, Sendai)

Fig. 8.37 Damaged cladding (Miyagino, Sendai)

Fig. 8.38 Fallen cladding (Miyagino, Sendai)

Fig. 8.39 Fallen cladding (Ishinomaki)

Fig. 8.40 Fallen and twisted cladding (Koriyama)

Fig. 8.41 Damage to cladding covering a column (Wakabayashi, Sendai)

Fig. 8.42 Damaged cladding in a mechanical penthouse (Koriyama)

Fig. 8.43 Damaged cladding (Sukagawa)

Fig. 8.44 Damage to windows (Miyagino, Sendai)

8.3 Damage Caused by Tsunami

In areas that were attacked by high tsunami waves, severe damage was observed such as failure of the column base that led to overturning and dislocation of the building, and extreme distortion of structural members and connections. In instances where the external claddings were washed away, the load produced by tsunami was reduced to cause little damage to the structural system. In the following, tsunami damage is described for different districts.

8.3.1 Port of Ishinomaki (Reported Inundation Height: 5 m [1])

8.3.1.1 Office Building A (Figs. 8.45 and 8.46)

X-dir.: five spans, Y-dir.: two spans, number of stories: one
X-dir.: moment frame, Y-dir.: moment frame
Columns: square HSS's, beams: W-shapes, external finish: ALC panels
Observations: internal and external finishes washed away. No structural damage

Fig. 8.45 Office building A: external view

Fig. 8.46 Office building A: inside view

Fig. 8.47 Office building B: external view showing noticeable residual drift

Fig. 8.48 Office building B: net section fracture of a single-angle brace

8.3.1.2 Office Building B (Figs. 8.47 and 8.48)

X-dir.: five spans, Y-dir.: one span, number of stories: two
X-dir.: braced frame, Y-dir.: moment frame
Columns: W-shapes, beams: W-shapes, braces: angles (1 F), round bars (2 F), external finish: ALC panels
Observations: extensive structural damage. Foundation exposed due to scouring, braces fractured

Fig. 8.49 Port facility A: external view

Fig. 8.50 Port facility A: inside view

8.3.1.3 Port Facility A (Figs. 8.49, 8.50, and 8.51)

X-dir.: braced frame, Y-dir.: moment frame

Columns: W-shapes, beams: W-shapes, braces: round bars, column base: exposed type

Observations: external cladding partly lost. Residual drift of 1/400 in the north and east directions

Fig. 8.51 Port facility A: column base rusted presumably due to proximity to seawater

Fig. 8.52 Port facility B: foundation exposed after scouring

8.3.1.4 Port Facility B (Fig. 8.52)

X-dir.: braced frame, Y-dir.: moment frame, pile foundation
Columns: W-shapes, beams: W-shapes, braces: round bars
Observations: piles exposed after scouring

Fig. 8.53 Port facility C: external view

Fig. 8.54 Port facility C: fractured anchor bolt

8.3.1.5 Port Facility C (Figs. 8.53 and 8.54)

X-dir.: braced frame, Y-dir.: moment frame
Columns: W-shapes, beams: W-shapes, braces: angles, column base: exposed type
Observations: two spans closest to the shore collapsed. Fracture of anchor bolts

Fig. 8.55 Building A: external view showing large tilt

8.3.2 Onagawa Town (Reported Inundation Height: 15 m [1])

8.3.2.1 Building A (Figs. 8.55 and 8.56)

X-dir.: three spans, Y-dir.: one span, number of stories: three
Y-dir.: moment frame
Columns: W-shapes, beams: W-shapes, external finish: corrugated metal sheets
Observations: building tilted after foundation scouring

8.3.2.2 Building B (Figs. 8.57 and 8.58)

Number of stories: three (partially four)
X-dir.: moment frame, Y-dir.: moment frame
Columns: square HSS's, beams: W-shapes, external finish: ALC panels
Observations: building dislocated about 15 m and lain sideways after all piles failed due to pullout

Fig. 8.56 Building A: foundation scour

Fig. 8.57 Building B: building collapsed towards left

Fig. 8.58 Building B: view of foundation showing fracture piles

Fig. 8.59 Building C: external view

8.3.2.3 Building C (Figs. 8.59 and 8.60)

X-dir.: four spans, Y-dir.: two spans, number of stories: two
X-dir.: moment frame, Y-dir.: moment frame
Columns: square HSS's, beams: W-shapes
Observations: internal and external finishes washed away

Fig. 8.60 Building C: contents washed away

8.3.2.4 Building D (Figs. 8.61 and 8.62)

Columns: square HSS's, beams: W-shapes
Observations: collapsed after extensive connection failure

8.3.2.5 Building E (Figs. 8.63, 8.64, and 8.65)

X-dir.: moment frame, Y-dir.: moment frame
Columns: older built-up square HSS's, beams: W-shapes, column bases: exposed type
Observations: internal and external finishes washed away, damage in column bases

Fig. 8.61 Building D: external view

Fig. 8.62 Building D: failed connection

Fig. 8.63 Building E: external view

Fig. 8.64 Building E: column base lifted and anchor bolts deformed

Fig. 8.65 Building E: beam-to-column connections at corner column

Fig. 8.66 Building F: external view

8.3.2.6 Building F (Figs. 8.66, 8.67, and 8.68)

Columns: square HSS's, beams: W-shapes, column bases: exposed type

Observations: collapsed. Fracture of beam-to-column connections, anchor bolts, and horizontal braces

Fig. 8.67 Building F: beam-to-column connection

Fig. 8.68 Building F: column base

Fig. 8.69 Warehouse A: external view

Fig. 8.70 Warehouse A: damage to column base

8.3.2.7 Warehouse A (Figs. 8.69, 8.70, and 8.71)

Use: refrigeration
X-dir.: braced frame, Y-dir.: moment frame
Observations: residual drift exceeding 1/20. East frame collapsed

Fig. 8.71 Warehouse A: inside view

8.3.2.8 Other Damage (Figs. 8.72, 8.73, and 8.74)

8.3.3 Shiogama City (Reported Inundation Height: 4 m [1])

8.3.3.1 Warehouse B (Figs. 8.75 and 8.76)

Columns: W-shapes, beams: W-shapes, braces: round bars, column bases: exposed type, external finish: slate tiles over metal lath

Observations: foundation subsided due to tsunami and liquefaction

Fig. 8.72 Fractured column base plate; structure was washed away

Fig. 8.73 Cladding and roof washed away

Fig. 8.74 Cladding and roof washed away

Fig. 8.75 Warehouse B: external view; right end subsided

Fig. 8.76 Warehouse B: view of front side; far-right end subsided

8.3.4 Miyagino, Sendai City (Reported Inundation Height: 8 m [1])

8.3.4.1 Office Building C (Figs. 8.77, 8.78, and 8.79)

Columns: square HSS's, beams: W-shapes, column bases: exposed type
Observations: underground piping exposed after scouring

8.3.4.2 Office Building D (Figs. 8.80 and 8.81)

X-dir.: moment frame, Y-dir.: moment frame
Columns: square HSS's, beams: W-shapes, column bases: exposed type
Observations: first floor cladding washed away. No structural damage

8.3.5 Minami-Sanriku Town (Reported Inundation Height: 13–15 m [2])

8.3.5.1 Office Building E (Figs. 8.82 and 8.83)

X-dir.: moment frame, Y-dir.: moment frame
Observations: evidence of beam yielding. No residual drift

Fig. 8.77 Office building C: external view

Fig. 8.78 Office building C: severe damage to cladding

Fig. 8.79 Office building C: under-ground piping exposed after scouring

Fig. 8.80 Office building D: cladding washed away in first story; minimal structural damage

Fig. 8.81 Office building D: nonstructural damage in first story

Fig. 8.82 Office building E: majority of nonstructural elements washed away; minor structural damage

Fig. 8.83 Office building E: yielding of beam near beam-to-column connection

8.3.5.2 Office Building F (Figs. 8.84 and 8.85)

X-dir.: moment frame, Y-dir.: moment frame

Observations: evidence of beam yielding. Entire building submerged during tsunami attack

8.3.5.3 Store A (Figs. 8.86 and 8.87)

X-dir.: moment frame, Y-dir.: moment frame

Observations: evidence of beam yielding. No residual drift

8.3.5.4 Store B (Figs. 8.88 and 8.89)

X-dir.: moment frame, Y-dir.: moment frame

Observations: evidence of beam yielding. Residual drift of 1/200 in first story. Cracking of concrete wrapping the column base

Fig. 8.84 Office building F: external view

Fig. 8.85 Office building F: slight yielding of beam near beam-to-column connection

Fig. 8.86 Store A: external view

Fig. 8.87 Store A: slight yielding of beam and column

Fig. 8.88 Store B: external view

Fig. 8.89 Store B: slight yielding of beam near beam-to-column connection

Fig. 8.90 Store C: nonstructural elements washed away

8.3.5.5 Store C (Fig. 8.90)

X-dir.: X-braced frame, Y-dir.: moment frame
Observations: brace buckling. No residual drift

8.3.5.6 Factory A (Figs. 8.91 and 8.92)

X-dir.: moment frame, Y-dir.: moment frame
Observations: chord member of roof truss buckled presumably due to impact of debris. Part of structure demolished and cleaned by the time of visit

8.3.5.7 Gymnasium A (Figs. 8.93, 8.94, and 8.95)

X-dir.: moment frame, Y-dir.: moment frame
Observations: wall closest to shore line deformed out of plane. Roof truss buckled and collapsed. Anchor bolts in column base fractured. Bolts joining roof truss members fractured. Adjacent 3-story R/C school building completely submerged during tsunami attack

Fig. 8.91 Factory A: external view

Fig. 8.92 Factory A: buckled chord members in roof truss

Fig. 8.93 Elementary school gymnasium A: external view

Fig. 8.94 Elementary school gymnasium A: damaged column base

Fig. 8.95 Elementary school gymnasium A: severely damaged roof

8.3.6 Kesennuma City (Reported Inundation Height: 4–10 m [2])

8.3.6.1 Warehouse C (Kawaguchi District) (Figs. 8.96 and 8.97)

X-dir.: moment frame, Y-dir.: X-braced frame
Observations: collapsed and severely deformed

8.3.6.2 Warehouse D (Figs. 8.98 and 8.99)

X-dir.: moment frame, Y-dir.: moment frame
Observations: external cladding washed away in first story. No structural damage. No residual drift

8.3.6.3 Residence A (Nagaiso-Ushirozawa District) (Figs. 8.100 and 8.101)

X-dir.: one span braced, Y-dir.: moment frame
Observations: residual drift in Y-direction of 1/33. Round-bar brace fractured presumably due to tsunami. Foundation scour and damage

Fig. 8.96 Warehouse C: external view

Fig. 8.97 Warehouse C: deformed columns and beams

Fig. 8.98 Warehouse D: external view

Fig. 8.99 Warehouse D: roof bracing

Fig. 8.100 Residence A: external view

Fig. 8.101 Residence A: damaged column base

Fig. 8.102 Store D: external view

Fig. 8.103 Store D: fractured beam-to-column connections

8.3.6.4 Store D (Figs. 8.102 and 8.103)

X-dir.: moment frame, Y-dir.: moment frame

Observations: severe damage caused by impact of debris. Damage concentrated in single-story segment. (Residual drift of 1/160 in Y-direction. Anchor bolts fractured. Fracture in beam-to-column connections. Ceiling braces fractured.) No damage in two-story segment

Fig. 8.104 Traditional hotel A: external view

Fig. 8.105 Traditional hotel A: close-up view of collapsed second story

8.3.6.5 Hotel A (Niihama District) (Figs. 8.104 and 8.105)

X-dir.: moment frame, Y-dir.: moment frame

Observations: second story collapsed after weld fracture between through-diaphragm plate and column

Fig. 8.106 Store E: external view

Fig. 8.107 Store E: distorted columns

8.3.6.6 Store E (Figs. 8.106, 8.107, 8.108, and 8.109)

Single story
X-dir.: moment frame, Y-dir.: moment frame
Observations: plastic hinging and fracture at top and bottom of column. Residual drift of 1/5 in Y-direction

Fig. 8.108 Store E: plastic deformation at top of first-story column

Fig. 8.109 Store E: damaged column base

Fig. 8.110 Factory B: external view

Fig. 8.111 Factory B: cracking of reinforced concrete encasing a column base

8.3.6.7 Factory B (Figs. 8.110, 8.111, and 8.112)

X-dir.: unknown, Y-dir.: moment frame

Observations: residual deformation of 1/10 in lower segment with cracking of concrete encasing the column bases, yielding of panel zone, and local buckling of beam flanges

Fig. 8.112 Factory B: yielding in beam-to-column connection

Fig. 8.113 Factory C: majority of cladding washed away in first story

8.3.6.8 Factory C (Figs. 8.113 and 8.114)

X-dir.: X-braced frame, Y-dir.: moment frame

Observations: bolt fracture in bracing connections, panel zone yielding, residual drift of 1/200 in both X and Y directions

Fig. 8.114 Factory C: yielding of column panel zone

8.3.6.9 Factory D (Figs. 8.115 and 8.116)

X-dir.: inverted V-braced frame, Y-dir.: moment frame
Observations: buckling and fracture of braces, bending of column base plates

8.3.6.10 Factory E (Figs. 8.117 and 8.118)

X-dir.: braced frame, Y-dir.: gable frame
Observations: collapsed in Y-direction. Yielding in column bases and panel zones and gable frame beams

8.3.6.11 Factory F (Figs. 8.119 and 8.120)

X-dir.: X-braced frame, Y-dir.: gable frame
Observations: rotation of foundation accompanied by collapse in Y-direction. Lateral-torsional buckling of gable frame beams

Fig. 8.115 Factory D: external view

Fig. 8.116 Factory D: net section fracture of a single-angle brace at the column base

Fig. 8.117 Factory E: building leaning towards right

Fig. 8.118 Factory E: yielding of joint region

Fig. 8.119 Factory F: external view

Fig. 8.120 Factory F: collapsed column

Fig. 8.121 Office building G: external view

8.3.7 Rikuzen-Takata City (Reported Inundation Height: 12–16 m [2])

8.3.7.1 Office Building G (Figs. 8.121 and 8.122)

X-dir.: X-braced frame, Y-dir.: moment frame
Observations: out-of-plane bending failure of ALC panels. Residual drift in Y-direction of 1/50–1/33

8.3.7.2 Store F (Figs. 8.123 and 8.124)

X-dir.: moment frame, Y-dir.: moment frame
Observations: no residual drift. Torsional deformation of beams in frame facing the shore line

Fig. 8.122 Office building G: damage to ALC panels

Fig. 8.123 Store F: damage to cladding

Fig. 8.124 Store F: torsional deformation of beam

Fig. 8.125 Store G: damage to cladding

8.3.7.3 Store G (Figs. 8.125 and 8.126)

X-dir.: moment frame, Y-dir.: moment frame

Observations: residual drift of 1/200. Damage to second-floor concrete slab. Suspected cause is lifting force produced by air pocket in the first story

Fig. 8.126 Store G: deformed metal floor slab

Fig. 8.127 Factory G: cladding washed away

8.3.7.4 Factory G (Figs. 8.127, 8.128, and 8.129)

X-dir.: X-braced frame, Y-dir.: moment frame

Observations: lateral-torsional buckling of beam. Yielding of column base. Severe deformation of column panel zones

Fig. 8.128 Factory G: local flange buckling of column near base

Fig. 8.129 Factory G: large deformation of column panel zone

8.3.7.5 Gymnasium B (Figs. 8.130 and 8.131)

X-dir.: X-braced frame, Y-dir.: gable frame

Observations: first story collapsed in X-direction. Building displaced in X-direction by 20 m

Fig. 8.130 High-school gymnasium B: external view of collapsed building

Fig. 8.131 High-school gymnasium B: building displaced by 20 m

8.3.7.6 Gymnasium C (Figs. 8.132, 8.133, 8.134, and 8.135)

X-dir.: X-braced frame, Y-dir.: moment frame, roof: space truss

Observations: severe deformation of frame facing the shore line. Buckling of X-braces. Buckling of chord members in roof truss

Fig. 8.132 Gymnasium C: external view

Fig. 8.133 Gymnasium C: severely distorted columns

Fig. 8.134 Gymnasium C: distorted roof trusses

Fig. 8.135 Gymnasium C: buckled braces

Fig. 8.136 Cladding washed away; minimal structural damage

Fig. 8.137 Office building: anchor bolts fractured; building leaning on next building

8.3.8 Otsuchi Town (Reported Inundation Height: 10–15 m [2]) (Figs. 8.136, 8.137, 8.138, 8.139, and 8.140).

8.3.9 Kuji City (Reported Inundation Height at Kuji Port: 8–9 m [1])

8.3.9.1 Warehouse E (Figs. 8.141 and 8.142)

X-dir.: moment frame, Y-dir.: moment frame
Columns: W-shapes, beams: W-shapes
Observations: no structural damage. Internal and external finish washed away

Fig. 8.138 Fish processing factory: damage to cladding

Fig. 8.139 Office building: a structural frame surrounded by debris

8.3.9.2 Factory H (Figs. 8.143 and 8.144)

X-dir.: moment frame, Y-dir.: X-braced frame

Columns and beams: built-up W-shapes, braces: round bars, external finish: corrugated metal sheets

Observations: evidence of debris impact on shore side wall

Fig. 8.140 Office building: cladding washed away from first to third story

Fig. 8.141 Warehouse E: nonstructural elements washed away

Fig. 8.142 Warehouse E: minimal structural damage; no residual drift

Fig. 8.143 Factory H: cladding remained

Fig. 8.144 Factory H: damage to concrete encasing column base

Fig. 8.145 Factory I: external view

8.3.9.3 Factory I (Figs. 8.145 and 8.146)

X-dir.: one span, Y-dir.: three spans, number of stories: two
X-dir.: moment frame, Y-dir.: moment frame
Columns: square HSS's, beams: W-shapes, external finish: corrugated metal sheets
Observations: internal and external finishes washed away. No residual drift

Fig. 8.146 Factory I: little structural damage

Fig. 8.147 Factory J: fallen roof

8.3.9.4 Factory J (Figs. 8.147 and 8.148)

X-dir.: one span, Y-dir.: four spans, number of stories: one
Observations: corner column collapsed

Fig. 8.148 Factory J: leaning columns

Fig. 8.149 Hotel B: external view

8.3.10 Miyako City (Reported Inundation Height at Taro Port: 13.4 m [2])

8.3.10.1 Hotel B (Figs. 8.149 and 8.150)

X-dir.: one span, Y-dir.: six spans, number of stories: six
X-dir.: moment frame, Y-dir.: moment frame

Fig. 8.150 Hotel B: interior damage seen at first story

Columns: square HSS's, beams: W-shapes, column base: concrete encased, external finish: ALC panels

Observations: internal and external finish washed away in first to third stories. Little residual drift

8.3.10.2 Ice Making Factory A (Figs. 8.151 and 8.152)

X-dir.: one span, Y-dir.: two spans, number of stories: four

Columns: square HSS's (concrete-encased in first story), beams: W-shapes, external finish: extruded cement panels

Observations: internal and external finishes washed away in first to third stories. Little residual drift

8.3.11 Kamaishi City

8.3.11.1 Office Building H (Figs. 8.153 and 8.154)

X-dir.: four spans, Y-dir.: two spans, number of stories: two

X-dir.: moment frame, Y-dir.: moment frame

Columns: square HSS's, beams: W-shapes, column bases: exposed type, external finish: extruded cement panels

Observations: internal and external finishes washed away. Little residual drift. Foundation exposed after scouring

Fig. 8.151 Ice making factory A: exterior view

Fig. 8.152 Ice making factory A: damage to cladding

Fig. 8.153 Office building H: external view

Fig. 8.154 Office building H: foundation exposed after scouring

Fig. 8.155 Factory K: external view

8.3.12 Ofunato City

8.3.12.1 Factory K (Figs. 8.155 and 8.156)

X-dir.: two spans, Y-dir.: three spans, number of stories: two
Columns: older built-up square HSS's (1 F), W-shapes (2 F), beams: W-shapes, column bases: exposed type, external finish: slate tiles
Observations: internal and external finishes washed away. First story deformed towards shoreline. Round bar braces fractured.

8.4 Damage Caused by Ground Deformation

In areas with poor soil conditions, ground deformation caused structural damage. In some cases the foundation mounted on piles was undamaged but the surrounding soil subsided. Liquefaction caused differential subsidence that tilted the building (Figs. 8.157, 8.158, and 8.159).

Fig. 8.156 Factory K: first story leaning towards inland

Fig. 8.157 Shopping complex: ground subsided near footing foundation (Rifu)

Fig. 8.158 Office building: two right-end spans subsided due to liquefaction (Iwanuma)

Fig. 8.159 Subsided corner of same building shown in Fig. 8.158 (Iwanuma)

8.5 Damage Caused by Fire

Damage in some buildings was exacerbated by fire that initiated after the earthquake and tsunami. The cause of fire is unknown (Figs. 8.160, 8.161, 8.162, and 8.163).

Fig. 8.160 Warehouse (Miyagino, Sendai)

Fig. 8.161 Building (Miyagino, Sendai)

8.6 Summary

Preliminary observations are summarized in the following:

1. The ground motion caused limited structural damage to buildings constructed after major update in the Building Standard Law was implemented in 1981. However, older buildings constructed prior to 1981 saw notable damage caused

Fig. 8.162 Same building shown in Fig. 8.161 (Miyagino, Sendai)

Fig. 8.163 Delivery station (Natori)

by ground motion. Nonstructural damage to internal and external finishes and ceilings was extensive regardless of construction age.
2. A large proportion of industrial and commercial facilities in the tsunami affected area was constructed in steel. Widespread damage was seen in these steel buildings.
3. In areas attacked by violent tsunami, some buildings saw limited structural damage because their internal and external finishes were immediately washed away. Many buildings were damaged by debris impact.
4. In areas attacked by less violent tsunami, steel buildings saw varying degrees of nonstructural damage depending on the tsunami height. However, the majority of buildings saw limited structural damage.

Acknowledgments This material is based on the damage reconnaissance reports prepared by the Steel Structures Committee of the Architectural Institute of Japan. Contribution of the damage reconnaissance team members, listed below, are greatly appreciated: Susumu Kuwahara, Seiji Mukaide, Akinobu Takada (Osaka Univ.); Keiichiro Suita, Yuji Koetaka, Yuichi Sato, Masanobu Sakashita (Kyoto Univ.); Naoto Yamada (JFE Shoji Construction Materials Sales Co.); Yoshiya Taniguchi, Sho Watanabe (Osaka City Univ.); Taichiro Okazaki, Taisuke Muraki, Hironori Otomo (Hokkaido Univ.); Tsuyoshi Tanaka, Hisashi Namba, Kenzo Taga (Kobe Univ.); Jun Kawaguchi, Yoshito Tomioka, Norihisa Hirabayashi, Seiko Tsuge, Bu-Sung Kong (Mie Univ.); Takamasa Yamamoto (Toyota National College of Tech.); and Hideki Idota, Atsushi Sato (Nagoya Institute of Tech.).

References

1. Port and Airport Research Institute (2011) Damage assessment of ports in Tohoku region (Preliminary field survey report), Attachment 2: Reconnaissance results of respective ports, 23 March 2011
2. The 2011 Tohoku Earthquake Tsunami Joint Survey Group (2011) The 2011 off the Pacific coast of Tohoku Earthquake tsunami information. http://www.coastal.jp/tsunami2011/. Accessed 20 May 2011

Chapter 9
Damage to Non-structural Elements

Tsuyoshi Seike, Akira Natori, Ryohei Kumagai, and Toru Eguchi

Abstract In this chapter, the damages of non-structural elements are reported as follows: roofing tile and exterior walls installed by wet joint in wooden houses, damage of ceilings (especially ceilings of large space like a hall, ceiling of general building and ceiling of low-rise commercial building) and damage of exterior outer walls, claddings and openings (especially (1) damage on the exterior tiles of RC buildings, (2) damage on the lath sheet of steel buildings, (3) damage on the ALC panels of steel buildings, (4) damage on the glass screens, (5) damages on window glasses and (6) damages on other outer wall parts). The investigation area is as follows: Fukushima and Miyagi in Tohoku region, Kanagawa, Tokyo, Saitama, Chiba, Tochigi, Ibaraki and Tokyo in Kanto region and Shizuoka in Chubu region.

Keywords Ceiling • Expansion joints • Exterior walls • Interior finishes • Roofing tiles

9.1 Overview of Investigation

Regarding regional tendencies, in Kanto region, many damages were found in wide area-damages and fall of ceilings, damages of exterior walls and mortar finished lath sheets, damages of fixed glass windows and damages of roofing tiles of wooden houses

T. Seike (✉)
Graduate school of Frontier Sciences, The University of Tokyo,
5-1-5- Kashiwano-ha, Kashiwa, Chiba 277-8563, Japan
e-mail: seike@k.u-tokto.ac.jp

A. Natori
Faculty of Human Life Design, Toyo University, Tokyo, Japan

R. Kumagai
Faculty of Engineering, Tokyo University of Science, Tokyo, Japan

T. Eguchi
Faculty of Urban Innovation, Yokohama National University, Tokyo, Japan

Architectural Institute of Japan (ed.), *Preliminary Reconnaissance Report of the 2011 Tohoku-Chiho Taiheiyo-Oki Earthquake*, Geotechnical, Geological and Earthquake Engineering 23, DOI 10.1007/978-4-431-54097-7_9, © Springer Japan 2012

in Kanagawa, Tokyo, Saitama, Chiba, Tochigi and Ibaraki. In Tohoku region, there were many damages in wide area as well. Some areas have many damages and some are less and this tendency is same as Kanto region. In Fukushima prefecture, the damages in Koriyama city located inland were greater than other area. In Miyagi prefecture, the damages in coastal area were greater than city center in Sendai City for example. This tendency may be caused by strength of ground in each area. In Tokai region in Fujinomiya city, damages were found only in old and less seismic strength buildings and in some elements of steel frame buildings which are easily damaged by earthquakes.

9.1.1 Introduction

This chapter outlines the damage to non-structural elements such as roofing tiles of wooden houses, exterior walls, non-structural walls, ceilings, interior finishes, and expansion joints. Damage to non-structural elements was observed over a very wide area. Because damaged non-structural elements were removed or replacement within days after the earthquake, it was not possible to obtain an even view of the damage across all areas. While this report is not intended as a comprehensive view of the damage, common patterns and key aspects as noted from the first investigation are summarized for future reference.

9.1.2 Regional Tendency

In Tohoku region, there was much damage in a widespread area as well. In Fukushima prefecture, the damage in Koriyama city was greater than any other area. In Miyagi prefecture, the damage in the coastal area was greater than the central area in Sendai City. This tendency may be caused by the variation in ground strength in these areas.

In Kanto region, much damage was found in a widespread area; damage and collapse of ceilings, damage to exterior walls and mortar finished lath sheets, damage to fixed glass windows, and damage to roofing tiles were all found in the Prefectures of Kanagawa, Tokyo, Saitama, Chiba, Tochigi and Ibaraki. In general it is difficult to analyze why these elements failed as damage could be linked to weak ground or inappropriate building structures and construction methods. This report does not cover whole area since damage is quite widespread.

9.1.3 Seismic Design Provisions for Non-structural Elements

There are two important design standards for nonstructural elements. One is the detailing rules prescribed in the "*Standards of Roofing, Exterior Finishes and Exterior Curtain Walls*" (originally public notice No. 109 by the Ministry of Construction dated Jan. 29, 1971, latest revision public notice No. 1348 by the Ministry of Construction dated May 23, 2000) that supplement the Building Standard Law Enforcement Order Article 39 paragraph 2.

The other is the "*Standard of Comprehensive Seismic Design Plan of Government Buildings.*" Published in 1996, this recommendation provides a target story drift limit for building structures and restricts the materials and installation methods for non-structural elements that are suitable for the specific target story drift limit.

Other provisions include the "*Technical Advice*" by the Housing Bureau of Ministry of Land, Infrastructure, Transport, and Tourism, which applies to ceilings, and the "*Recommendations for Seismic Design and Construction of Nonstructural Elements*" published by AIJ originally in 1985 and revised in 2003, which detail the design and construction method for each type of non-structural element.

9.1.4 Overview of Damage

The following four were the most commonly observed damage types:

1. Damage to roofing tiles and exterior walls in wooden houses
2. Damage to ceilings
3. Damage to exterior walls, claddings, and openings
4. Damage to other non-structural elements

9.2 Damage to Roofing Tiles and Exterior Walls in Wooden Houses

Over a widespread area, wooden houses suffered extensive damage to their exterior walls and roofing tiles. Damage to roofing tiles can be classified into damage to the ridge and damage to other segments. Damage to the ridge generally involved failure of the clay bonding the ridge tiles to the sheathing. In some cases, the entire ridge had failed. There were cases where the majority of roof tiles fell down (Fig. 9.1).

Fig. 9.1 Damage to roofing tiles

Fig. 9.2 Damage to plaster-finished wall

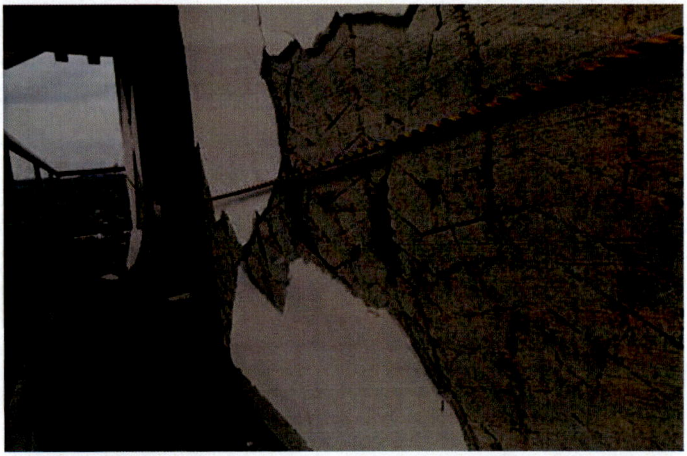

Fig. 9.3 Damage to plaster-finished wall

Failure of exterior walls installed using wet construction was another pattern of damage to wooden houses observed over a widespread area. Cracking in traditional mud walls and lime plaster walls was likely due to their poor ability to accommodate deformation of the structure (Figs. 9.2 and 9.3). Damage was also found in exterior walls finished with mortar on metal lathing (Fig. 9.4).

Fig. 9.4 Damage to mortar finish on wood

9.3 Damage to Ceilings

Since the 2005 Miyagi-Oki earthquake caused the entire ceiling to fall in a swimming pool, numerous guidelines had been made to increase the safety of ceilings particularly in buildings with a large open space. However, according to reports, ceilings fell even in relatively new buildings in spite of these warnings.

Damage to ceilings was widespread. Unfortunately, the investigation reported herein is limited because access to the interior was difficult for the majority of buildings. Furthermore, because the vast majority of ceiling damage is concealed and never made public, the extent of damage remains unknown.
Damage to ceilings is categorized into the following three building types:

1. Ceilings of large space, such as a hall or auditorium
2. Ceilings of office buildings
3. Ceilings of low-rise commercial buildings

9.3.1 Ceilings of Large Space

Damage in the Tohoku region included the falling of ceiling panels in auditoriums (Fig. 9.5), the falling of ceiling panels, lighting equipment, insulation, etc. in gymnasiums (Fig. 9.6), and damage to joints between the ducts and the ceiling panels in a school cafeteria.

In Kanto region, there was human impact due to the fall of hall ceilings in Tokyo. In addition, there was the damage in which the large amount of ceilings of a music hall in Kanagawa Prefecture fell. The damage to sports facilities was reported as well (Figs. 9.7 and 9.8). In addition, some of the ceilings in an indoor pool fell down (Fig. 9.9) and some of the ceilings fell down in a new airport terminal.

Fig. 9.5 Damage to ceiling panels in a concert hall

Fig. 9.6 Damage to ceiling panels in a martial arts gymnasium

Fig. 9.7 Damage to ceiling panels in a gymnasium

Fig. 9.8 Fall of ceiling panels in a gymnasium

Fig. 9.9 Damage to ceiling panels in an indoor pool

Fig. 9.10 Fall of ceiling panels in an office building

9.3.2 Ceilings of Office Buildings

For general buildings, the ceilings were often damaged. Ceilings on the first floor in high-rise buildings frequently fell in Tokyo (Fig. 9.10). Almost of this damage was seen at the edge, near columns, which means that the panels could not follow the deformation of a building frame.

9.3.3 Ceilings of Low-Rise Commercial Buildings

Low-rise commercial steel structures have been frequently constructed all over Japan. Most ceilings in this type of structure were severely damaged. The main damage was in the falling and breaking of ceiling panels connected to light-gauge steel, especially near the connection with a wall (Figs. 9.11 and 9.12). The exterior wall made of ALC panels, as shown in Fig. 9.13, exhibited peculiar damage. This is possibly due to interference between exterior wall and light-gauge steel of the ceiling.

Fig. 9.11 Damage to ceiling panels

Fig. 9.12 Damage to ceiling panels

Fig. 9.13 Damage to exterior wall due to the damaged internal ceiling

9.4 Damage to Exterior Walls, Claddings and Openings

In many areas, there is damage to exterior walls, claddings, and openings. Typical types of damage are listed as follows, though they are not the entire extent of damage:

1. Damage to exterior tiles of RC buildings
2. Damage to lath sheets of steel buildings
3. Damage to ALC panels of steel buildings
4. Damage to glass screens
5. Damage to window glasses
6. Damage to other exterior wall parts

9.4.1 Damage to Exterior Tiles of RC Buildings

In some cases, the tiles on RC and SRC buildings are damaged. This kind of damage is found in many areas from Kanagawa Prefecture to Tohoku area (Figs. 9.14, 9.15, 9.16, and 9.17).

Fig. 9.14 Damage to exterior tiles around openings

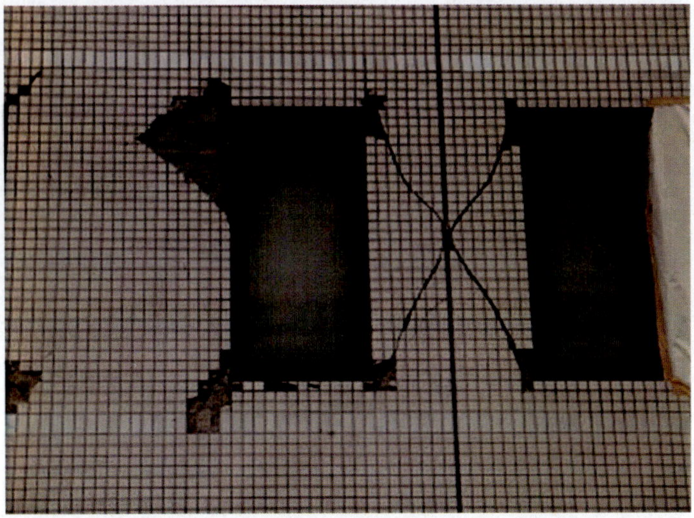

Fig. 9.15 Damage to exterior tiles associated with shear failure of wall

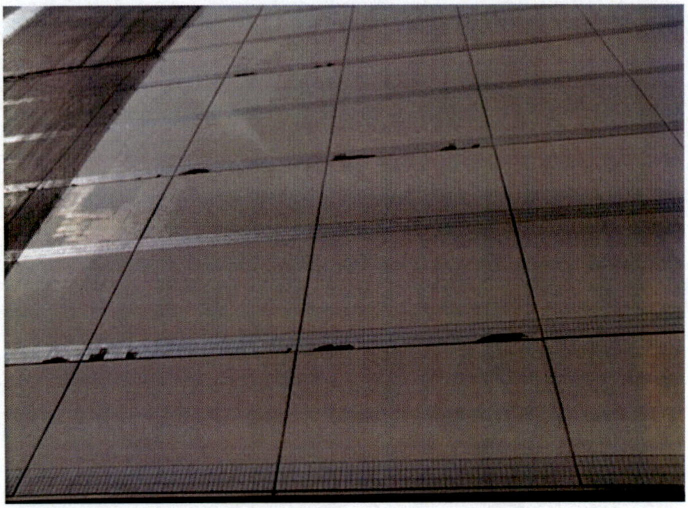

Fig. 9.16 Damage to exterior tiles at joints of precast wall

Fig. 9.17 Damaged tiles between the base and wall

9.4.2 Damage to Lath Sheets of Steel Buildings

A "lath sheet" is used with mortar finish on the substrate made of it attached on the square-corrugated galvanized steel plate in constructing exterior walls of steel frame. This method of construction was primarily used on steel buildings from 1960s to 1970s. The damage was found in a widespread area, similar to the Niigata-Ken-Chuetsu Earthquake in 2004 (Figs. 9.18, 9.19, and 9.20).

Fig. 9.18 Damaged mortar finish on lath sheets

Fig. 9.19 Detail of damaged mortar finish on lath sheets

9.4.3 Damage to ALC Panels of Steel Buildings

Autoclaved lightweight concrete (ALC) panels also experienced damage in a widespread area. Rocking walls, a seismically safer option for ALC panels, have been implemented since 2002. In this investigation, most of the damaged walls were installed with the old construction methods (Figs. 9.21, 9.22, and 9.23).

Fig. 9.20 Fallen walls with mortar finishing lath sheets

Fig. 9.21 Damaged ALC panels

9.4.4 Damage to Glass Screens

In the glass screen construction, there are two typical types of construction methods. One is a ribbed glass system which is used in car showrooms, and the other is two new systems, the DPG and MPG systems, which uses large toughened glass. In this earthquake, many of ribbed glass systems are damaged and most of the newer systems are not (Figs. 9.24, 9.25, and 9.26).

Fig. 9.22 Damaged ALC panels

Fig. 9.23 Damaged ALC panels

Fig. 9.24 Damaged glass screen

Fig. 9.25 Damaged glass screen

Fig. 9.26 Damaged glass screen

Fig. 9.27 Damage to fixed window glasses

9.4.5 Damage to Window Glasses

Although putty-glazed fixed sash window glasses were banned by the "*Notification No.109 of the Ministry of Construction/1971*" after numerous damage was reported during the 1978 Miyagi-Ken-Oki earthquake, cracked fixed window glasses were still widely observed (Figs. 9.27 and 9.28).

Fig. 9.28 Damage to window glasses in a commercial building

Fig. 9.29 Damage to *horizontal* composite windows

Horizontal composite windows was damaged due to large inter-storey displacement (Fig. 9.29). Furthermore, partial damage on glass curtain walls were also reported (Fig. 9.30). However, glass curtain walls of skyscrapers appeared to be intact (Fig. 9.31).

Fig. 9.30 Damaged glass curtain walls

Fig. 9.31 Glass curtain wall without damage

Fig. 9.32 Damaged precast concrete panels

9.4.6 Damage to Other Exterior Wall Parts

In addition to damage observed above, there were falling-off of non-curtain wall type precast concrete panels in Sendai (Fig. 9.32). There was also damage in the form of falling-off of glass blocks in Fujinomiya (Fig. 9.33). Cracks on exterior walls, such as steel structure siding boards, were also observed (Fig. 9.34).

9.5 Damage to Other Non-structural Elements

Damage to other non-structural elements is as follows:

1. Damage to interior elements
2. Damage to smoke preventive hanging glasses
3. Damage to panels under eaves
4. Damage to expansion joints
5. Damage to workpieces

9.5.1 Damage to Interior Elements

Regarding the interior part of the building with the exception of ceilings, damage to interior walls, doors, and floor finishes were seen in a widespread area (Figs. 9.35 and 9.36). However, in quite a few interior investigations, the damage was reported but the reason of the damage is unclear.

Fig. 9.33 Damage to all glass blocks at curved corner on the second floor

Fig. 9.34 Cracks on siding boards of exterior wall

Fig. 9.35 Damage to interior walls in a staircase

Fig. 9.36 Damage to floor finish in a staircase

9.5.2 Smoke Preventive Hanging Glasses

Smoke preventive hanging glasses are generally employed for commercial buildings. Similar to damage seen in the past earthquakes, damage to smoke preventive hanging glasses was seen in a widespread area (Figs. 9.37 and 9.38).

Fig. 9.37 Crack of smoke preventive hanging glass in a commercial building

Fig. 9.38 Damage to smoke preventive hanging glass and panel

9.5.3 Damage to Panels Under Eaves

In steel gymnasiums and low-rise commercial buildings, damage to panels under eaves was occasionally seen (Figs. 9.39 and 9.40).

Fig. 9.39 Damage to panels under eaves of commercial building

Fig. 9.40 Damage to panels under eaves of gymnasium

9.5.4 Damage to Expansion Joints

In lots of cases, damage to expansion joints was seen in school buildings, commercial buildings between parking areas, and factories (Figs. 9.41 and 9.42).

Fig. 9.41 Damaged expansion joint

Fig. 9.42 Damaged expansion joint

Fig. 9.43 Damaged ALC signboard on a store

Fig. 9.44 Damaged parapet

9.5.5 Damage to Workpieces

Workpieces such as signboards saw extensive damage in many cases (Figs. 9.43 and 9.44).

Block and masonry walls (Figs. 9.45 and 9.46) were damaged in many cases as well.

Fig. 9.45 Falling down of block wall

Fig. 9.46 Damaged masonry wall

9.6 Further Note

This section summarizes the concerning damage to non-structural element and gives an explanation of post-event situation in most structures. However, it was difficult to grasp overall damage in each area right after the earthquake since they had been removed within a few days. Even though it seems the non-structural elements have little damage in some cases, further inspection could reveal damage in other areas such as part of non-structural connection with ceiling panels. As further investigations continue, the process causing non-structural damage becomes more clear as more cracks and damage are discovered.

Acknowledgement This material is based on the damage reconnaissance reports prepared by the members of Kanto chapter and the Tohoku chapter of the Architectural Institute of Japan. Contribution of the damage reconnaissance team members, listed below, are greatly appreciated: Shigeki SAITO (Center For Better Living), Masaaki OHNUMA (Tohoku Institute of Technology), the members of College of Engineering, Nihon University.

Chapter 10
Damage to Soils and Foundation

Kohji Tokimatsu, Shuji Tamura, Hiroko Suzuki, and Kota Katsumata

Abstract An overview of the geotechnical aspects of the building damage in the 2011 Off the Pacific Coast of Tohoku earthquake is presented based on field reconnaissance made after the quake. It is shown that: (1) extensive soil liquefaction occurred along the coast of Tokyo Bay and around the Tone River floodplain. Liquefaction primarily occurred within relatively new reclaimed area, with large ground settlement up to 60 cm, accompanied by settlement/tilting of wooden and reinforced concrete buildings supported on spread foundations; (2) numerous houses in Sendai's hilly residential areas constructed with cut-and fill methods were badly damaged not only by simple collapse of retaining walls, but also by slope failures of fill; and (3) several pile-supported buildings tilted and settled not only in the Tohoku region but also in the Kanto plain, implying damage to pile foundations.

Keywords Foundation • Ground subsidence • Landside • Soil liquefaction

10.1 Introduction

In this chapter, an overview of my group's findings on soils and foundations damage occurring during the 2011 earthquake off the Pacific coast are reported. The survey [1] was carried out in the Tone River basin and in the northeastern region, in mid March and in early April, respectively, to analyze the damage state, including the effects of aftershocks.

K. Tokimatsu (✉) • H. Suzuki • K. Katsumata
Tokyo Institute of Technology, Tokyo, Japan
e-mail: kohji@o.cc.titech.ac.jp; hsuzuki@arch.titech.ac.jp;
katsumata.k.aa@m.titech.ac.jp

S. Tamura
Kyoto University, Kyoto, Japan
e-mail: tamura@sds.dpri.kyoto-u.ac.jp

Architectural Institute of Japan (ed.), *Preliminary Reconnaissance Report of the 2011 Tohoku-Chiho Taiheiyo-Oki Earthquake*, Geotechnical, Geological and Earthquake Engineering 23, DOI 10.1007/978-4-431-54097-7_10, © Springer Japan 2012

10.2 Liquefaction Damage in Tokyo Bay

10.2.1 Soil and Ground Motion Characteristics

Figure 10.1 shows the sites that have been reported liquefaction in the Tokyo Bay reclaimed land and in the Chiba Prefecture. A relationship between reclamation age [2] and liquefaction [3–5] for reclaimed land site is shown. It is confirmed that prominent liquefaction occurred only in reclaimed land. Figure 10.2 shows the relationship between liquefaction point and the depth of alluvial soil in Urayasu, Chiba Prefecture, and Tokyo Metropolis [6, 7]. It is interesting to note that significant liquefaction damage is caused in the case of soils at the depth of at least 35–40 m or deeper.

There are strong motion time histories during the main shock in the Tokyo Bay at two points where the liquefaction was observed, K-NET Inage (CHB024) station [8] and K-NET Tatsumi (TKY017) station. In addition, liquefaction was observed in the north of the former coastline of Urayasu, in the neighborhood of the K-NET Urayasu (CHB008) station.

Figure 10.3 shows the acceleration record (100 s, i.e., including the main motion) for the Inage K-NET (CHB024) station NS and EW, maximum accelerations, equal to

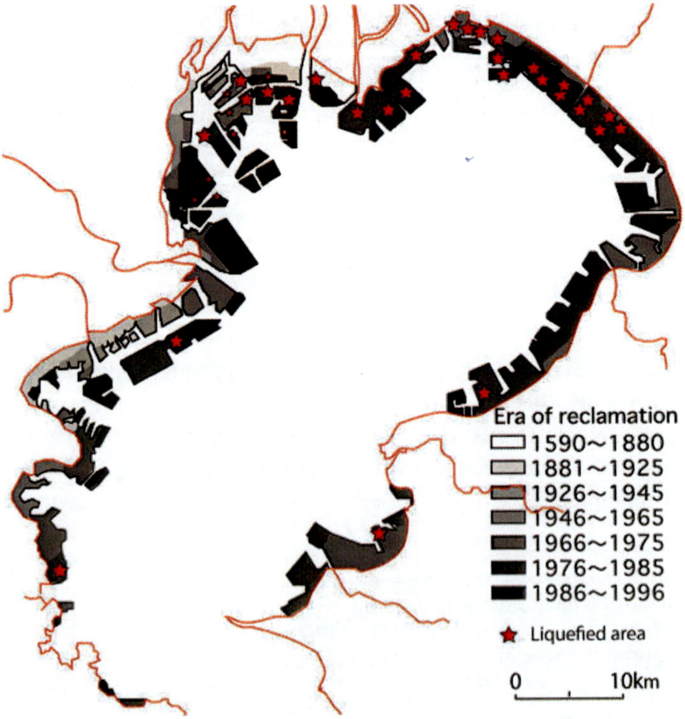

Fig. 10.1 Relationship between reclamation age and liquefaction for reclaimed land site [1]

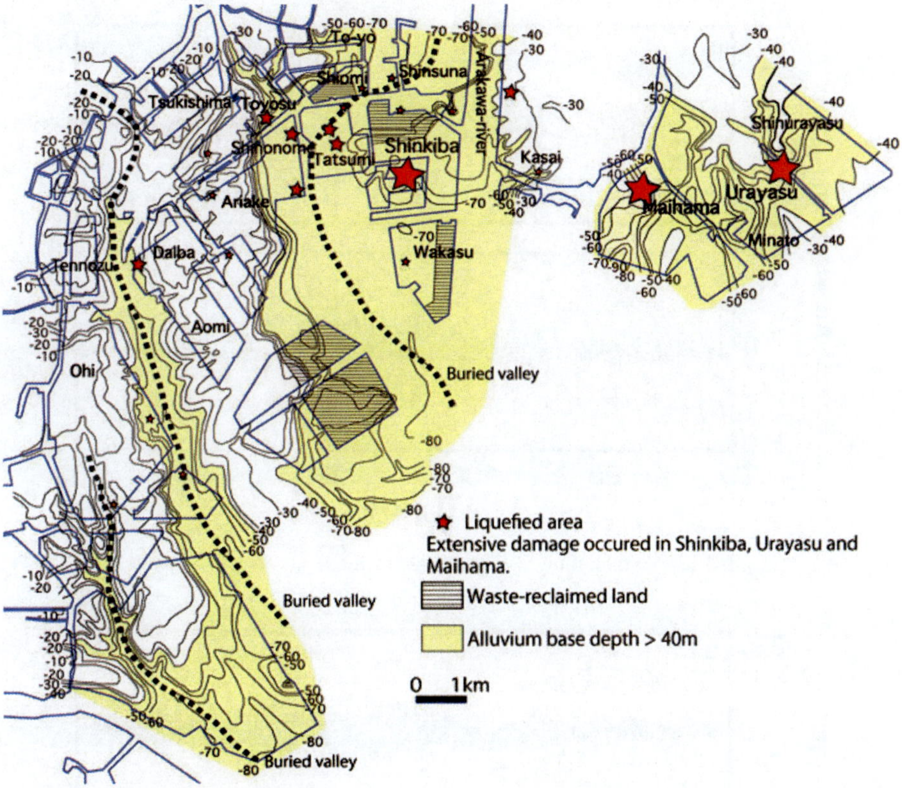

Fig. 10.2 Relationship between depth and soil liquefaction for alluvium sites [1]

2.34, 2.03 m/s^2, respectively, are recorded around 120 s, with spike-shaped waveform which suggests the possibility of cyclic mobility due to liquefaction. A trend is remarkable, i.e. a long-period peak become more important with time, up to around about 110–140 s. Then, it is possible to conclude that, by repeating for the next 30 s, lead to a gradual soil liquefaction. Figure 10.4 shows for the K-NET Urayasu station at which no liquefaction was observed. Compared with the Inage K-NET station, it is not evident that the predominant period changed in a similar fashion. In addition, the comparison of Figs. 10.3 and 10.4 reveals that the liquefaction behavior was not primarily in Urayasu while Inage was completely liquefied between 110 and 140 s.

10.2.2 Damage in Urayasu

Extensive damage to lifelines, such as roads and bridges, and to wooden houses due to liquefaction of reclaimed land was observed in the coast of Tokyo Bay; Kiba in Koto Ward, Urayasu City, Ichikawa City, Narashino City, and in Mihama-ward in Chiba City. In the following section, damage in the Urayasu area is summarized.

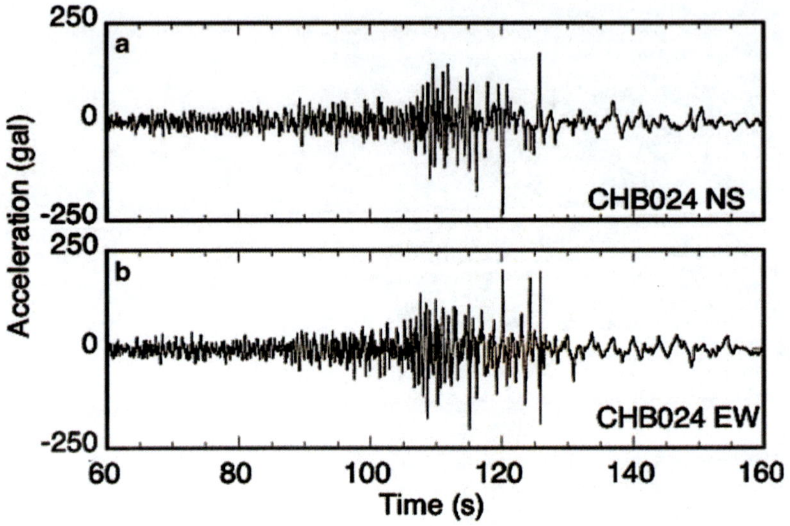

Fig. 10.3 Recorded acceleration time-history at K-NET Inage

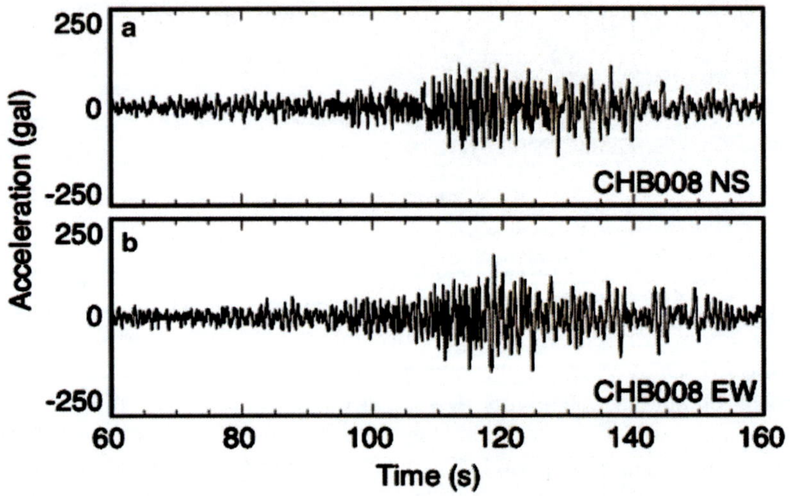

Fig. 10.4 Recorded acceleration time-history for K-NET Urayasu

Figure 10.5 shows the map of Urayasu, Chiba Prefecture, with the reclamation age for each district [9]. Reclamation was started from about 1964 outside the old shoreline breakwater. There are many completed projects such as residential areas, commercial facilities, public facilities, and a finished business district in the first reclaiming work until 1975. The second phase of the reclamation project was completed by 1980; high-rise condominiums, universities, hotels, warehouses, etc. are present. The sea sand, dredged from the offshore Urayasu, is mainly used for these reclaimed lands.

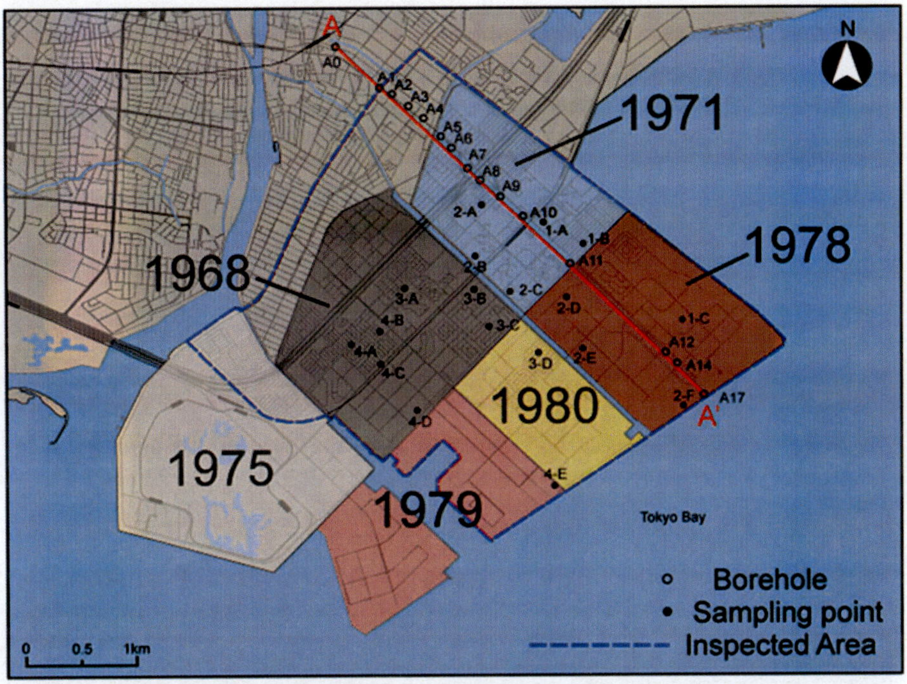

Fig. 10.5 Age of reclaimed land [1]

After the 1987 Chiba earthquake (M6.7) that occurred on December 17, liquefaction damage was reported at Kairaku 1-chome, Mihama 3-chome, and a part of Irifune 4-chome [10]. This time, liquefaction damage was not observed in the area at north-east of the old coastline, in the vicinity of Metro Urayasu Station which includes the location of K-NET Urayasu Station.

The phenomena recognized during damage investigation are reported in the following:

- In many reclaimed locations no liquefaction was observed including the property of Disneyland, where soil was subjected to stabilization. Hence such soil stabilization is proved to be effective to the base acceleration of 2.0 m/s^2.
- In the area where liquefaction occurred, sand boils (Fig. 10.6), ground subsidence, inclination and settlement of buildings with spread foundation (Figs. 10.7 and 10.8), damage of piping due to the difference in level between buildings with pile foundation and surrounding buildings (Fig. 10.9), uplift of underground structure (manhole, emergent water tank and underground parking) (Fig. 10.10), damage to water and sewage, cave-in of roads, and falling of the poles, were observed to varying degrees. But little structural damage due to vibration was observed.
- Inclination and settlement of the foundation did not lead to structural damage of the upper building because they usually adopt mat foundations or stiff foundations to mitigate the effects of settlement and liquefaction.

Fig. 10.6 Sand boils and buried cycle

Fig. 10.7 Tilted building

- In some cases, building experienced relatively great settlement, probably due to high ground contact pressure caused by underground story for flood control or use of reinforced concrete for the first story.
- For adjacent buildings, they usually tilted to the direction such that they should meet together at upper part of the building, because the superposition of the effect of two buildings increases ground contact pressure.

Fig. 10.8 Large building settlement

Fig. 10.9 Ground subsidence around the building with pile foundation

- Some pile foundations of buildings under the construction suffered damage due to liquefaction.

Intensity of liquefaction is shown in a color-coded map (Fig. 10.11) with three levels of damage (red, yellow, green) based on the extent of land subsidence, inclination and settlement of buildings, and degree of damage to pavement.

Fig. 10.10 Uplift of manhole

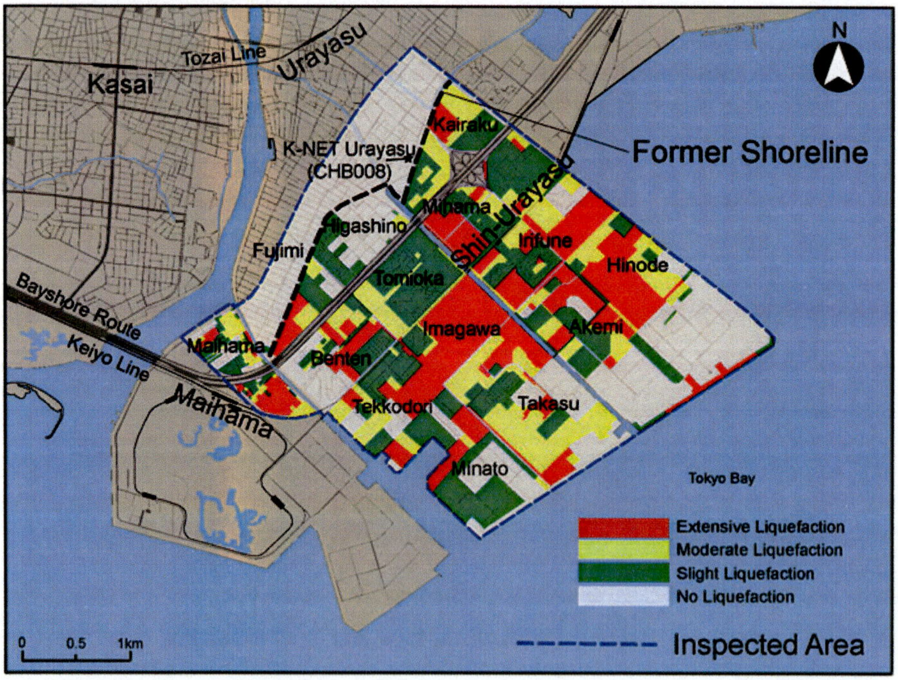

Fig. 10.11 Map liquefaction damage (preliminary) [1]

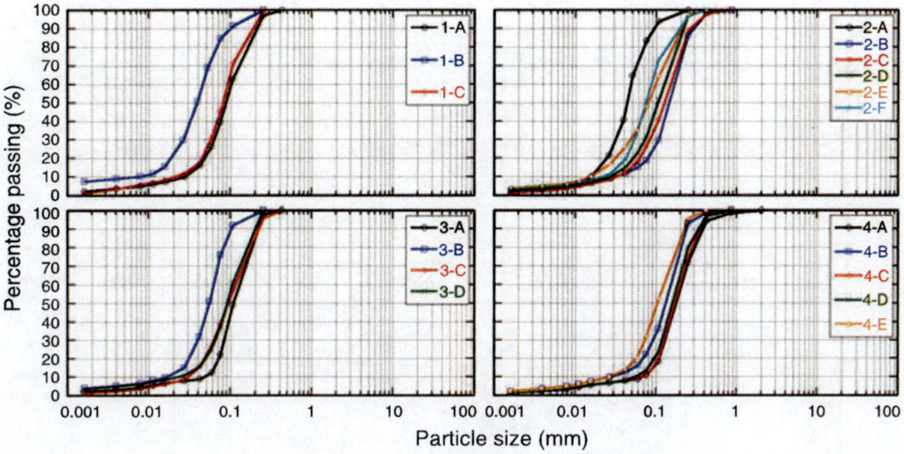

Fig. 10.12 Grain size distribution in sand volcanoes [1]

It is revealed that the liquefaction occurred only in the district north to the coast in 1964. The area of liquefaction is wider and includes the area of liquefaction during the 1987 Chiba earthquake, where liquefaction occurred again this time. The extent of damage caused by liquefaction differs depending on the location in the reclaimed area.

10.2.3 Relationship Between Soil and Liquefaction Damage in Urayasu

Figure 10.12 shows the grain size distribution by district of the sand volcanoes sampled at the locations shown in Fig. 10.5. In all the samples, the fine fraction content was high at approximately 15–70%, and it was considered to be nonplastic fine sand and silty sand. This corresponds to the filling sand layer at a depth of 10 m below sea and lower, so it is thought to be filling sand which underwent liquefaction during the earthquake.

Figure 10.13 shows the artificial fill and sand layer for each district, gathered from Chiba Prefecture data and our surveys. The depth distribution of N values is shown in gray, and the mean is shown in red. It also shows that the damage in Akemi and Hinode is significantly different in the northwest and southeast in the respective cases. The figure shows that the N value of the sand layer is very low in Tomioka-Imagawa, and Akemi-Hinode (northwest). In the vicinity around Urayasu Station which is not reclaimed land and in Akemi-Hinode (southeast) which is reclaimed but with the highest elevation, the N value is high. In addition, the thickness of the artificial fill and sand layer differs in each region, and it is thick in Maihama, Mihama-Irifune, Takasu, and Akemi-Hinode.

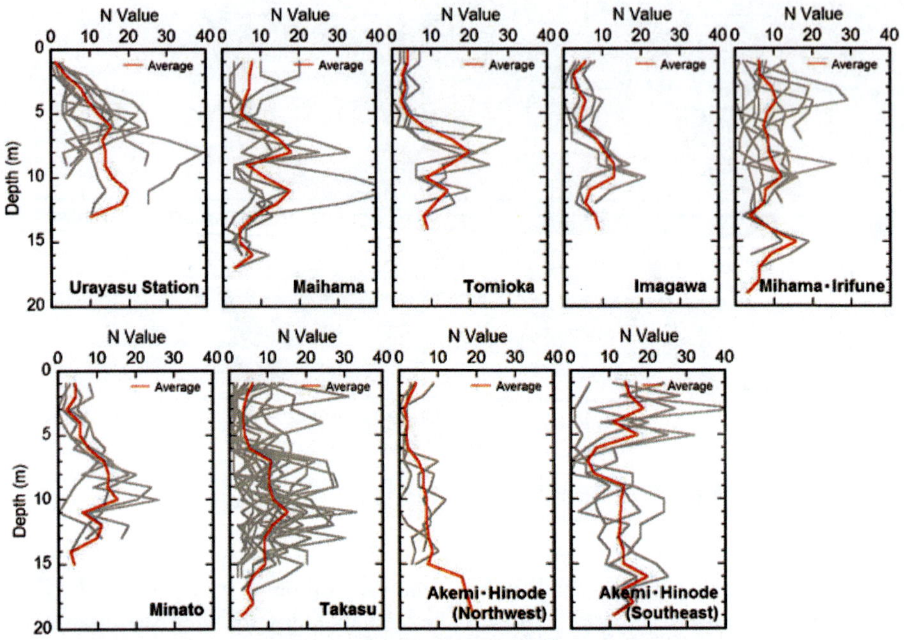

Fig. 10.13 Depth distribution of N values in each district [1]

When the results above are compared with the liquefaction damage, the following tendency was observed. The elevation is low inland of the pre-1964 coastline, therefore the groundwater level is shallow. Nevertheless, liquefaction was not observed. Since the N value of this district is higher than districts with young landfill where liquefaction occurred, so-called aging may have had an impact on the degree of liquefaction.

In Akemi-Hinode (southeast), the N value is relatively high, and since the liquefaction damage is minor, differences in the filling material and method may have had an effect on the damage. Furthermore, as the elevation of this district is relatively high, the height of the elevation may affect the degree of liquefaction. This can be attributed to the fact that the groundwater level is correspondingly lower in relation to the higher elevation, and the silty sand layer below the groundwater level will have undergone compression and so on. In addition, since this is the most recently developed area, measures against liquefaction may have been taken in many places, with preloading for accelerated consolidation of the ground (overconsolidation of the liquefaction layer).

10.2.4 Liquefaction Damage and Liquefaction Prediction

Figure 10.14 shows the result of liquefaction assessment with the method from Recommendation for the Design of Building Foundations [11] using the mean N

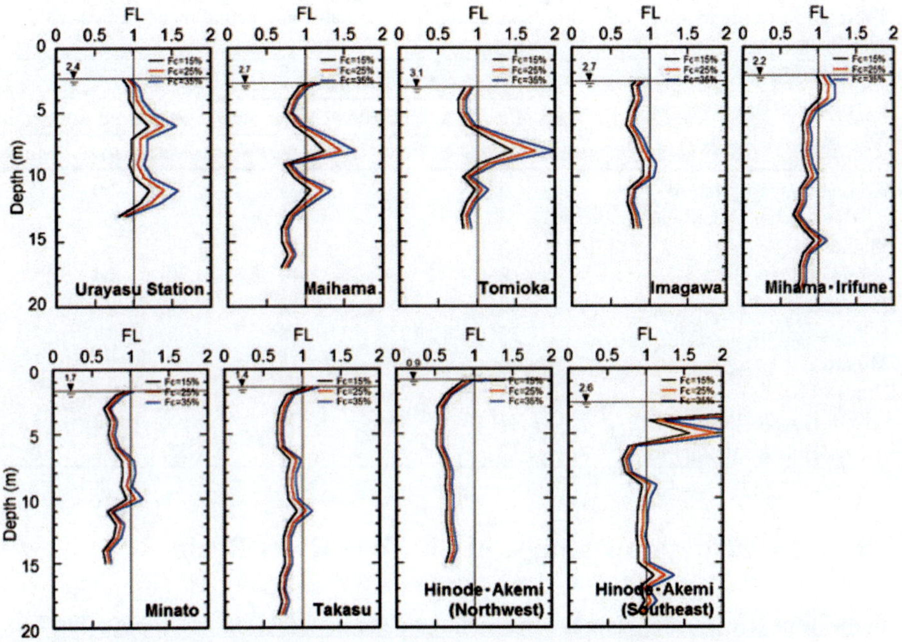

Fig. 10.14 Liquefaction safety factor in each district [1]

value for each area in Fig. 10.13, with acceleration of 2.0 m/s^2 and magnitude of 9.0. The respective average was taken as the groundwater level of each area, and three cases were considered using a fine fraction content of 15%, 25%, and 35%. The FL value at most depths around Urayasu Station where no liquefaction damage was seen and in the Akemi-Hinode (southeast) district where liquefaction damage was minor was 1 or more, but in other districts, it was less than 1 in many cases. In Mihama-Irifune, Takasu, and Akemi-Hinode (northwest) in particular, there are continuous layers with an FL value lower than 1 at a depth up to 20 m. These results agree to the state of damage.

Table 10.1 compares the inferred amount of ground subsidence in relation to the distribution of N values for each area shown in Fig. 10.13 using the method from Recommendation for the Design of Building Foundations [11] with the measured maximum, average and minimum values for each district (relative subsidence of pile foundation buildings and the surrounding ground). The inferred subsistence, when a fine fraction content of 25% was assumed, was 6 cm around Urayasu Station, 11 cm in Akemi-Hinode (southeast), 16–33 cm in other districts where liquefaction was severe, with the highest values in Akemi-Hinode (northwest). There was no significant discrepancy with the actual measured values. While it will be necessary to reconsider and clarify the fine fraction content at each location and each depth, the current design guidelines can be considered to provide a reasonably accurate estimate of the potential for liquefaction and the degree of damage.

Table 10.1 Comparison of the inferred subsidence mean value and actual measurement value in each district [1]

	Estimated value (cm)			Measured value (cm)		
	Fc = 15%	Fc = 25%	Fc = 35%			
	Av.	Av.	Av.	Max.	Av.	Min.
Around Urayasu Station (Nekozane, Todaijima, Kitasakae)	9	6	5	0	0	0
Maihama	25	18	14	–	–	–
Tomioka	18	13	10	30	26	15
Imagawa	23	16	12	50	22	5
Mihama-Irifune	32	23	18	45	19	7
Minato	26	19	15	60	22	5
Takasu	38	28	23	50	23	2
Akemi-Hinode (northwest)	44	33	27	65	32	3
Akemi-Hinode (southeast)	17	11	9	15	8	2

10.3 Liquefaction Damage in the Tone River Basin

In the Tone River basin, damage to housing due to liquefaction was reported in Kuki City and Satte City in Saitama Prefecture, Chiba Prefecture and Ibaraki Prefecture (Fig. 10.15). The following is a report on the damage in Katori City in Chiba Prefecture and Itako City and Kamisu City in Ibaraki Prefecture.

10.3.1 Katori City

In the Sawara district of Katori City, waterways leading to the Tone River flow in a network across the town. When the town is compared with old maps, most of the town and waterways correspond to old river courses or wetlands. The liquefaction damage was especially pronounced in reclaimed areas including these waterways. Various types of damage were observed; housing with spread foundations tilted and subsided, the ground around pile foundation buildings subsided, underground structures rose up, and roads and pedestrian walkways experienced subsidence and became uneven. Around the waterways, liquefaction caused lateral soil movement and in addition to the damage mentioned above as shown in Fig. 10.16. The river walls and ground behind them were pushed in towards the river, making the river narrower, and the river bottom protruded. The bank fill subsided significantly, moved horizontally, and caused damage to the bridges over the waterway.

10.3.2 Itako City

In the Hinode district of Itako City, many sand volcanoes occurred causing housing and buildings with spread foundations to tilt and subside, the ground around pile

Fig. 10.15 Liquefaction areas in the Tone River basin (more in [12])

Fig. 10.16 Lateral spreading towards the river [1]

Fig. 10.17 Damage to a pedestrian walkway due to liquefaction [1]

foundation buildings to subside, underground structures to rise up, roads and pedestrian walkways to experience subsidence and become uneven (Fig. 10.17), and electricity poles to lean. Immediately after the earthquake, water and sewage were disrupted throughout the Hinode district. Damage from liquefaction tended to be greater in the south of the district where the Hitachitone River flows. The resulting ground subsidence was about 40–50 cm in the south around the Itako Purification Center, and about 10 cm or less in the northern part.

Old maps show that the Hinode district corresponds to the Uchinasakaura reclaimed land (reclaimed between 1934 and 1949 year). In the past, as reclaimed land, it was used as rice paddy, but was later developed for buildings. It suffered enormous damage from liquefaction. Liquefaction damage in the Hinode district was also reported in the survey by Wakamatsu [10] following the East Chiba earthquake of 1987, but in this earthquake, the scale was enormous over a wider area. Therefore it is inferred that the impact on lifelines such as water and sewage was more serious.

10.3.3 Kamisu City and Kashima City

In the Fukashiba and Horiwari districts, many sand volcanoes occurred causing housing with spread foundations to tilt and subside (Fig. 10.18), the ground around pile foundation buildings to subside, underground structures to rise up, and roads and pedestrian walkways to experience subsidence and become uneven. In Fukashiba

Fig. 10.18 A leaning building

district, there were sand volcanoes with a thickness of about 50 cm, some of which even buried the external units of air conditioning systems. In addition, there was a tendency for adjacent buildings to lean inwards towards the center, and the central part to subside significantly. However, houses at the margin of banked earth were sometimes seen to be leaning significantly outwards where liquefaction caused the embankment to collapse. In addition, in the northern part of the Horiwari district, culverts were pushed upwards and only houses along the road were badly damaged by subsidence of about 50 cm in relation to the road surface and adjacent houses. In contrast, the damage was minor in the southern part of the district. At all of these places where liquefaction occurred, there were many puddles to be seen at the time of the survey, suggesting that the groundwater level is very shallow.

In the Wanigawa district of Kamisu City and the north of Horiwari, the area of the Wanigawa land reclamation was reclaimed from the Wani River between 1928 and 1941 and used as rice paddy according to the Ibaraki prefectural government office. The area was later developed for housing and other purposes, and it is thought to have experienced liquefaction in the recent earthquake. However, the southern part of Horiwari where damage from liquefaction was minor is known to have been used as a coniferous forest.

Furthermore, in the Fukashiba district, areas that were used as rice paddy and were later developed for housing suffered significant damage, but there was little damage along the old highway and in long-established villages. A stone monument in Fukashiba records that the whole area where liquefaction damage occurred was once improved with soil brought from elsewhere, which suggests the possibility that this may have contributed significantly to the severe liquefaction damage.

10.4 Damage to Foundations in the Tohoku District

10.4.1 Low-Lying Areas of Sendai City

10.4.1.1 K-NET Sendai

Around K-NET Sendai (Miyagi Fire Department MYG013) in Nigatake Wakabayashi ward, Sendai City, sand volcanoes were observed, and subsidence of about 3 cm occurred in the ground around the pile foundations. This happened immediately after the main earthquake. No structural damage to the building was found. At this site, maximum NS and EW acceleration of 15.15 and 9.77 m/s^2 was recorded at the time of the earthquake. Figure 10.19 shows a time history of acceleration for the recent earthquake. A spiked waveform can be seen at around 90 s, suggesting the possibility of cyclic mobility due to liquefaction.

10.4.1.2 Damage to Pile Foundations in Sendai City

In the western part of Oroshimachihigashi, sand volcanoes occurred, the ground around pile foundation buildings subsided, underground structures rose up, and roads and pedestrian walkways experienced subsidence and became uneven. Ground subsidence of around 10–20 cm was observed around buildings with pile foundations, and at least two pile foundation buildings were observed to be leaning noticeably. Building damage thought to be due to damage to pile foundations was observed in

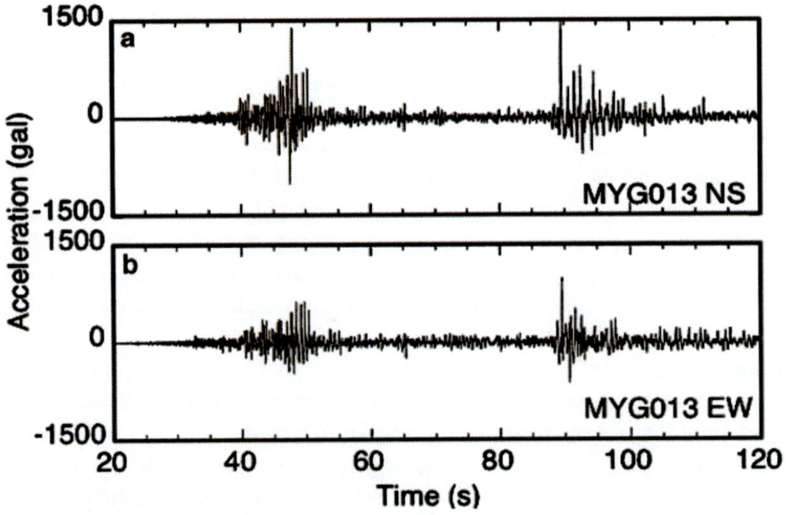

Fig. 10.19 Acceleration record at K-NET Sendai

Fig. 10.20 A leaning pile foundation building

Fukumuro, Miyagino-ku (Fig. 10.20). This building is a 14-story structure built using steel reinforced concrete and pile foundations, completed in 1976. Two buildings are arranged in an L-shape with expansion joints, and a national road runs along the horizontal line of the "L" (south). In the 1978 Miyagi earthquake, shear fracture damage to nonstructural walls was reported [11]. In the recent earthquake, subsidence of the foundations on the south side of the south building resulted in a lean of about 1° towards the south. Although large cracks appeared in nonstructural wall throughout the building, no major damage to the main structure was observed. Ground subsidence of around 10 cm was observed around neighboring buildings with pile foundations.

10.4.2 Damage in Hilly Areas in Sendai City

In Sendai and its suburbs, there are many hilly areas that were developed for housing by cutting and filling [13]. The respective thickness is more than 30 m at maximum.

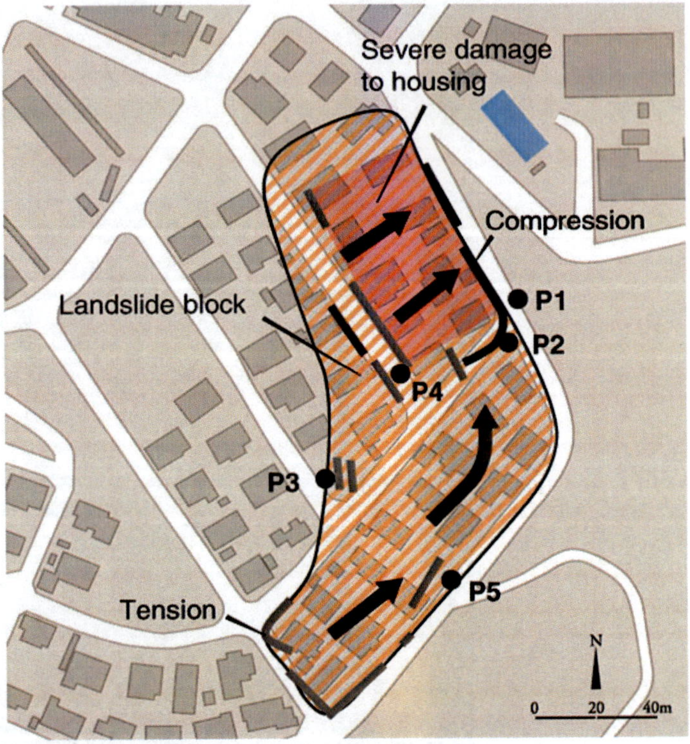

Fig. 10.21 Damaged area in Oritate 5-chome [1]

According to reports by the Architectural Institute of Japan, the Japan Society of Civil Engineers and other organizations, damage to buildings due to ground deformation at housing sites was reported in Izumi-ku (Takamori, Kitatakamori, Kitayamanaka, Chomeigaoka, Kamo, Kuromatsu, Nankodai), Miyagino-ku (Tsurugaya), Aoba-ku (Sakuragaoka, Futabagaoka, Asahigaoka, Seikaen, Oritate), Taihaku-ku (Midorigaoka, Aoyama), Natori (Sogodai), Shiroishi and other places. The following is a report on the damage in Oritate, Aoba-ku and Aoyama, Taihaku-ku.

10.4.2.1 Oritate 5-Chome, Aoba Ward

The Oritate housing complex was developed in the latter half of the 1960s, and the lots were sold in the early 1970s. No damage was recorded in the report by the Architectural Institute of Japan and the Tohoku Chapter of the Japan Society of Civil Engineers after the 1978 Miyagi earthquake [13, 14].

Figure 10.21 shows the area of damage in Oritate 5-chome. In this area, many retaining walls were damaged. As Fig. 10.22 (P1 in Fig. 10.21) shows, at the bottom of the slope, the retaining wall has been destroyed as if pushed outwards by its

Fig. 10.22 Damage to retaining walls

Fig. 10.23 Large cracks

backfill. Some of the upper retaining walls were newer than the lower ones, and some may have undergone repair. However near the upper part of the slope (P3), there were large cracks in the site as shown Fig. 10.23. At the top of the slope, there were also places with tensile cracks in the retaining wall. This means that at these points, tensile force was produced in the ground. The locations of compression and

Fig. 10.24 Damage to house

tension in the ground described above are shown in Fig. 10.21. The shaded area is thought to have experienced landslides. When compared with an old 1/25,000 scale map from around 1964, the landslide zone corresponds roughly to the former topography of a valley. In the recent earthquake, the whole earth filling of the valley is thought to have moved. The road that was straight before the earthquake has become curved at the site of the landslide (P4).

The houses that straddled the block where the landslide occurred and the cut earth were damaged near the boundary (P5). The serious damage to the houses (Fig. 10.24) was concentrated in the hatched area in Fig. 10.21 at the edge of the block where the landslide occurred. All of the plots were carried from the mountain side (left) to the valley side (right), and the plots were destroyed by subsidence towards the valley side (right). There appeared to be no reinforcement in the footing foundations. Ground deformation of the edge of the landslide block had significant horizontal and vertical components which are thought to have increased the damage to the buildings.

10.4.2.2 Aoyama 2-Chome, Taihaku Ward

The Aoyama housing complex was developed in the latter half of the 1960s. According to the report by the Architectural Institute of Japan on the 1978 Miyagi earthquake [15], there were cracks and bulges in retaining walls, some of which collapsed. Residents reported that foundations of houses that were damaged in the Miyagi earthquake were damaged again in this earthquake.

Figure 10.25 shows the area of damage in Aoyama 2-chome. At the top of the shaded part, large cracks occurred (P1 in Fig. 10.25). The depth of these cracks

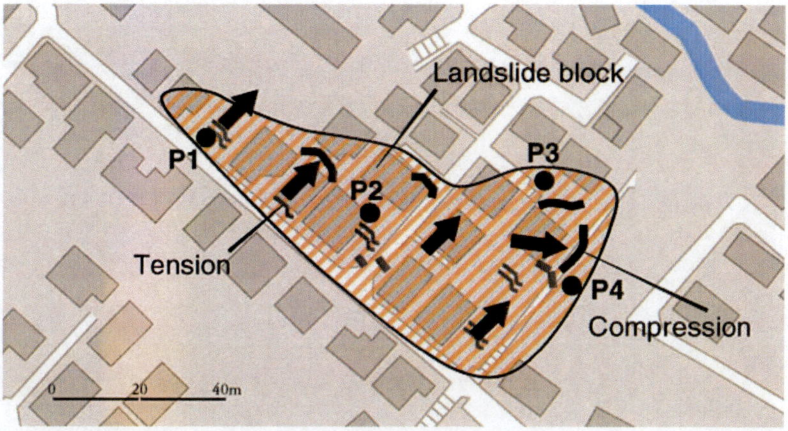

Fig. 10.25 Damaged area in Aoyama 2-chome [1]

Fig. 10.26 Damage to house foundations

reached 70 cm. As a result, at location P2 on the valley side, the retaining wall moved about 1 m towards the valley causing the soil to fall off, leaving the foundation hanging in the air as shown in Fig. 10.26. In addition, as Fig. 10.27 (P3 in Fig. 10.25) shows, at the bottom of the slope, the building plot has been destroyed as if pushed outwards. The locations of compression and tension in the ground described above are shown in Fig. 10.25. The shaded area is thought to have experienced landslide.

When compared with an old 1/25,000 scale map from around 1964, the landslide zone corresponds roughly to the former topography of soil cliffs. The landslide is thought

Fig. 10.27 Damage to a house

Fig. 10.28 Damage to a house

to have occurred in the widened embankment which was built on these cliffs. According to residents, the groundwater level in this area was very shallow at 1 m below ground. This shallow groundwater level is also thought to be a factor in the landslide.

As Fig. 10.28 shows, the building in the block below the landslide (P4) is severely damaged in general. In Aoyama 2-chome, there was also major ground movement in the block above the landslide, and many houses in this block were severely damaged.

Some of the houses that were completely destroyed had been reinforced against earthquakes above ground. This suggests that when reinforcing houses against earthquakes, it is also necessary to consider not only the part above ground but also the underlying ground and foundations.

According to testimony from residents, the ground cracked and the retaining walls bulged with the initial earthquake on March 11, and the cracks spread and the walls became more deformed with subsequent aftershocks. In particular, with the aftershock on April 7, the ground was deformed further, and the foundations made a grating sound. The ground may be undergoing progressive failure.

10.5 Conclusion

The Great East Japan Earthquake in 2011 caused liquefaction on the shores of Tokyo Bay, the Tone River basin, and the Tohoku region. A survey of the damage found the following:

1. The places where liquefaction occurred around Tokyo Bay and the Tone River basin were mostly relatively new reclaimed land. In some places, severe sand volcanoes and ground subsidence of about 50 cm caused wooden and reinforced concrete buildings with spread foundations to subside and lean over, buried structures to rise out of the ground, roads to subside, and other damage. It also caused big gaps to open between the pile foundation buildings and the ground around them, although it caused no structural damage to the buildings themselves. Buildings with rigid spread foundations such as slab foundations did not suffer structural damage above ground, even if subsidence caused them to lean. Damage was observed to some pile foundations under construction.
2. The degree of liquefaction differed even within the same districts, and there is a possibility that the age, method and material of the filling, whether ground improvement was implemented, the thickness of the filling layer, elevation, depth of bedrock and so on have a significant impact on the degree of liquefaction.
3. The fine fraction content of sampled sand volcanoes was comparatively high, and it is possible that sand including a large fine fraction underwent liquefaction.
4. Past methods of assessing liquefaction, including prediction of subsidence, agreed to the actual state of damage, but it is necessary to understand the information about the ground in more detail and consider its validity. In addition, it is necessary to consider approaches to predict liquefaction involving houses where fundamental countermeasures are difficult, and to consider measures and methods of recovery.
5. In Sendai, damage to several buildings with pile foundations such as subsidence and leaning was observed, suggesting damage to their foundations. Ground subsidence of around 10 cm was observed around these buildings, including the effects of sand volcanoes and liquefaction.

6. In the Oritate and Aoyama districts of Sendai City, damage to houses was likely to have caused not simply by damage to retaining walls but also by landslide of the embanked ground under the houses. Future damage to housing land cannot be prevented by localized reinforcement of retaining walls, and measures against landslides on the scale of major public engineering work is required.

Acknowledgments Tokyo Institute of Technology Professor Saburoh Midorikawa and Associate Professor Jiro Takemura provided invaluable information. We used K-NET data of NIED. We greatly appreciate their support and cooperation.

References

1. Tokimatsu K, Tamura S, Suzuki H, Katsumata K (2011) Damage to soils in the 2011 Tohoku Pacific earthquake. Res Rep Earthquake Eng 118:21–47 (in Japanese)
2. Kaizuka S (1993) Geographical and geological and features and water of Tokyo Bay. Tsukiji-shokan, Tokyo (in Japanese)
3. Chibaken Kankyo Kenkyu Center (2011) Soil liquefaction in Chiba Prefecture in the Tohoku Pacific earthquake, http://www.wit.pref.chiba.lg.jp/_sui_chi/chishitu/touhoku/touhoku.html. Accessed 27 Oct 2011
4. Ministry of Land, Infrastructure, Transportation and Tourism (2011) Report on Great East Japan earthquake, http://www.mlit.go.jp/common/000139083.pdf. Accessed 27 Oct 2011
5. Koto Ward (2011) Damage in Koto Ward in the 2011 Tohoku Pacific earthquake, http://www.city.koto.lg.jp/seikatsu/bosai/58976/58570.html. Accessed 27 Oct 2011
6. Urayasu City: Project of reclaimed land, http://www.city.urayasu.chiba.jp/menu2863.html. Accessed 27 Oct 2011
7. Bureau of Port and Harbor, Tokyo Metropolitan Government (2001) Geographical and geological features of Tokyo Port (in Japanese)
8. National Research Institute for Earth Science and Disaster Prevention (2011) Kyoshin Network K-NET, http://www.k-net.bosai.go.jp/k-net/. Accessed 27 Oct 2011
9. Urayasu City: Geographical and geological features of Urayasu city, http://www.city.urayasu.chiba.jp/menu3185.html. Accessed 27 Oct 2011
10. Wakamatsu K (1991) Maps for historic liquefaction sites in Japan. Tokai University Press, Kanagawa (in Japanese)
11. Architectural Institute of Japan (2001) Recommendations for design of building foundations, Maruzen (in Japanese)
12. Kanto Regional Development Bureau, Ministry of Land, Infrastructure, Transportation and Tourism (2011) Investigation on soli liquefaction in Kanto region in the 2011 Tohoku Pacific earthquake. http://www.ktr.mlit.go.jp/bousai/bousai00000061.html. Accessed 27 Oct 2011 (in Japanese)
13. Architectural Institute of Japan (1980) Report of the 1978 Miyagiken-oki earthquake (in Japanese)
14. Japan Society of Civil Engineering (1980) Report of the 1978 Miyagiken-oki earthquake (in Japanese)
15. Fukkenn Gijutu Consultant Co., Ltd. (2008) Map of developed residential sites, http://www.fgc.jp/solution/technical/kouzou01/index.html. Accessed 27 Oct 2011 (in Japanese)

Chapter 11
Summary

Hisahiro Hiraishi

Abstract We have prepared a preliminary reconnaissance report pertaining to the magnitude 9.0 earthquake that occurred off Miyagi Prefecture on 11 March 2011 (the Tohoku-Chiho Taiheiyo-Oki earthquake). Presented herein is a summary of this report.

11.1 Outline of Earthquake

1. On March 11, 2011, one of the strongest earthquakes that strike Japan in recorded history occurred at a depth of 24 km off the coast of Miyagi Prefecture. This subduction-zone earthquake, which had a magnitude of 9.0, resulted from thrust faulting on the boundary between the Pacific plate and the North America plate (on which rests eastern Japan). This event is officially called the 2011 Tohoku-Chiho Taiheiyo-Oki Earthquake in Japanese or the 2011 Off the Pacific Coast of Tohoku earthquake in English.
2. Direct factors triggering this earthquake were presumably a magnitude 7.3 earthquake that occurred west of the 11 March earthquake's epicenter (i.e., also off the coast of Miyagi Prefecture) on 16 August 2005 and a magnitude 7.3 earthquake that occurred just two days beforehand (9 March 2011) off the Sanriku coast. The 11 March earthquake produced a substantial 25–30 m slip along the west side of the Japan Trench axis all the way from the sea off the Sanriku coast in the north down to the sea off Ibaraki Prefecture in the south.
3. One feature of the 11 March earthquake was that its seismic waves came from at least two clusters with the second cluster even stronger than the first. The maximum acceleration of the second cluster was recorded at approximately 100 s

H. Hiraishi (✉)
Meiji University, Tokyo, Japan
e-mail: hiraishi@isc.meiji.ac.jp

Architectural Institute of Japan (ed.), *Preliminary Reconnaissance Report of the 2011 Tohoku-Chiho Taiheiyo-Oki Earthquake*, Geotechnical, Geological and Earthquake Engineering 23, DOI 10.1007/978-4-431-54097-7_11, © Springer Japan 2012

after the starting time of motion. Only one of the two major clusters was widely felt in the Kanto region.

4. The maximum reading on the Japan Meteorological Agency's seismic intensity scale, a reading of 7, was recorded at only one location—the city of Kurihara, in Miyagi Prefecture—while 6-upper readings were recorded at 40 locations in four prefectures. Strong motion records having peak acceleration in excess of 1,000 Gal were recorded at 18 locations, although all had principal frequencies of 5 Hz or above. Pseudo-velocity response spectra also showed principal frequencies of around 1 and 5 Hz, and peak velocity exceeded 100 cm/s at only one measurement location, in Kurihara. A comparison with a strong motion record for the Great Hanshin (Kobe) Earthquake reveals that although the 11 March earthquake lasted for a very long time, relatively speaking, its ground motion did not have a dominant period in the intermediate range of around 1 s, which could lead to particularly heavy damage.
5. The 11 March earthquake presumably had some influence on two events that followed, specifically a magnitude 6.7 earthquake that occurred in northern Nagano Prefecture on 12 March and a magnitude 6.4 earthquake that occurred in eastern Shizuoka Prefecture on 15 March. Although readings of 6-upper on the JMA seismic intensity scale were recorded locally in both prefectures, the seismic intensity of these earthquakes was generally low in surrounding areas, and the principal frequencies of their shocks were approximately 5 Hz.

11.2 Topography and Geology

1. This series of earthquakes caused extensive damage over much of eastern Japan. Damage was particularly heavy in Iwate, Miyagi, and Fukushima Prefectures, all in the Tohoku region as well as Ibaraki and Chiba Prefectures, both in the Kanto region. The geological and geomorphological characteristics of these areas are diverse; included are basins, hills, alluvial plains, coastal areas, and even landfill.
2. Trending north–south along the western edge of the three Tohoku Prefectures (Iwate, Miyagi, and Fukushima) is the Ou Range, composed primarily of Cretaceous granitic and Neogene rock. In Iwate Prefecture, together with the Kitakami Mountains, which trend north–south along the eastern edge of the prefecture, the Ou Range forms a basin where cities such as Morioka, Hanamaki, and Kitakami are located. In Miyagi Prefecture, an alluvial plain (the Sendai Plain) extends throughout the center of the prefecture. On it are located Sendai, Natori, Ishinomaki, and numerous other cities. In Fukushima, the Ou Range, together with the Abukuma Mountains running roughly parallel to the east, forms the Naka-dori, or central valley, within which the cities of Fukushima, Koriyama, and Sukagawa, among others, are located. The coastline runs to the east of the Abukuma Mountains, forming the Hama-dori, or coastal region, on which sit the cities of Soma and Iwaki.

3. The northern part of Ibaraki Prefecture contains several mountainous regions, most notably the Abukuma and Yamizo mountains, while the southern part, comprising the Hitachi Upland and the downstream basin of the Tonegawa River, is notable for a considerable degree of urban development, including the cities of Mito and Tsukuba. The Hitachi Upland itself consists of numerous plateaus, the terraces of which tend to be covered with a thick layer of Kanto loam over relatively soft sand and gravel strata. Chiba Prefecture likewise can be generally classified into a northern part, specifically the Shimousa Upland and its peripheral lowlands, and a southern part where hills predominate. The Shimousa Upland itself comprises the Narita Group (marine strata formed during the Pleistocene epoch) and, on top of that, layers of Kanto loam. Numerous cities are located on the upland, including Noda, Funabashi, and Narita. The northwestern section of Chiba Prefecture runs along the northern side of Tokyo Bay, where the ground is naturally low and, in many cases, presently covered by landfill. Indeed, more than 70% of the city of Urayasu is built on landfill.

11.3 Outline of Tsunamis

1. Based on readings from GPS wave buoys placed in waters 100–300 m deep roughly 10–20 km off the Pacific coast, the first tsunami surge was recorded from the area off northern Iwate Prefecture (Kuji) down to the area off Fukushima Prefecture (Onahama) at 14:50, immediately after the occurrence of the earthquake (14:46). The largest surge was recorded from 15:12 to 15:19, approximately 30 min after event. The tsunami height was 2.6–6.7 m, with the maximum value recorded off southern Iwate Prefecture (Kamaishi).
2. Waveforms recorded off southern Iwate Prefecture (Kamaishi) reveal that the first surge was particularly high, with the height gradually decreasing in subsequent surges. Even the seventh surge, which occurred 6 hours after the first one, was still higher than the largest tsunami generated by the 2010 Chile earthquake.
3. The tsunami height was generally 8–9 m off the coast from Miyako (Iwate Prefecture) down to Soma (Fukushima Prefecture). The inundation height tended to progressively increase from Misawa toward the south, exceeding 10 m around Kuji and reaching roughly 10–15 m along the Sanriku coast from northern Iwate Prefecture to Oshika Peninsula, Miyagi Prefecture. Inundation was relatively low along Matsushima Bay at 5 m or less but nonetheless exceeded 10 m along the coast from Sendai Bay down to Soma. Extremely high tsunami run-up heights were recorded within those regions, most notably 40.5 m at Miyako, Iwate Prefecture.
4. As for the Kanto region, the first surge (height: 1.8 m) arrived at Oarai (Ibaraki Prefecture) at 15:15, with maximum tsunami height (4.2 m) recorded at 16:52. At Choshi (Chiba Prefecture), the first surge (0.4 m) arrived at 15:13, with maximum height (2.4 m) recorded at 17:22. Note that the largest surge reached Ibaraki and Chiba Prefectures approximately 2–2.5 hours after the earthquake. Tsunami height, at approximately 3–5 m off Ibaraki Prefecture, 3–4 m off northern Chiba

Prefecture (Kujukuri), and 3 m off southern Chiba Prefecture (outer Pacific coast of the Boso Peninsula), tended to decrease from Ibaraki down south to Chiba. However, high tsunamis were recorded locally within those two prefectures, including 7.2 m off Hirakata, Kita Ibaraki (Ibaraki Prefecture) and 7.6 m off Iioka, Asahi (Chiba Prefecture).

5. The tsunami surges produced by the main earthquake were the largest ever recorded in Japan. This event caused widespread destruction along not only the Sanriku coast, which has been frequently struck by earthquake-induced tsunamis in the past and also has a complicated geomorphology (including a sawtooth (ria) coastline) that tends to amplify the effects of a tsunami, but also along the southern coast, which heretofore has largely escaped extensive damage, from Sendai Bay down to Fukushima, northern Ibaraki Prefecture, and Kujukuri and the outer Boso Peninsula of Chiba Prefecture.

11.4 Damage Statistics

1. There are 22,801 people listed as dead or missing as of 30 June 2011. The toll is highest in Miyagi Prefecture at 13,803 people, followed by Iwate Prefecture at 6,942, Fukushima Prefecture at 1,754, Ibaraki Prefecture at 25, and Chiba Prefecture at 22. Fatalities were also reported in the Prefectures of Hokkaido, Aomori, Yamagata, Tochigi, Saitama, Tokyo, and Kanagawa. There were seven fatalities in Tokyo but less than five in each of the others. Nationwide, a total of 5,565 people are listed as injured.
2. The proportion of dead/missing relative to population for specific communities runs highest for Otsu-Cho at 11.3% and Onagawa-Cho at 10.3%, among communities in Iwate and Miyagi Prefectures, respectively. The tsunami event presumably accounted for the majority of fatalities. According to a survey conducted on 9 April by the Metropolitan Police Office and later reported by Asahi Shimbun, elderly people—those 65 years of age or older—accounted for the majority of fatalities at 55.4%, followed by those in the 40–64 age bracket at 27.9%, the 19–39 age bracket at 10.0%, the 7–18 age bracket at 3.9%, and the 0–6 age bracket at 2.8%.
3. As of 30 June, the number of buildings listed as totally destroyed is 105,940. Miyagi Prefecture accounts for the majority, at 65,492 buildings, followed by Iwate Prefecture at 20,998, Fukushima Prefecture at 15,897, Ibaraki Prefecture at 2,163, Chiba Prefecture at 771, Aomori Prefecture at 306, and all other prefectures at a total of less than 100. Nationwide, 107,855 buildings are listed as partially destroyed and 426,405 buildings are listed as damaged.
4. A total of 95,227 buildings were examined within a series of emergency risk assessments conducted by various prefectural governments soon after the 2011 Tohoku-Chiho Taiheiyo-Oki earthquake, wherein 11,557 were assessed as unsafe and 23,149 as requiring caution. Miyagi Prefecture had the most buildings judged unsafe, at 5,088, followed by Fukushima Prefecture at 3,314, Ibaraki Prefecture at 1,561, Chiba Prefecture at 677, Tochigi Prefecture at 676, Tokyo at 59, Gunma

Prefecture at 30, and Kanagawa Prefecture at 4. Similarly, a total of 2,318 buildings were examined after the 12 March earthquake in northern Nagano Prefecture, of which 297 buildings were judged unsafe in Nagano and 78 in Niigata. In addition, a total of 513 buildings were examined after the 15 March earthquake in eastern Shizuoka Prefecture on 15 March, whereupon 13 buildings in that prefecture were judged unsafe.

5. With lifeline systems cut by the earthquake and dwellings swept away by tsunamis, many people immediately fled to evacuation centers. There were approximately 470,000 evacuees nationwide as of 14 March, of which roughly 410,000 had gone to approximately 2,000 evacuation centers located within the three Tohoku Prefectures of Iwate, Miyagi, and Fukushima. The number of evacuees gradually diminished as lifeline systems were restored and temporary housing was built. Nonetheless, there remained as of 11 May, exactly 2 months after the earthquake, a total of 115,000 evacuees (evacuation center occupants) nationwide, of which 94,000 were within the three aforementioned Tohoku Prefectures.
6. As of 25 June, total damages were approximately ¥16.9 trillion, of which buildings and other structures account for roughly ¥10.4 trillion, lifeline systems for ¥1.3 trillion, societal infrastructure for ¥2.2 trillion, agricultural and fisheries related facilities for ¥1.9 trillion, and "other" for ¥1.1 trillion. Damage was especially heavy in the Tohoku region, breaking down by prefecture as roughly ¥6.4 trillion in Miyagi Prefecture, ¥4.3 trillion in Iwate Prefecture, and ¥3.1 trillion in Fukushima Prefecture (all three numbers are as of 13 May). Note that these figures, respectively, correspond to 11.9% of the total asset base of Miyagi Prefecture and 12.6% of that of Iwate Prefecture.

11.5 Damage to Wooden Houses

1. Substantial vibrational damage to buildings and structures, including collapse or other heavy damage to houses, occurred over a wide swath from Tohoku down to northern Kanto, although in only a fairly limited number of clusters.
2. Collapse and other heavy damage affecting wooden houses, which tend to have long natural period and little seismic protection, were extensive in areas having soft ground (along river beds, etc.). Also reported was damage to earthen wall storehouses or other buildings not only in areas where houses in general suffered severe damage, but also in areas where they suffered little damage.
3. Slight damage, typically to roof tiles (especially fallen ridge tiles) and exterior wall coverings (peeling, etc.) was observed over a wide area extending from Tohoku to Kanto.
4. Particularly conspicuous in the aftermath of this earthquake are (1) collapse or other severe damage to houses and other structures as a result of landslides or failure of retaining wall on slopes, especially on land developed for residential use and (2) tilting or sinking of entire structures as a result of the liquefaction of sandy soils. Furthermore, many of the areas that experienced significant vibra-

tional stress are characterized by soft soil, and in several cases it is not entirely clear whether damage to structures there should be attributed to vibration or to ground deformation.

5. As for tsunami damage, many wood buildings were swept away within areas struck particularly hard. There were cases, however, of wood buildings that, because they were shielded from direct tsunami impacts by relatively large surviving structures, remained in place. There were also cases where low-rise wooden dwellings presumably having excellent structural specifications remained in place despite the lack of a shielding building, although some did suffer severe damage to their walls or frames. The use of hold-down bolts appeared not to be an important factor in determining whether a wood building was swept away.

11.6 Damage to Reinforced Concrete Buildings

1. Although not great in number, some reinforced concrete buildings in need of rebuilding/repair because of collapse or other severe damage were observed. More numerous were buildings that despite suffering only moderate or slight damage were nonetheless unfit for further utilization. Many of the damaged buildings predate 1981 (current seismic standards were implemented in June 1981), and a good number of the buildings that suffered particularly severe damage were built before 1971 and had not been seismically reinforced afterwards (shear design modifications in construction code for reinforced concrete buildings were implemented in June 1971). The large majority of buildings built after the current earthquake standards took effect (along with buildings built before the standards took effect but subsequently reinforced) suffered only slight damage. Several cases were observed in which buildings that were undergoing retrofitting, but not fully reinforced, suffered considerable damage to their unreinforced portions.
2. Also observed were (1) cases of damage apparently attributable to amplification of ground motion by geological or geomorphological effects and (2) cases of foundation tilting or sinking apparently attributable to pile damage or soil collapse.
3. With regards to member damage, shear failure and axial collapse of long and short columns were observed. Such damage was apparently due to the following factors: an insufficiency of shear reinforcement bars, a concentration of seismic forces, high torsional forces within long spans, or the use of piloti columns. Also noted were damage to and collapse of concrete junction areas within buildings having a hybrid structure of reinforced concrete in some parts and steel or wood in others.
4. Many buildings were rendered unusable, despite a lack of serious damage to structural members, by damage to nonstructural elements (mullion walls, etc.) or cave-ins of ceiling material. This phenomenon was also apparent in relatively new buildings built after the new seismic-strength standards took effect. As for

tsunami-induced damage, little damage was observed to structural members in school buildings but considerable damage to nonstructural members and finishing. Many such buildings could be difficult to repair.

5. Steel-reinforced concrete buildings typically remained close to their original condition even in areas where the large majority of wooden structures had been swept away by the tsunami. However, in Onagawa and certain other areas some smaller buildings had tipped over upon being completely submerged. Also observed were buildings that, while remaining in place, nonetheless suffered wall damage caused by perpendicular strikes by wave surges or drifting objects.

11.7 Damage to Steel-Reinforced Concrete Buildings

1. Almost all damage to steel-reinforced concrete buildings was to nonstructural members, whereas little damage to structural members was observed. However, some buildings designed under the old seismic design codes suffered severe structural damage. Story collapse in the middle floors of steel-reinforced concrete buildings, which was one of the characteristics of the 1995 Hyogo-ken Nanbu earthquake, was not observed in structures damaged by the 2011 earthquakes.
2. Notable types of structural damage apparent in steel-reinforced concrete buildings within the city of Sendai include flexural cracking or shear cracking of columns, shear failure of column bases, flexural cracking or shear cracking of beams, flexural yielding of beams (crushing of concrete at the ends of beams), shear failure or bond-splitting failure of coupling beams (including beams with openings), cracking of reinforced concrete braces, shear cracking of shear walls or concrete crushing at its base, flexural failure of multi-story shear walls, severe damage to portions of elevator shafts, damage to pile foundations, and damage to rooftop structures.
3. As for nonstructural members, the following damage was observed: shear cracking and shear failure of nonstructural walls (spandrel walls, wing walls, and stud walls), horizontal cracking of concrete construction joints at stairwells, jammed (unopenable or unclosable) doors, tile peeling, expansion joint damage, damage or collapse of ceiling panels, damage or collapse of autoclaved lightweight concrete (ALC) panels, damage to connecting corridor concrete, damage to glass blocks, damage to hand railings, and damage to building peripheries due to ground subsidence.

11.8 Damage to Reinforced Concrete Boxed Wall Buildings and Masonry Structures

1. A total of 634 public apartment buildings having three structural types were examined within the city of Sendai. Of the 496 buildings having a reinforced concrete boxed wall (WRC) structure, 97.4% were found to be undamaged or to

have only slight damage. Among the remaining 2.6%, only one building had moderate damage and no buildings had severe damage (the others had minor damage). All of the 47 buildings having a precast prestressed reinforced concrete boxed wall (WPCaPS) structure in Sendai were found to have suffered no or slight damage. However, there was one WPCaPS building in Natori which suffered severe damage to its foundation structure. Of the 91 buildings having a boxed wall of thin ribbed panel structure, 72.5% were found to have suffered no or slight damage. Among the remaining 27.5%, 11 buildings were moderately damaged and 1 building was severely damaged. Common to all three structural types is that structural damage was apparently promoted by ground deformation in the vicinity of the building.

2. In the above investigation in Sendai, only one of the WRC buildings examined was within an area exposed to a seismic intensity of 7, and it suffered slight damage. In areas where coexisting WRC, WPCaPS, and ribbed panel buildings were together exposed to a seismic intensity of 6 upper, WRC and WPCaPS buildings suffered approximately the same degree of damage, whereas ribbed panel buildings suffered relatively higher damage. This is probably because 71 out of the 91 ribbed panel buildings were constructed before the new earthquake-resistance standards took effect in 1981.
3. WRC and WPCaPS buildings within coastal areas in Miyagi Prefecture, where tsunami damage was particularly severe, did not encounter any major structural damage, although they did suffer some damage to balconies and the like. In addition, there were ribbed panel buildings in which the panels themselves were damaged.
4. Along the Miyagi coast, some reinforced hollow concrete block masonry buildings were observed that had fallen over or collapsed as a result of the tsunami, or tilted as a result of soil washing out from under the foundation. Partial collapse of nonbearing concrete block masonry walls was also observed. In an investigation of areas in Sendai away from the coastline, no notable damage was observed to the followings: reinforced hollow concrete block masonry buildings, reinforced fully grouted concrete masonry buildings, or nonbearing concrete block masonry walls.
5. Masonry structures in northern Miyagi Prefecture and a part of Iwate Prefecture were also investigated. With regards to stone buildings, large cracks in stone walls, outward collapse of gables, tsunami-induced collapse, and other damage were observed. On the other hand, some seismically retrofitted stone buildings were found to be entirely undamaged.
6. An investigation of concrete block garden walls in Miyagi Prefecture revealed the following: collapse due to ground deformation, collapse of unreinforced portions due to illegal construction; tip-over and collapse due to insufficient anchoring of vertical bars into foundations or due to insufficient embedding of the foundation into the ground, tilting or collapse due to rebar corrosion, major cracking at joints along the horizontal top rebar, and tsunami-induced tip-over and collapse. An investigation in Natori City showed that 16% of the concrete block garden walls examined had tipped. Severe damage to non-reinforced stone garden walls was observed.

11.9 Damage to Steel Buildings

1. Steel buildings presumably built after the current seismic design code took effect suffered little structural damage due to ground motion, although some damage was observed in nonstructural elements (interior and exterior finishing, ceilings, etc.). Structural damage was apparent in buildings presumably built before the current code came into effect. Damage induced by ground motion to such buildings included damage to beam-to-column connections, buckling/joint deformation/ fracture of diagonal bracing, cracking of column base concrete, and plastic elongation/fracture of anchor bolts. This pattern of damage is similar to what has been reported from earlier earthquakes.
2. Among damage to nonstructural members are many reported cases of ceiling or exterior finishing damage or collapse within steel buildings having relatively long spans (gymnasiums, factories, etc.). This is particularly the case for dry-construction (prefabricated) ceiling frames that utilize light-gauge steel supports.
3. With regards to tsunami damage, some buildings whose column bases fractured were washed away with hardly a trace. Some buildings, while free of column base fracture, tilted or collapsed because of failed joints or members. Some buildings had their exterior finishing washed away, even though their structural skeletons remained in place. The type and extent of damage varied considerably depending on tsunami inundation height and other factors. In the areas struck by extremely large tsunami there were several buildings that we could see little damage to their primary structural members because their exterior/interior finishing were washed away. Yet, even in such cases, structural members were often damaged by the impact of debris carried by the tsunami. In areas less severely struck by the tsunami, we could see varying degrees of nonstructural damage in steel buildings, depending on the tsunami height. However, the majority of buildings suffered from limited structural damage.

11.10 Damage to Nonstructural Elements

1. As for wooden housing and similar structures, damage to roof tiles was observed over a wide geographical area, and damage to ridge tiles was almost always accompanied by damage to clay tile roofing. Damage to wet-applied outer walls was also observed in several cases.
2. The following damage was also noted: collapse of ceiling panels in large halls; collapse of ceiling panels, lighting fixtures, insulation, and other similar elements in gymnasiums; and damage to tie-in points between ductwork and ceiling panels within cafeteria kitchens. Such ceiling panel damage occurred even within buildings erected after technical guidance on that issue was issued and was even apparent in a relatively new airport terminal building, in which a portion of the ceiling collapsed. Many ceiling collapse and the like were also observed in

general buildings. Particularly common was the collapse of ceiling panels from ceiling frames consisting of lightweight steel supports (joists). Similarly, within relatively large, steel-framed commercial facilities having upper parking floors, wide-scale ceiling collapse was often observed in lower storefront floors. This type of damage was evident even in relatively distant locales such as Chiba Prefecture and, as with other types of nonstructural damage, tended to increase in severity with increasing proximity to the epicenter (a tendency apparent within Tochigi Prefecture, Ibaraki Prefecture, and the three Prefectures in Tohoku District). The primary forms of damage were slippage of ceiling panels from lightweight steel support frames, collapse of the ceiling panels themselves, and deformation/collapse of the support frames themselves.

3. Widely observed was damage to external tile walls of reinforced concrete buildings; lath sheets of steel buildings; ALC panels, glass screens, and window glass of steel buildings; and external walls, external finishing, and openings of other types of buildings. Other damage includes that to interior glass, hanging smoke barriers, extended eave walls, expansion joints, and miscellaneous works.

11.11 Damage to Soils and Foundations

1. Ground subsidence, presumably due to diastrophism (deformation of the Earth's crust, especially folding and faulting), was measured at 76 cm at Ofunato (Iwate Prefecture) and 56 cm at Kesennuma (Miyagi Prefecture). Subsidence was also quite notable around the present mouth of the Kitakami River in Ishinomaki (and also around the river's historical mouth). Parts of the central business and residential districts of that city are now at sea level (SL +0 m).
2. Extensive liquefaction was observed along the coast of Tokyo Bay and around the Tonegawa River floodplain. Liquefaction primarily occurred within relatively new landfill, with numerous sand boils and large (approximately 50 cm) ground subsidence, leading to settlement/tilting of wooden and reinforced concrete buildings supported on spread foundations, uplift of underground structures (pipelines, etc.), and collapse of roadways. Nonetheless, no structural damage was observed to the superstructures of pile-supported buildings. In addition, no cases of structural damage were apparent in the superstructures of buildings built on spread foundations (including mat foundations) having a high degree of rigidity, despite their settlement and/or tilting.
3. Within emergency risk assessments, a total of 886 houses were judged to be dangerous in Miyagi Prefecture, 269 in Fukushima Prefecture, and 98 in Iwate Prefecture. Of those 886 in Miyagi Prefecture, the large majority—794 houses—were in the city of Sendai. Many of the damaged houses within residential areas were damaged not by simple collapse of retaining walls, but rather by slope failure.

4. Several cases of settlement and tilting of pile-supported buildings were observed in the city of Sendai, implying damage to the pile foundations themselves. Ground subsidence (approximately 10 cm), sometimes the apparent result of sand boils and liquefaction, was evident around these buildings.
5. Within Onagawa and Rikuzen-Takata, several steel and reinforced concrete structures were noted that, having suffered damage to their pile foundations, were knocked over by tsunami surges. Much of the pile damage was to the (1) joint between the pile cap and the pile and (2) the area around the pile head. The buildings suffering such damage were old; apparently their pile foundations were not designed to withstand earthquakes. We infer that the pile cap joints or the piles themselves were fragile and, as a result, damaged by the earthquake to some extent, making them unable to withstand tsunami wave pressure and buoyancy forces.

11.12 Damage to Historical Structures

1. There were 578 reported incidents of damage to designated cultural properties as of 7 June 2011 (note that two or more buildings within the same site are counted as one property). These include 115 properties in Ibaraki Prefecture, 82 in Miyagi, 78 in Tochigi, 61 in Gunma, 45 in Fukushima, 45 in Tokyo, and 31 in Iwate. Damage was also reported within six important preservation districts for groups of traditional buildings.
2. Not least because there are so many historical structures, a good number of registered tangible cultural properties—290 structures, to be exact—were reported to be damaged. This corresponds to approximately 15% of all such structures within a broad swath running from Tokyo north through 11 prefectures up to and including Aomori.
3. No less than 130 structures listed as national treasures or important cultural properties and located within Tokyo or 14 prefectures were reported to have suffered some of the heaviest damage they had ever encountered under natural circumstances. However, few such structures were heavily damaged by ground movement. Furthermore, the only damage caused by tsunamis was an inundated floor within one designated important cultural property.
4. The destructive forces of the 2011 earthquake do not extend that far beyond what has been encountered in the historical past. The following damage to historical wood structures was observed: (1) misalignment with (slippage on) foundations, (2) cracking, peeling, and caving of earthen walls (particularly within earthen wall storehouses), pertaining to the framework between the foundation and roof, (3) partial leaning, separation between attached *geya* penthouses and main dwellings, and pertaining to roofs, and (4) misalignment and falling of pantile roof tiles. Among damage to masonry structures were cases of serious structural cracking within their frames.

5. Generally speaking, little damage was apparent to structures that had undergone repair in recent years, particularly those which had also been structurally reinforced. On the other hand, the earthquakes acted to abruptly accelerate damage in structures which had sagged or tilted through the years and were due for repair. In some cases, seismic forces acted to tilt the framework to such an extent that the structure now appears on the verge of collapse.

Appendix A
Outline of Earthquake Provisions in the Japanese Building Codes

Masaomi Teshigawara

Abstract The outline of seismic provisions in the building code of Japan is introduced. They feature a two-phase design for earthquakes. The first phase design is for medium earthquake motions, and this is basically working stress design. The second phase design is intended to give protection to buildings in case of severe ground shaking. It requires the checking of several aspects of the building. These include story drift, vertical stiffness distribution, horizontal eccentricity and ultimate lateral load carrying capacity. Both phases of the design are reviewed in detail. Some provisions are discussed in the light of recent earthquake damage in Japan. This paper is revised one written by Aoyama [1]. That is to say, this appendix is heavily quoted from the paper by Aoyama [1], and revised to include several amendments in the building code of Japan after 2000.

A.1 Introduction

The seismic design of Japanese buildings is featured by a two-phase design for earthquakes. The first phase design for earthquakes aims at the safety and reparability of buildings during medium earthquake motions. The second phase design for earthquakes is added to give safety against severe ground shaking.

The history of seismic design in Japanese building code started in 1924 when the Urban Building Law was revised, as a consequence of the disaster of great Kanto earthquake of 1923. This adopted a set of structural provisions including a seismic coefficient of 0.1. After the World War 2 the Building Standard Law replaced the Urban Building Law with much more elaborate provisions for various aspects of

M. Teshigawara (✉)
Nagoya University, Nagoya, Japan
e-mail: teshi@corot.nuac.nagoya-u.ac.jp

Architectural Institute of Japan (ed.), *Preliminary Reconnaissance Report of the 2011 Tohoku-Chiho Taiheiyo-Oki Earthquake*, Geotechnical, Geological and Earthquake Engineering 23, DOI 10.1007/978-4-431-54097-7, structural design. The standard value of seismic coefficient was raised to 0.2. The essential feature of seismic design was, however, unchanged as this increase in seismic loading was accompanied by comparable increase in the allowable stresses for various materials.

Both the Urban Buildings Law and the Building Standard Law specified only loadings and allowable stresses, and certain minimum requirements for the detailing of members. Details of structural design, such as methods of structural analysis and the proportioning of members, are specified in the Structural Standards issued by the Architectural Institute of Japan (AIJ), and "Commentary on the Structural Calculation based on the Revised Enforcement Order, Building Standard Law (in Japanese)", supervised by Housing Bureau and National Institute of Land Infrastructure Management in Ministry of Land, Infrastructure, Transport, and Tourism (MLIT), and Japan Conference of Building Administration, 2007. These Standards, prepared separately for each structural material, have served as the supplements to the Law. They have been revised more frequently to adapt new knowledge and to provide for new materials as they developed.

A particularly important event regarding seismic design was the 1968 Tokachi Oki earthquake which caused significant damage to modern buildings designed in accordance with building regulations. Various actions were undertaken as a consequence of this event. A partial revision of the Building Standard Law incorporating ultimate strength design in shear of reinforced concrete, the establishment of review procedure of existing buildings for seismic safety, were some of the changes.

The following year, another important event took place, the 1978 Miyagi-ken Oki earthquake. Damage was as severe as in the 1968 Tokachi Oki earthquake. It also demonstrated the more complicated characteristics of urban disaster in the city of Sendai with more than 600,000 populations.

In July, 1980 a revision of the Enforcement Order of the Building Standard Law was released. It was also announced that this order, together with supplementary documents, would be enforced from the first of June, 1981. The second phase design for earthquakes is added in this time to give safety against severe ground shaking.

After the 1995 Hyogo-ken Nanbu earthquake, another seismic design method, "Response and Limit Deformation", is introduced in 2000. This utilizes the linear response spectrum, in which earthquake motion is defined on the engineering bedrock whose shear wave velocity is not less than 400 m/s, and amplification of subsoil in construction site is considered.

In this chapter a simple review of the revised seismic design method in 1981 is attempted.

A.2 Types of Construction in Japan

Figure A.1 shows types of building construction and the commonly employed number of story for each type of construction. Traditionally, Japanese houses have been built in timber, and they are one or two story high. They are still very common.

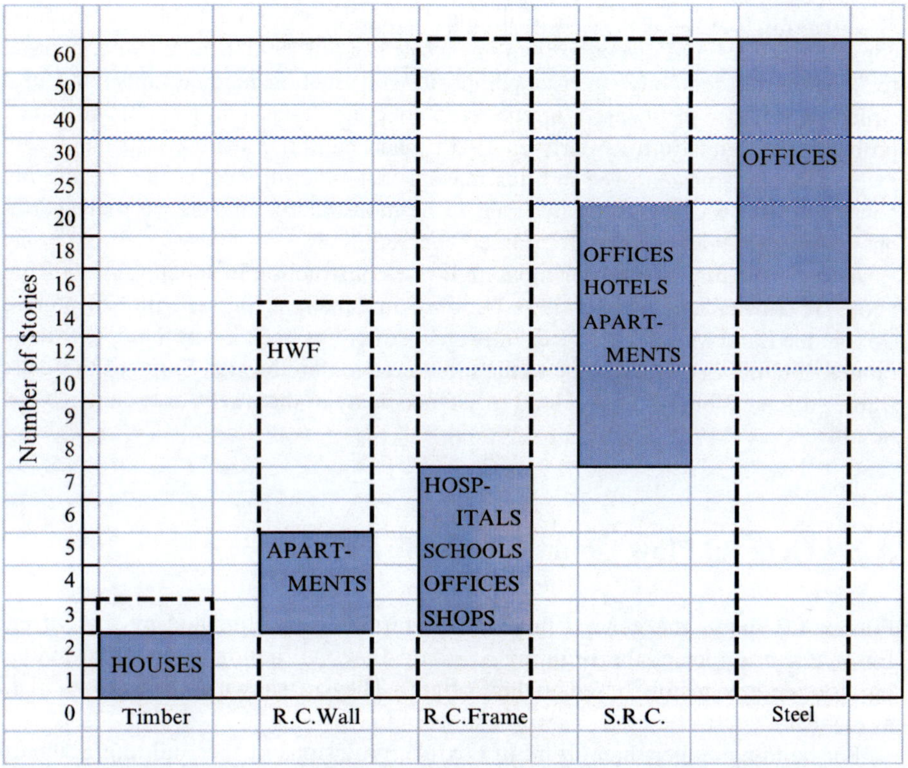

Fig. A.1 Types of construction in Japan

Recently, three story timber houses can be built. Masonry houses are scarce. Although provisions exist in Japan for the construction of reinforced concrete block masonry apartment houses up to three stories high, this construction is not shown in Fig. A.1.

The more common type of construction for apartment houses uses reinforced concrete "wall" structures. For these structures the building code allows a much simpler method of design than for ordinary reinforced concrete frame construction.

This wall construction is in the majority of low-rise apartment houses, ranging from three to five stories. High rise framed wall construction (HFW) which consists of wall-column and wall-beam in the longitudinal direction is also used for higher buildings up to 15 stories.

Ordinary reinforced concrete (RC) frame structures, with or without shear walls, represent the most common type of construction for various types of buildings, such as shops, offices, schools, and hospitals, ranging from three to seven stories. It may also be used for lower or higher buildings. At present the tallest RC frame structure is a 56 storied apartment building in Tokyo.

However, in ordinary cases Building Officials recommend the use of SRC construction for buildings taller than seven stories until 1990. SRC, an abbreviation for steel-reinforced-concrete, is a composite construction method consisting of a structural steel frame encased in reinforced concrete. It has a long history in Japan, probably evolving from as early as 1920, when buildings were constructed with exterior steel frames encased in brick masonry and interior steel frames encased in concrete. SRC was generally believed to be more ductile and hence more earthquake-resistant than ordinary reinforced concrete.

A steel structure is used for most high-rise construction in Japan ranging from about 15 stories up to the tallest building in Japan, at present the 70 storied Yokohama Land Mark Tower. Recently steel structures became more popular in all ranges of buildings, mainly due to the rapid erection on the site. Some of these are frame structures but for lateral load resistance many of them rely, at least partly, on bracing.

A.3 General Flow Chart of Seismic Design

Figure A.2 shows the general flow of structural design stipulated by the current Enforcement Order of the Building Standard Law. All buildings are first divided into four groups, mainly based on their heights. They are shown in the boxes marked as (1)–(4).

For buildings taller than 60 m, in box (4), provisions of the Building Standard Law do not apply directly. These high-rise buildings are to be designed by the "special study", usually incorporating time-history, non-linear response analyses. The design is then subjected to the technical review by the High-rise Building Structure Review Committee in examination organization entrusted the Minister of Land, Infrastructure, Transport, and Tourism, such as the Building Center of Japan. Upon its recommendation, a special approval of the structural design is issued by the MLIT.

For buildings, not exceeding 60 m in height, the basic intent of the general flow in Fig. A.2 is to make a two-phase design. This means that an additional design phase, hereafter called the second phase design for earthquakes, follows the working stress design, including the seismic design, hereafter called the first phase design for strong earthquakes which can occur several times during the life time of the building. The second phase design is intended mainly for severe or extraordinary earthquakes which could occur once in the life time of the building.

The application of the two-phase design is shown in Fig. A.2 for three different groups of buildings up to 60 m in height, in boxes (1)–(3).

For all of these buildings, working stress design is carried out first, box (5), including the first phase seismic design. As explained later, this is an allowable stress design for permanent and temporary loadings, also taking ultimate strength into account. Major changes of the code in 1981 relevant to this phase of the design is the method of seismic force evaluation.

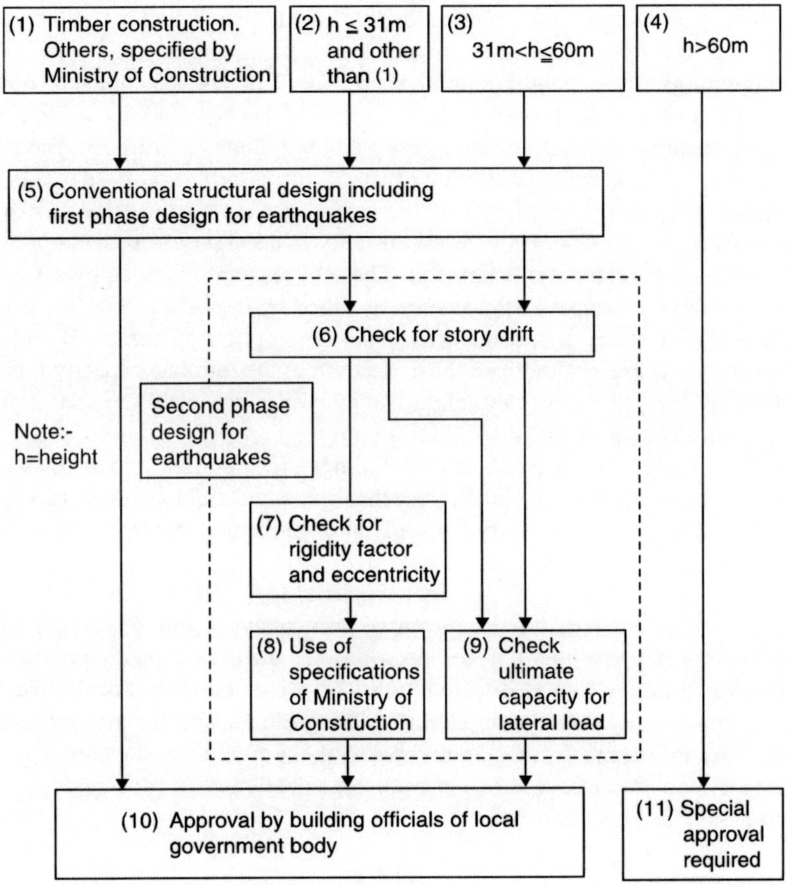

Fig. A.2 General flow chart of seismic design

Box (1) is for buildings of the most prevalent construction type in Japan. A detailed description of these buildings is given later. They include low-rise buildings of reinforced concrete with a generous amount of shear walls. For these buildings there is ample experience in Japan in seismic design and also evidence of seismic behavior. The first phase design, basically unchanged from the rules of the former building code, should be successful in providing sufficient seismic resistance for these buildings to withstand severe earthquakes. Hence the second phase design need not be applied to these buildings.

For buildings of boxes (2) and (3), the second phase design follows. The most important step in the second phase design is the evaluation of ultimate capacity for lateral load in box (9). However, other considerations listed in boxes (6)–(8), are also included in the second phase design.

The evaluation of story drift, box (6), is intended to eliminate soft structures which might experience excessively large lateral deflection under the seismic loading. In the case of ordinary reinforced concrete buildings this check will never be critical. However, the results of calculations are needed later for items given in box (7) or (9).

For buildings whose height is up to 31 m, box (2), there is a choice of flow into boxes (7) and (8), or into box (9).

Box (7) requires a check for the rigidity factor and the eccentricity. The rigidity factor refers to the vertical distribution of lateral stiffness. The purpose of this check is to eliminate buildings with one or more soft stories among the other stories, such as the soft first story. Checking for eccentricity is necessary to provide protection against excessive torsional deformation. These checks are followed by satisfying a set of additional minimum requirements specified by the MLIT, box (8), to ensure certain levels of strength and ductility. Thus, the option of boxes (7) and (8) is intended to allow exemption from the evaluation of the ultimate capacity for lateral load for buildings up to 31 m in height if they have a reasonably regular structural system conforming to the additional minimum requirements.

However, there is no such choice for buildings exceeding 31 m in height, box (3). These must be evaluated and checked for the ultimate lateral load carrying capacity. This route is the most straightforward application of the philosophy of the two-phase design.

The purpose of the check for ultimate capacity for lateral load is to evaluate the actual strength of the structure by means of limit analysis, and to ascertain that the response deformations would indeed lie within the ductility capacity provided.

The structural design and drawings must be presented to the local government body for approval by the corresponding building officials of the city, town, or prefecture, as conforming to the requirements of the Building Standard Law. This approval system became open to private organizations after the revision of the Building Standard Law in 2000.

A.4 First Phase Design for Earthquakes

A.4.1 Load Combinations and Design Method

The first phase design for earthquakes, as shown in box (5) of Fig. A.2, is a part of working stress design as prescribed by the Enforcement Order of the Building Standard Law, which considers in usual circumstances five kinds of loading.

These are: dead load, D; live load, L; snow load, S; wind forces, W; and seismic force, E. For permanent loading, the following load combinations are considered:

$$F = D + L \tag{A.1a}$$

$$F = D + L + S \text{ (in designated snowy areas)} \tag{A.1b}$$

For short time (temporary) loading, the following load combinations are considered:

$$F = D + L + S \tag{A.2}$$

$$F = D + L + W \tag{A.3a}$$

$$F = D + L + S + W \text{ (in designated snowy areas)} \tag{A.3b}$$

$$F = D + L + E \tag{A.4a}$$

$$F = D + L + + S + E \text{ (in designated snowy areas)} \tag{A.4b}$$

In the case of reinforced concrete buildings, Eqs. (A.2), (A.3a) or (A.3b) will never be critical because of the dominance of dead load, hence only Eqs. (A.1a) and (A.4a), or (A.1b) and (A.4b), are used.

The design is still based on the working stress design method. However, for short time loading, the allowable stress for steel is taken as its specified yield strength, and that for concrete is taken at two-thirds of its specified compressive strength. This means that in most cases where the tensile strength of the steel is a dominant factor in flexural calculation, the resulting section would be very similar to that obtained from the use of the ultimate strength design method. The design for shear is based on an empirical equation derived from the ultimate shear strength. The design shear force is taken as the smaller of the shear force associated with the flexural yielding of the member (in case of columns yielding of the column at one end and the yielding of beams at the other end of the column may be assumed) or the shear force calculated using factored earthquake load.

The load factor for this case should not be less than 1.5.

A.4.2 Evaluation of Seismic Force

The chief revision in 1981 of the first phase design for earthquakes was in the method of evaluating lateral seismic force, Q. This is to be calculated as the seismic shear at an i-th level of a building with the following equation:

$$Q_i = C_i \sum_{i=i}^{n} W_i \tag{A.5}$$

$$C_i = Z R_t \mathrm{A}_i C_0 \tag{A.6}$$

where

Q_i = seismic shear force at i-th story
W_i = weight of i-th story (This includes dead load plus reduced live load and, if located in the designated snowy zone, reduced snow load.)
n = number of stories
C_i = shear coefficient at i-th story
Z = seismic zone factor

R_t = vibration characteristics factor
A_i = vertical distribution factor
C_0 = standard shear coefficient

Before 1981, lateral seismic force at i-th story, P_i, was evaluated by the following equation:

$$P_i = 0.2W_iZ\ (\text{h is not higher than 16m})$$
$$= 0.2 + 0.01 \times \text{int}\left(\frac{(h-16)}{4}\right)(\text{h} > 16\text{m})$$

where W_i, Z are the same as above.

A.4.3 The Seismic Zone Factor

The seismic zone factor Z is shown in Fig. A.3. A similar seismic zoning has been in effect since 1952. The zoning in Fig. A.3 was published in 1978. It is based on the most recent assessment of seismicity over Japan at that time. As seen in this figure, the value of Z=0.7 is the smallest, and is applicable only to Okinawa Islands.

The large cities such as Tokyo, Osaka or Nagoya are within the zone A where Z=1.0.

A.4.4 The Vibration Characteristics Factor

The vibration characteristics factor R_t in Eq. (A.6) is a function of natural period, T and the type of subsoil. It is evaluated from the following equations:

$$R_t = 1 \text{ when } T \le T_c$$

$$R_t = 1.0 - 2.0\left(\frac{T}{T_c} - 1\right)^2 \text{ when } T_c < T \le 2T_c \tag{A.7}$$

$$R_t = 1.6\frac{T_c}{T} \text{ when } T > 2T_c \tag{A.8}$$

where T is the fundamental natural period in seconds, and T_c is the critical period in seconds, determined according to the type of subsoil.

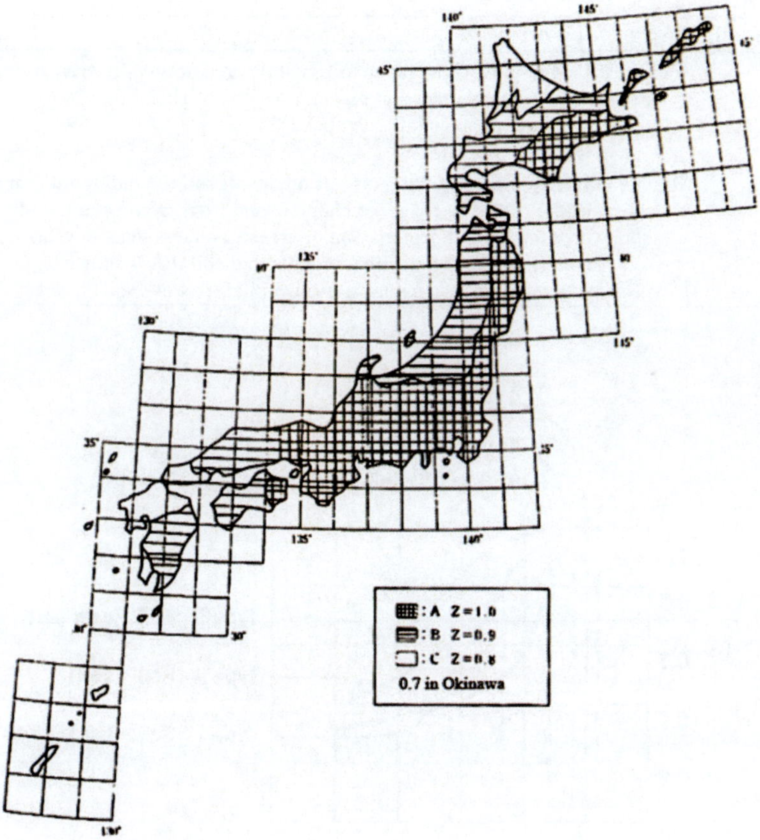

Fig. A.3 Seismic zone factor

The fundamental natural period is to be calculated by the following expression:

$$T = (0.02 + 0.01\alpha)h$$

where,

h = height of the building in meters
α = ratio of the height of story, consisting of steel columns and girders, to the entire height h

This means that T=0.03 h for steel structures and T=0.02 h for concrete structures.

The critical period T_c is given in Table A.1. The resulting value of R_t seen in Eqs. (A.7) and (A.8) is shown in Fig. A.4.

Table A.1 Values of *Tc*

Type of soil	Description of soil	T_c
Type I	Rock, stiff sand or gravel, and other soils mainly consisting of tertiary or older layers or any other soil that is shown by a special study to possess a natural period similar to above soils	0.4
Type II	Other than type I or type III	0.6
Type III	Alluvium mainly consisting of organic or other soft soil (including fill if any) whose depth is 30 m or greater, reclaimed land from swamps or muddy shoal where the ground depth is 3 m or greater and less than 30 years have passed since the reclamation, or any other soil that is shown by special study to possess a natural period similar to above soils	0.8

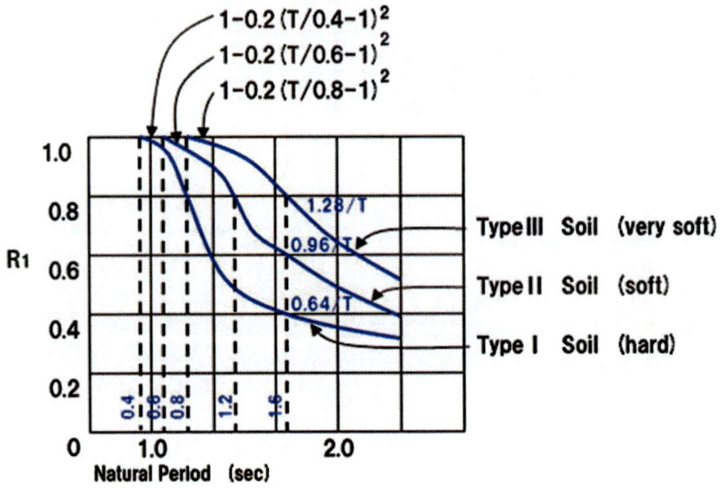

Fig. A.4 Vibration characteristics factor R_t

When the period of vibration is more accurately evaluated by a suitable method, or the vibration characteristics coefficient R_t is evaluated by a special study considering structural behavior during earthquakes, such as soil–structure interaction, the value of R_t may be taken less than the value given by Eqs. (A.7) and (A.8), yet limited to be 3/4 of the value given by these equations.

A.4.5 Story Shear Force Distribution Factor Along Building Height A_i

The vertical distribution factor A_i in Eq. (A.6) specifies the distribution of lateral seismic forces in the vertical direction, and is calculated by the following expression:

Fig. A.5 The vertical distribution factor A_i

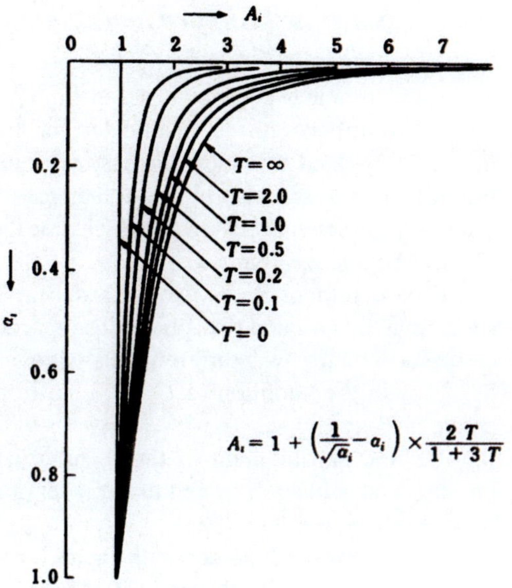

$$A_i = 1 + \left(\frac{1}{\sqrt{\alpha_i}} - \alpha_i \right) \frac{2T}{1+3T} \tag{A.9}$$

where,
α_i = non dimensional weight (or height)by the following expression:

$$\alpha_i = \frac{\sum_{i=i}^{n} W_i}{\sum_{i=1}^{n} W_i} \tag{A.10}$$

where,

W_i = weight of i-th story
n = number of stories

As seen in Fig. A.5, the vertical distribution factor is close to a uniform value for shorter periods. Larger lateral force is assigned to the upper part of a building with long period.

When the vertical distribution of seismic force is evaluated by a special study considering the dynamic characteristics such as spectral modal analysis, Eqs. (A.9) and (A.10) need not be applied.

A.4.6 Standard Shear Coefficient

The standard shear coefficient C_0 in Eq. (A.6) has been determined to be not less than 0.2 for the first phase seismic design. An exception is the case of wooden buildings in designated soft subsoil areas, when the value of C_0 must be not less than 0.3. Equation (A.6) will also be used later for ultimate capacity for lateral load in the second phase seismic design, in which case the value of the standard shear coefficient C_0 must be not less than 1.0.

In the commentary of the revised Building Standard Law [4], it is explicitly stated that the adopted two-phase design procedure for earthquakes can be regarded as the design for two different intensities of earthquake motion. The first phase design with the adoption of $C_0=0.2$, is from the experience of 1923 Kanto earthquake in Tokyo.

After the enforcement of the seismic design in 1924, practically all Japanese buildings have been designed to the level of protection corresponding to the design seismic coefficient $C_0=0.2$.

Experiences earthquakes with Japan Meteorological Agency (JMA) intensity 4 and 5 have shown that the majority of these buildings have behaved satisfactorily, and were almost without damage. The purpose of the first phase design is now to protect buildings in case of earthquakes which can occur several times during the life of the building. Such earthquake motions may be considered as having seismic intensity 5 on the JMA intensity scale, with the maximum acceleration of 80–100 cm/s^2. Buildings are expected to respond to earthquakes of this level without loss of function. This design objective is assumed to be achieved by the first phase design.

The second phase design is intended to ensure safety against an earthquake which could occur once in the life time of the building. Such earthquake motion may be as strong as that of 1923 Kanto earthquake in Tokyo, whose seismic intensity of 6 upper or even 7 in terms of JMA intensity scale with the maximum acceleration of 300–400 cm/s^2. The traditional seismic design, as prescribed by the previous Enforcement Order before 1981, did not include direct evaluation of safety against such earthquake motions.

It was expected that buildings designed under seismic coefficient of $C_0=0.2$ would safely survive severe earthquakes as a result of built-in over strength and ductility. Whether the structure possessed adequate levels of over strength and sufficient ductility was not expressly required to be confirmed.

Recent experiences in major earthquakes, such as 1968 Tokachi Oki earthquake or the 1978 Miyagi-ken Oki earthquake, have shown in fact that the majority of Japanese buildings had adequate over strength and ductility to survive without any damage or with minor damage. However, about 10 % of the affected buildings suffered appreciable damage and several buildings among these reached the stage of collapse.

The second phase design procedure is intended to ensure that serious damage will not occur, and this will be discussed later.

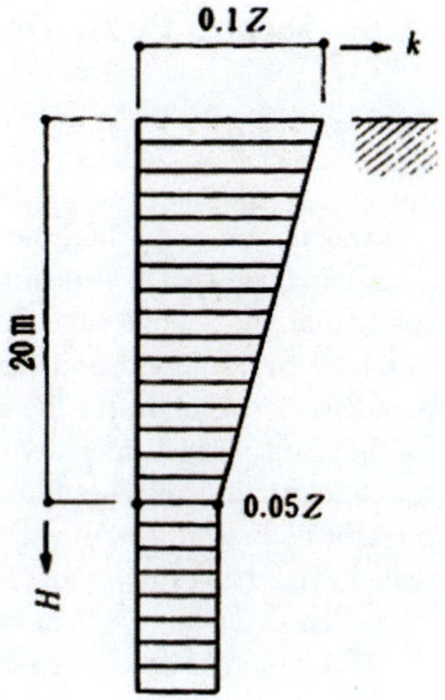

Fig. A.6 Seismic coefficient for basement

A.4.7 *Seismic Force Acting in the Basement*

The Building Standard Law also contains provisions for the seismic force to be considered at basements. Unlike the seismic shear force for the upper portion of buildings, the seismic force in the basement is calculated as the inertia force acting at the basement of the building directly, by multiplying the sum of the total dead and live loads for the basement, W_B, by the following seismic coefficient:

$$k \geq 0.1\left(1-\frac{H}{40}\right)Z \qquad \text{(A.11)}$$

where,

k=horizontal seismic coefficient
H=depth from the ground level in m measured to the bottom of the basement, but it shall not be taken more than 20 m
Z=seismic zone factor as in Eq. (A.6)

The story shear force Q_B at any basement level may then be calculated as follows:

$$Q_B = Q' + kW_B \qquad \text{(A.12)}$$

where

Q'=the portion of seismic story shear force in the adjacent upper story that is carried by columns and shear walls directly above the basement being considered
W_B=weight of the basement story considered

Figure A.6 shows the distribution of seismic coefficient by Eq. (A.11).

A.5 Second Phase Design for Earthquakes

A.5.1 Outline of the Second Phase Design

As was discussed earlier, the two phase design procedure in the Building Standard Law can be regarded as design for two different intensities of earthquake motion. In particular, the second phase design was introduced as the direct and explicit evaluation of overstrength and ductility. This identifies whether the available ductility capacity is sufficient for the ductility demand in case of severe earthquake motions of the class of JMA intensity scale of 6 upper or 7, considering also the over-strength of the structure provided. Such evaluation can be made only when the object is clearly defined. Structural configuration and dimension of members and in some cases even the details, must be given. In this sense the second phase design may be regarded as analysis, rather than design.

The principal objective of the first phase design may be said to create the object for this analytical procedure. Thus the essential part of the second phase design is a check for ultimate lateral load carrying capacity, i.e., box (9) in Fig. A.2. However, it has been pointed out that several other checks are necessary as indicated by recent earthquakes in Japan. They are thus included in the second phase design as shown in boxes (6)–(8) in Fig. A.2.

A.5.2 Story Drift

The story drift limitation is to be investigated first, as shown by the box (6) in Fig. A.2. The story drift δ_i under the action of design seismic shear force of Eq. (A.5) is calculated by the elastic analysis, and the story deformation angle R_i is calculated as:

$$R_i = \frac{\delta_i}{h_i} \tag{A.13}$$

where h_i is the story height. The story deformation angle R_i should not be greater than 1/200.

If nonstructural elements are used that can sustain greater structural deformation, or if they are made of deformable materials, R_i can exceed the above limit. However, in no case should R_i exceed 1/120. This story drift limitation was introduced in view of increasing earthquake damage to architectural parts of buildings, particularly of steel buildings.

It should be noted that the calculated story drift results from the action of seismic shear force given by Eq. (A.5), i.e. for the first phase design for earthquakes. Under the action of severe earthquakes assumed in the second phase design, the story drift will become much larger than that predicted by Eq. (A.13).

For reinforced concrete buildings the story drift limitation will be seldom critical because of the large initial stiffness of the structure. Nevertheless this check cannot be omitted as the results are used in the subsequent steps of second phase design.

A.5.3 Rigidity Factor and Eccentricity Factor

One of the most effective ways in the design of earthquake-resistant buildings is to provide a well-balanced, regular structural layout for the building in the early stage of the design. Buildings with non-uniform rigidity along the building height, for example having a soft first story, or buildings with horizontal eccentricities, for example having an eccentric service core, should be designed with particular care. They are certainly more difficult to design than regularly shaped buildings. As a result of these requirements the structure is likely to become more costly and yet the benefits in terms of behavior remain doubtful. Whenever possible, such irregular buildings should be avoided.

Damage observed after the 1968 Tokachi Oki earthquake or the 1978 Miyagi-ken Oki earthquake revealed in particular the vulnerability of irregular buildings and indicated the superior performance of regular buildings.

Consequently, it was intended to provide some relaxation in the design of regular buildings in the flow chart of Fig. A.2. For buildings whose height does not exceed 31 m, [box (2)] it was made possible to bypass box (9) by ensuring the requirements of boxes (7) and (8).

The rigidity factor is defined in Fig. A.7. The reciprocal of the story deformation angle R_i is expressed by r_{si}, and then the rigidity factor of the i-th story R_{si} is defined as the ratio of r_{si} and the average value, i.e.,

$$R_{si} = \frac{r_{si}}{r_{sa}} \tag{A.14}$$

where

$$r_{si} = \frac{1}{R_i}$$

and

$$r_{sa} = \frac{\sum_{i=1}^{n} r_{si}}{n}$$

R_{si} is less than 1.0 in a story whose rigidity is less than the average rigidity of stories. The Building Standard Law now requires to meet Eq. (A.15) for any stories.

$$R_{si} \geq 0.6 \tag{A.15}$$

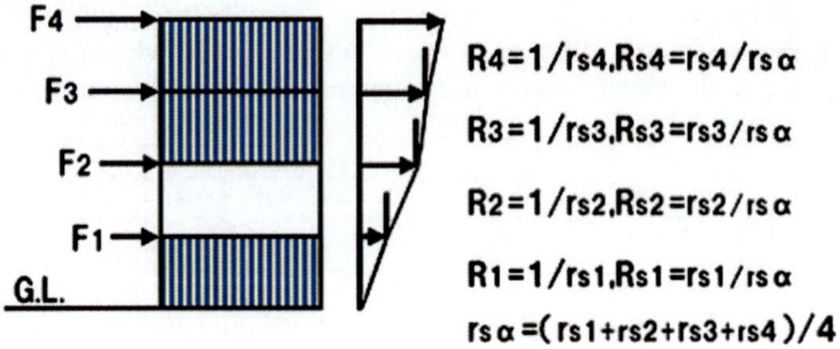

Fig. A.7 Rigidity factor

If there is a story which does not satisfy Eq. (A.15), the structure has to be checked for ultimate lateral load carrying capacity, i.e. it must follow the flow into box (9) in Fig. A.2.

The definition of eccentricity factor is as shown in Fig. A.8. G is the centre of gravity of the total mass above the story considered. R is the centre of rigidity, or centre of rotation under the action of torsional moment. Eccentricity distances associated with loadings x and y-direction, e_x and e_y, respectively, are measured from R as shown in this figure. Then the eccentricity factors R_{ex}, R_{ey} are defined as follows:

$$R_{ex} = \frac{e_x}{r_{ex}} \tag{A.16a}$$

$$R_{ey} = \frac{e_y}{r_{ey}} \tag{A.16b}$$

where

r_{ex}, r_{ey} = elastic radii defined as follows:

$$r_{ex} = \frac{\sqrt{rotational\ stiffness}}{translational\ stiffness\ in\ x} \tag{A.17a}$$

$$r_{ey} = \frac{\sqrt{rotational\ stiffness}}{translational\ stiffness\ in\ y} \tag{A.17b}$$

The terms of rotational and translational stiffness will be defined subsequently.

The Building Standard Law specifies

$$R_{ex} \leq 0.15 \quad \text{and} \quad R_{ey} \leq 0.15$$

in all stories. If there is any story which does not satisfy Eqs. (A.17a) and (A.17b), the building has to be checked for the ultimate lateral load carrying capacity, i.e., should go to box (9) in Fig. A.2.

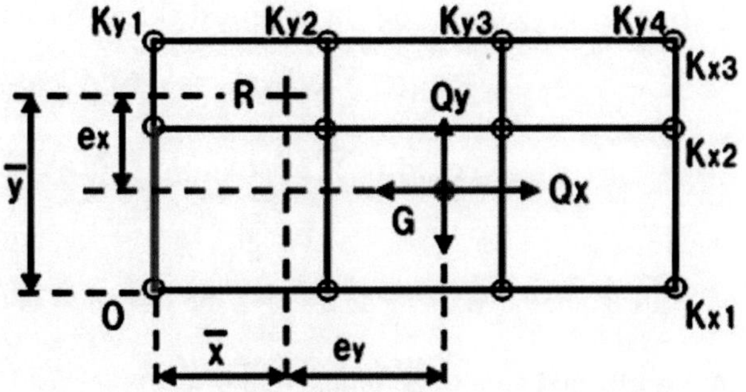

Fig. A.8 Principal quantities to determine story stiffness properties

A.5.4 Specified Minimum Requirements

A building which passed the checks for the rigidity and eccentricity factors is considered to be a regular building. It is likely to be seismically safe. Hence they can be exempted of the time consuming check for ultimate lateral load carrying capacity as long as they can also satisfy a set of minimum requirements specified by the MLIT.

For steel structures not exceeding 31 m in height, the following conditions must be satisfied in the structural calculations.

1. If the building includes stories (except the basement) that are braced to carry horizontal forces, the force in the members in each of those stories due to the design seismic force must be increased by the factor β as follows:

$$\text{When } \beta \le \frac{5}{7} \quad \beta = 1 + 0.7\beta$$

$$\beta > \frac{5}{7} \quad \beta = 1.5$$

   where, β is the ratio of horizontal force carried by the braces to the entire story shear in the story being considered.
2. It must be shown, when necessary, that columns, girders, and their connections, do not lose their load carrying capacity rapidly due to local buckling or fracture.

Thus for steel structures the minimum requirements result in assurance of increased strength and ductile behavior of braces, and ductile behavior of frames when the frames carry some portion of lateral loads.

For reinforced concrete and steel reinforced concrete structures not exceeding 31 m in height, the minimum requirements consist of satisfying any one of the following three conditions.

1. In each story the following empirical equations must be satisfied:

$$\text{For RC} \sum 2.5\alpha A_w + \sum 0.7\alpha A_c \geq 0.75ZWA_i \quad \text{(A.18a)}$$

$$\text{For SRC} \sum 2.5\alpha A_w + \sum 1.0\alpha A_c \geq 0.75ZWA_i \quad \text{(A.18b)}$$

where

A_w = horizontal area of shear walls in the direction of the seismic forces considered in mm^2
A_c = horizontal area of columns in mm^2
Z = seismic zone factor given in Eq. (A.6)
W = weight of the portion of the building that is carried by the story considered in N
A_i = vertical distribution factor given in Eq. (A.6)

α = modification factor of concrete strength, $\sqrt{\frac{F_c}{20}}$

This expression is similar to the one used in specifying buildings for which second phase design for earthquakes is not required [box (1) in Fig. A.2]. This aspect will be discussed later in detail, but in essence Eqs. (A.18a) and (A.18b) specifies the condition for buildings with relatively large amount of shear wall area and hence with a relatively high ultimate lateral load carrying capacity.

2. In each story the following empirical equations must be satisfied:

$$\text{For RC} \sum 1.8\alpha A_w + \sum 1.8\alpha A_c \geq ZWA_i \quad \text{(A.19a)}$$

$$\text{For SRC} \sum 2.0\alpha A_w + \sum 20.\alpha A_c \geq ZWA_i \quad \text{(A.19b)}$$

where A_w, A_c, Z, W, A_i, α are as defined for Eqs. (A.18a) and (A.18b).

This expression will also be reexamined in connection with the conditions for which second phase design for earthquakes is not required. Equations (A.19a) and (A.19b) intend to distinguish buildings with many columns with wing walls and hence with relatively high ultimate lateral load carrying capacity and ductility. Columns with wing walls are extensively used in Japan. The wing walls significantly contribute to the strength and stiffness for earthquake actions.

3. All columns and girders must be designed so that premature shear failure is prevented. This is to ensure energy dissipating capacity for the frames.

In this case no requirements are made for the amount and behavior of shear walls.

A.5.5 Check for Ultimate Lateral Load Carrying Capacity

The most important step in the second phase design for earthquakes is to check the ultimate lateral load carrying capacity, i.e. box (9) in Fig. A.2.

First, member strengths are evaluated, based on the material strength and geometry of sections obtained from the results of the first phase design. Then the ultimate lateral load carrying capacity is calculated by any method, including incremental nonlinear analysis, limit analysis, and simplified methods suitable for approximate hand calculation. In any case, an appropriate vertical distribution of horizontal forces, A_i is assumed, and the each story shear, associated with the formation of collapse mechanism, is found.

It is required that the ultimate lateral load carrying capacity in each story thus found must exceed the required shear force Q_{un} given below:

$$Q_{un} = D_s \cdot F_{es} \cdot Q_{ud} \tag{A.20}$$

where,

Q_{ud} = standard seismic shear in a story
D_s = structural characteristics factor
F_{es} = a shape factor which considers the rigidity and the eccentricity factors

The standard seismic shear in the story, Q_{ud} is calculated using Eqs. (A.5) and (A.6). Thus this procedure is similar to the evaluation of seismic force in the first phase design for earthquakes, except that the standard shear coefficient C_0 in Eq. (A.6) is now taken to be not less than 1.0. Therefore, Q_{ud} is five times the design story shear obtained in the first phase design, unless the dynamic characteristics of the building, such as the natural period, vibration characteristics of the factor R_t, or the vertical distribution factor A_i, are recomputed.

The factors D_s and F_{es} are examined in the following two sections, respectively.

A.5.6 Structural Characteristics Factor

The structural characteristics factor D_s takes into account inelastic deformations and energy dissipation.

The second phase design is intended to ensure the safety against severe earthquakes of the order of 6 upper or 7 in terms of the JMA intensity scale. Although the building is expected to survive without collapse, some cracking or yielding is expected. Structures will have variable degrees of energy dissipating capacity, or ductility. The structural characteristics factor D_s is used to reduce the elastic response story shear according to the available ductility, as given in Tables A.2 and A.3 for steel and reinforced concrete structures, respectively.

Table A.2 D_s values for steel structure

Framing members	Structural type (a) Rigid frames or braced frames $\beta w \le 0.3$, or braced frames with very short braces	(b) Braced frames with very long or short braces	(c) Braced frames with long braces
(i) Most ductile	0.25	0.30	0.35
(ii) Very ductile	0.30	0.35	0.40
(iii) Ductile	0.35	0.40	0.45
(iv) Others	0.40	0.45	0.50

Note:
(a) β=load ratio carried by braces as defined in Sect. A.5.4
Very short braces correspond to $\lambda \le 500 \le \sqrt{fy}$, where λ is effective slenderness ratio of the bracing member considering the appropriate end conditions and fy is the yield strength in M_{pa}
(b and c) Short braces: $500/\sqrt{fy} < \lambda \ge 900/\sqrt{fy}$
Long braces: $900/\sqrt{fy} < \lambda \ge 2000/\sqrt{fy}$
Very long braces: $\lambda \ge 2000/\sqrt{fy}$
(i–iv) These depend on restrictions on width to thickness ratio, strength of joints and span length of beams. For details see [3]

Table A.3 D_s values for reinforced concrete structures (for steel reinforced concrete structures, subtract 0.05 from the tabulated value of D_s)

Framing members	Structural type (a) Rigid frames or v ery ductile shear wall with $\beta w \le 0.5$	(b) Very ductile or ductile shear wall with $\beta w \le 0.7$	(c) Very ductile or ductile shear wall with $\beta w \le 0.7$, or less ductile shear wall
(i) Most ductile	0.30	0.35	0.40
(ii) Very ductile	0.35	0.40	0.45
(iii) Ductile	0.40	0.45	0.50
(iv) Others	0.45	0.50	0.55

Note:
(a–c) β_w=ratio of load carried by shear walls to total story shear. The classification, very ductile, ductile, or less ductile shear walls depends mainly on the shear stress level at ultimate and on the mode of failure. For details see [3]
(i–iv) These depend on restrictions on length-to-depth ratio, axial force, axial reinforcement ratio, shear stress level at ultimate state, and the mode of failure. For details see [3]

A.5.7 Shape Factor

The shape factor F_{es} is intended to take into account their regularity of the structure expressed in terms of rigidity factor and eccentricity determined as follows:

$$F_{es} = F_s \cdot F_e \tag{A.21}$$

where

Fs=basic shape factor determined as a function of the rigidity factor Rs [Rs is given by Eq. (A.14)]

$$F_s = 1.0 \qquad \text{when } R_s \geq 0.6 \tag{A.22a}$$

$$F_s = 2.0 - \frac{R_s}{0.6} \qquad \text{when } R_s < 0.6 \tag{A.22b}$$

and

F_e = basic shape factor determined as a function of the eccentricity factor R_e [R_{ex} or R_{ey} are given by Eqs. (A.16a) and (A.16b)]

$$F_e = 1.0 \qquad \text{when } R_e \leq 0.15 \tag{A.23a}$$

$$F_e = 1 + \frac{0.5}{0.15}(R_e - 0.15) \quad \text{when } 0.15 < R_e < 0.3 \tag{A.23b}$$

$$F_e = 1.5 \qquad \text{when } R_e \geq 0.3 \tag{A.23c}$$

The shape factor F_{es} is 1.0 for values of rigidity factor and eccentricity factor permitted in case of buildings up to 31 m in height for which there is an exemption for checking the ultimate lateral load carrying capacity. When irregularity exceeds these limits, the shape factor necessitates the provision of higher ultimate lateral load carrying capacity, or in cases that the ultimate lateral load carrying capacity cannot be increased, lower structural characteristics factor D_s must be used.

Equations (A.22a), (A.22b), (A.23a) and (A.23b) show that the value of the shape factor is limited to 2.0 and 1.5, respectively. However, this should be interpreted that such an extraordinarily irregular structure cannot be handled by the shape factor concept. Any building with a story having a rigidity factor R_s less than 0.6, or eccentricity factor R_e greater than 0.3, should be redesigned for a better balance of mass and stiffness.

A.6 Buildings for Which the Second Phase Design Is Not Required

A.6.1 Specification of the MLIT

Box (l) in Fig. A.2 refers to buildings most popular in Japan. With ample experience in earthquakes, they are judged to be safe without being subjected to the second phase design for earthquakes.

Besides timber construction, they include the following, according to the specification of the MLIT [4, 5].

1. Masonry buildings with not more than three stories, excluding the basement.
2. Reinforced concrete block masonry buildings with not more than three stories, excluding the basement.

3. Steel buildings conforming to (a)–(f), or (g)–(l) below:

   (a) Not more than three stories, excluding the basement.
   (b) Not more than 13 m in height, and not more than 9 m at eaves height.
   (c) Horizontal distance between major vertical structural supports is not more than 6 m.
   (d) Total floor area is not more than 500 m^2.
   (e) Design seismic shear force in the first phase design is calculated with the standard shear coefficient taken not less than 0.3.
   (f) End connections and joints of braces carrying components of horizontal earthquake forces must not fracture when the bracing member yields.
   (g) Not more than two stories, excluding the basement.
   (h) Not more than 13 m in height, and not more than 9 m at eaves height.
   (i) Horizontal distance between major vertical structural supports is not more than 12 m.
   (j) Total floor area is not more than 500, or 3000 m^2 in case of flat.
   (k) Design seismic shear force in the first phase design is calculated with the standard shear coefficient taken not less than 0.3.
   (l) End connections and joints of braces carrying components of horizontal earthquake forces must not fracture when the bracing member yields.

4. Reinforced concrete buildings, steel reinforced concrete buildings, or buildings consisting in part of reinforced concrete and in part of steel reinforced concrete, conforming to the requirements (a) and (b) below:

   (a) Not more than 20 m in height.
   (b) Horizontal area of shear walls and columns in each story above ground shall satisfy the following equations.

$$\text{For RC} \sum 2.5\alpha A_w + \sum 0.7\alpha A_c \geq Z \cdot W \cdot Ai \tag{A.24a}$$

$$\text{For RC} \sum 2.5\alpha A_w + \sum 1.0\alpha A_c \geq Z \cdot W \cdot Ai \tag{A.24b}$$

   where, A_w, A_c, Z, W, A_i, α are as defined for Eqs. (A.18a) and (A.18b).

5. Buildings consisting of the mixture of two or more of the following constructions: timber, masonry, reinforced concrete block masonry and steel structures, or buildings consisting of any one or more of these above construction types and reinforced concrete or steel reinforced concrete construction, conforming to requirements (a)–(d) below:

   (a) Not more than three stories, excluding the basement.
   (b) Not more than 13 m in height, and not more than 9 m at eaves height.
   (c) Total floor area is not more than 500 m^2.
   (d) The story consisting of steel construction must conform to requirements (c), (e) and (f) of item 3 above.
   (e) The story consisting of reinforced concrete or steel reinforced concrete construction must conform to requirement (b) of item 4 above.

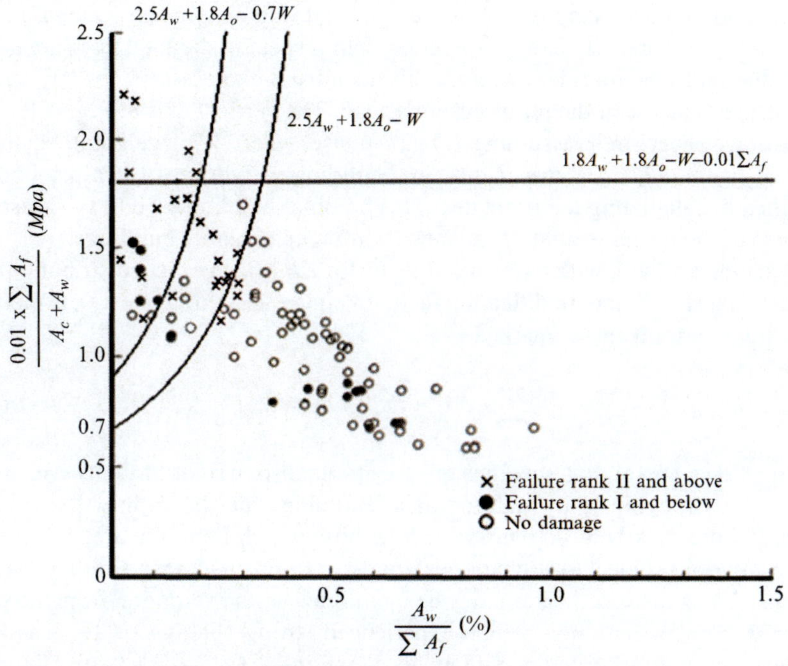

Fig. A.9 Wall area index and average shear stress in walls and columns [6]

6. Industrial (prefabricated) houses approved by the MLIT.
7. Other construction approved by the MLIT as having equivalent or higher safety against earthquakes as the items listed above.

A.6.2 Reinforced Concrete Buildings

Among the structural types listed in Sect. A.6.1, Item 4, defining reinforced concrete buildings that need not be subjected to the second phase design for earthquakes, is the most interesting and important from a technical point of view. The right hand side of Eqs. (A.24a) and (A.24b) may be regarded as the story shear due to 1.0 g response, or standard shear coefficient $C_o = 1.0$ and vibration characteristics factor $R_t = 1.0$. This would be the case for buildings whose height is not more than 20 m. Hence Eqs. (A.24a) and (A.24b) may be interpreted as a requirement that the lateral load carrying capacity, based on an average shear stress of 2.5 MPa for shear walls and 0.7 MPa for columns (1.0 MPa in case of SRC), should exceed the 1.0 g response story shear.

The meaning of Eqs. (A.24a) and (A.24b) may be explained with the aid of a "wall area index" vs "average shear stress" plot originated by Shiga [6]. Figure A.9 shows such a plot, also compiled by Shiga, with respect to damage observed in the 1978 Miyagi-ken Oki earthquake [2].

The horizontal axis in Fig. A.9 is the so-called wall area index, which is the ratio of the total wall area, A_w in each direction with respect to the total floor area, ΣA_f of the building. In the original literature this quantity is expressed in cm²/m² units, but here the percentage of the ratio is used instead. The vertical axis of Fig. A.9 records the average shear stress assuming 1.0 g response, where W is weight in N, and ΣA_w and ΣA_c were defined above. A nominal building weight of 10 kPa per floor was assumed in calculating the ordinates. This nominal building weight is a reasonable average of the actual weight of Japanese reinforced concrete buildings.

Equation (A.24a), with a seismic zone factor $Z = 1.0$, a vertical distribution factor $A_i = 1.0$ (first story), and modification factor of concrete strength $\alpha = 1.0$, is plotted on Fig. A.9 by a solid curve, marked

$$\sum 2.5A_w + \sum 0.7A_c = W \tag{A.25}$$

Equation (A.24a) thus specifies the area to the lower-right of the curve, and this corresponds roughly to the safe domain. Buildings that lie in this domain suffer slight (failure rank I) or no damage, while those which lie in the area of upper-left of the curve are subject to slight or no damage or to severe damage or they may even collapse. Thus the specification may be interpreted as an exemption from the second phase design for buildings which experienced strong shaking of 1968 and 1978 earthquakes, with recorded maximum accelerations of 0.25–0.3 g, with minor or no damage.

A.6.3 Specified Minimum Requirements

It was explained earlier that there was a set of minimum requirements specified by the MLIT, which must be met by any building that follows the flow along boxes (6)–(8) in Fig. A.2, instead of the flow along boxes (6) and (9). These are the minimum requirements for buildings to be designed without check for ultimate lateral load carrying capacity in the second phase design for earthquakes. For reinforced concrete buildings the rules consist of satisfying one of three conditions. The first two of these are essentially strength requirements and the last one is a ductility requirement.

The first of these three rules is to satisfy Eqs. (A.18a) and (A.18b). Comparing this with Eqs. (A.24a) and (A.24b), it can be seen that they are identical except for the right hand side of Eqs. (A.18a) and (A.18b), which is factored by 0.75. It is anticipated that some ductility is available in these buildings as indicated by this reduction of strength.

For the condition of $Z = 1.0$, $A_i = 1.0$, and $\alpha = 1.0$, Eq. (A.18a) for reinforced concrete is shown in Fig. A.9 by a dashed curve

$$\sum 2.5A_w + \sum 0.7A_c = 0.75W \tag{A.26}$$

The minimum requirement of Eq. (A.18a) implies that buildings which lie between the two curves in Fig. A.9 will be designed without check for ultimate lateral load carrying capacity. However, these structures must be checked for regularity and a minimum level of ductility. As seen there were several damaged buildings in this domain in the case of the 1978 earthquake. It is expected that response behavior will be improved by the regularity requirements expressed by the rigidity factor and the eccentricity factor limitation. In addition, these buildings should be designed to possess some ductility.

The second of the three conditions is to satisfy Eqs. (A.19a) and (A.19b). The meaning of this equation may be explained as follows [3].

Reinforced concrete buildings with reasonable regularity and with several shear walls with large openings or with columns with wing walls should be safe, provided that a certain level of ductility is maintained. Assuming that an average shear stress of 1.35 MPa is expected in columns and wing walls at the development of certain ductility, we may write

$$\sum 1.35A_w + \sum 1.35A_c \geq 0.75Z \cdot W \cdot A_i \tag{A.27}$$

from which the first expression in Eq. (A.19a) is derived. Equation (A.19b) for steel reinforced concrete is a modified form based on the engineering judgment.

For the condition of $Z=1.0$, $A_i=1.0$, and $\alpha=1.0$, Eq. (A.19a) for reinforced concrete is also shown in Fig. A.9 by a horizontal dashed line:

$$\sum 1.8A_w + \sum 1.8A_c = W \tag{A.28}$$

In the domain below this line, an even greater number of damaged buildings can be identified. Besides the improvement by the adoption of regular structural layout, care must be taken to ensure the availability of some ductility in the first phase design for earthquakes.

Acknowledgements The author thanks Emeritus Professor of Tokyo Univ. H. Aoyama with great appreciation, who allows for author to add several amendments in the building code of Japan after 2000 on his paper [1].

References

1. Aoyama H (1981) Outline of Earthquake Provisions in the recently revised Japanese Building Codes. Bull New Zeal Natl Soc Earthquake Eng 14(2):63–80
2. Nakano K, Ishiyama Y, Ohashi Y (1980) A proposal of a new aseismic design method for buildings in Japan. In: 7th WCEE Proceedings, vol 4, Istanbul, September 1980, pp 41–48
3. Umemura H (ed) (1979) New earthquake resistant design. Building Center of Japan, 344 pp, May 1979 (in Japanese)
4. MLIT (2007) Commentary on the Structural Calculation based on the Enforcement Order, Building Standard Law, supervised by Housing Bureau and National Institute of Land Infrastructure Management in Ministry of Land, Infrastructure, Transport, and Tourism, and Japan Conference of Building Administration, 173 pp (in Japanese)

5. MLIT (1980) Kampo (Official Notice, Government of Japan), No. 16043, 14 July 1980 (in Japanese)
6. Architectural Institute of Japan (1990) Lateral Loading Capacity of Seismic design of Buildings, Oct 1990 (in Japanese)

Appendix B
Design of Buildings for Tsunami Loads

Yoshiaki Nakano

Abstract The outline of the current guidelines for tsunami load in Japan is introduced. They are mainly based on the guidelines issued by the Japanese Cabinet Office in 2005, with some additions after the 2011 disaster. They feature a simple procedure to calculate loads imposed by tsunami under the assumption that the proper design inundation height of tsunami will be provided. In this appendix the guidelines named as "Interim Guidelines on the Structural Design of Tsunami Evacuation Building: a revision based on the lesson of the building damage by 2011 East Japan Tsunami Disaster" is introduced.

B.1 Introduction

For the structural design of tsunami evacuation buildings, the Japanese Cabinet Office issued in June 2005, "Guidelines for evacuation of buildings affected by tsunami" [1]. The guidelines was created by adopting the design method for tsunami load equations by the Building Center of Japan [2, 3] published in 2004. However, very few buildings have been designed based on these guidelines since then.

After the 2011 East Japan Earthquake and Tsunami Disaster, a joint team of the Institute of Industrial Science, The University of Tokyo, and the Building Research Institute extensively inspected the tsunami damaged buildings and have published a result of an inventory analysis of the damaged buildings [4] to scrutinize the former guidelines and has proposed some revisions. Housing Bureau and National Institute

Y. Nakano (✉)
Institute of Industrial Science, The University of Tokyo, Tokyo, Japan
e-mail: iisnak@iis.u-tokyo.ac.jp

Architectural Institute of Japan (ed.), *Preliminary Reconnaissance Report of the 2011 Tohoku-Chiho Taiheiyo-Oki Earthquake*, Geotechnical, Geological and Earthquake Engineering 23, DOI 10.1007/978-4-431-54097-7, © Springer Japan 2012

for Land and Infrastructure Management, MLIT adopted the proposal and issued interim guidelines based on the former guidelines by the Cabinet Office.

The relatively simple procedure to calculate loads imposed by tsunami is not changed but some design coefficients and additional cautions are added. This is currently the only standard for the design of tsunami evacuation buildings in Japan.

In the following sections, the guidelines named as "Interim Guidelines on the Structural Design of Tsunami Evacuation Building: a revision based on the lesson of the building damage by 2011 East Japan Tsunami Disaster" is introduced.

B.2 Scope and Application

Scope

This design guideline is applicable to the structural design of tsunami evacuation buildings and others to resist loads imposed by tsunami. In applications, the tsunami inundation depth for the design should be referred to the expected inundation depth shown in hazard maps designated by municipalities. However, the design inundation condition (area and inundation depth) can be specified in a particular area by a master plan of tsunami disaster mitigation made exclusively to a special local district.

Application to New buildings

If these provisions are to be applied to new buildings, they shall conform to the provisions shown here as well as to the Building Standard Laws, related laws and relevant enforcement orders.

Application to Existing Buildings

If these provisions are applied to existing buildings, and they do not conformed to the current Building Standard Laws, they shall conform to the enforcement orders No. 185, pursuant to No. 1, Paragraph 3 of Article 8 of the act on Promotion of Seismic Retrofitting of Buildings No. 123 or the Article 20 of the Building Standard Law as of June 1st, 1981 (see Appendix A).

B.3 List of Terms

Terms used in the design guidelines are defined as follows:

Design inundation depth: inundation depth of the tsunami or distance from inundation height to the ground level of a building site in meter.
Tsunami force: lateral force generated by tsunami acting on the building in kN.
Tsunami wave pressure: lateral pressure acting on the exposed surface of the building generated by tsunami in kN/m^2.
Tsunami wave force: lateral force acting on the building generated by tsunami in kN.
Buoyancy: upward force acting on the building generated by tsunami in kN.
Pressure-exposed surface: building surface directly subjected to the tsunami.
Pressure resistant components: building components that resist the tsunami direct pressure.
Non-pressure resistant components: building components that can potentially fail due to the tsunami direct pressure.
Structural system: entire structure that is able to transfer the tsunami force coming from the pressure-exposed surfaces to foundation.

B.4 Structural Design

In tsunami resisting building design, structural components should be classified as ones that are directly exposed to the tsunami pressure and ones that are not.

B.5 Calculation of Loads Imposed by Tsunami

Tsunami Wave Force Equation

Tsunami wave pressure q_z at height z is given by:

$$q_z = \rho g \left(ah - z\right) \tag{B.1}$$

where,

q_z: intensity of tsunami pressure at height z in kN/m^2,
ρ: density of water in t/m^3,
g: gravity acceleration in m/s^2,
h: design inundation depth in m,
z: location of acting pressure measured from the ground ($0 \le z \le ah$) in meter,
a: depth factor (=3.0). The value of a can be reduced if the building is located in the following condition (Fig. B.1).

	Condition	Value of a
(1)	The tsunami stream to the building is effectively blocked by other buildings and/or structures	2.0
(2)	In addition to the condition (1), the distance of the building to the coast (sea, river) is larger than 500 m	1.5

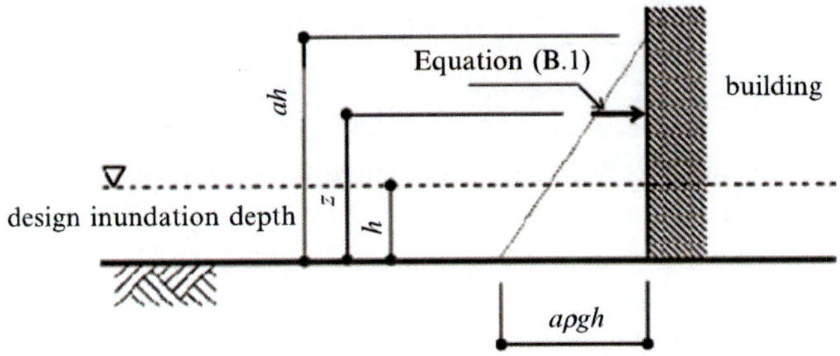

Fig. B.1 Tsunami wave pressure given by Eq. (B.1)

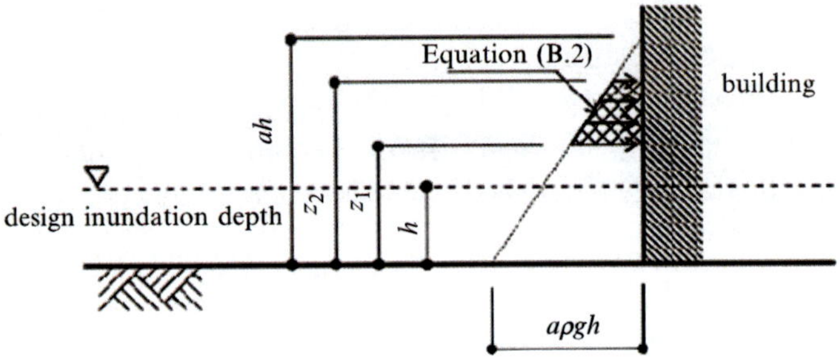

Fig. B.2 Tsunami wave force given by Eq. (B.2)

Calculation of Tsunami Wave Force

Tsunami wave force should be calculated by the following equation:

$$Q_z = \rho g \int_{z_1}^{z_2} (ah - z) B dz \tag{B.2}$$

where,

Q_z: tsunami wave force for structural design in kN,
B: width of pressure-exposed surface in m,
z_1: minimum height of pressure-exposed surface ($0 \le z_1 \le z_2$) in m, and
z_2: maximum height of pressure-exposed surface ($z_1 \le z_2 \le ah$) in m (Fig. B.2).

Lateral Load Reduction due to Openings

The area of openings (including the surface area of non-pressure resistant components designed to fail due to tsunami wave pressure) can be excluded from the area to calculate the tsunami wave force.

Piloti Structures (Structures with First Floor Supported Only by Columns)

The tsunami wave pressure on the part beneath a building supported by piloti (except on the pressure resistant components, such as columns and beams) can be neglected from the calculation of design tsunami wave pressure.

Direction of Lateral Load

Lateral force of tsunami is assumed to act on the structure in all horizontal directions.

If the direction of tsunami is clearly given by the shape of the coastline and the predicted inundation depth, this assumption does not have to be applied. Also, depending on the particular case, water backrush can be considered.

Calculation of Buoyancy

Buoyancy caused by the tsunami is calculated using the following formula:

$$Q_z = \rho g V$$

where, Q_z: buoyant force in kN, V: volume of the inundated part of the building in cubic meters.

The volume of the building can be reduced considering the inflow of water through the openings depending on the height of the openings and tsunami inundation depth.

Exception Based on Special Researches

The tsunami wave pressure obtained after special researches and investigations considering the condition of the location of the tsunami evacuation building can be used if it is more rational than those specified in this guideline.

B.6 Load Combinations

Following load combinations should be considered in tsunami resisting design of building structures.

$$G + P + 0.35S + T \text{ (Regions with heavy snow)} \tag{B.3}$$

$$G + P + T \text{ (Outside of regions with heavy snow)} \tag{B.4}$$

where, G: dead loads, P: live loads, S: snow load and T: force generated by tsunami.

The regions with heavy snow can be designated by municipalities in their jurisdiction as stated in the relevant articles in the Building Standard Law Enforcement Orders.

B.7 Design of Pressure-Exposed Surfaces

Design of Pressure Resistant Components

Ultimate strength and deformation capacity of supporting components should be sufficient to resist ultimate loads. Also, waterproof characteristics should be considered.

Design of Non-pressure Resistant Components

These members are allowed to fail if its failure does not cause damage to any structural member.

B.8 Design of Structural Members

Ultimate lateral strength of structural system in every horizontal direction at each story should be greater than lateral loads imposed by tsunami [Eq. (B.5)]

$$Q_{ui} \geq Q_i \tag{B.5}$$

where,

Q_{ui}: lateral strength of i-th floor, and

Q_i: lateral load caused by tsunami acting on the i-th floor.

Ultimate lateral strength should be calculated with respect to most critical load combination.

B.9 Design Against Sliding and Overturning

Structures including foundation members such as piles should be designed to be stable for overturning and to resist sliding due to tsunami wave pressure by considering the effects of gravity and buoyancy.

B.10 Miscellaneous

Local Scouring

Design of foundation shall consider tilting of buildings due to the soil washed away beneath the foundation by tsunami flow, in particular when no pile foundation is used.

Collision of Floating Debris

Impact load due to collision of floating debris shall be considered. Critical structural members shall be prevented from failure, or stability of the structural system shall be secured after failure of some structural members.

References

1. JCO/Task Committee under the Japanese Cabinet Office (2005) Design guidelines for tsunami shelters. http://www.bousai.go.jp/oshirase/h17/050610/tsunami_siryo2.pdf (in Japanese)
2. Okada T, Sugano T, Ishikawa T, Ogi T, Takai S, Hamabe T (2004) Structural design method of building to seismic sea wave, No. 1 preparatory examination, Building Letter. The Building Center of Japan, pp 7–13. October 2004 (in Japanese)
3. Okada T, Sugano T, Ishikawa T, Ogi T, Takai S, Hamabe T (2004) Structural design method of building to seismic sea wave, No. 2 design method (a draft), Building Letter. The Building Center of Japan, pp 1–8. November 2004 (in Japanese)
4. Institute of Industrial Science, The University of Tokyo (2011) Investigation on Building Code Development in Tsunami Hazardous Areas, granted by the Ministry of Land, Infrastructure, Transport and Tourism, Interim Report (July 2011) and Interim Report 2 (November 2011), Principal Investigator: Y Nakano, Professor of IIS, The University of Tokyo (in Japanese). http://www.nilim.go.jp/japanese/organization/kenchiku/iinkai/20110818pdf/siryou1.pdf; http://www.mlit.go.jp/report/press/house05_hh_000274.html

Author Index

Architectural Institute of Japan (ed.), *Preliminary Reconnaissance Report of the 2011 Tohoku-Chiho Taiheiyo-Oki Earthquake*, Geotechnical, Geological and Earthquake Engineering 23, DOI 10.1007/978-4-431-54097-7, © Springer Japan 2012

Subject Index

Architectural Institute of Japan (ed.), *Preliminary Reconnaissance Report of the 2011 Tohoku-Chiho Taiheiyo-Oki Earthquake*, Geotechnical, Geological and Earthquake Engineering 23, DOI 10.1007/978-4-431-54097-7, © Springer Japan 2012